Points, Lines,

&

Conic Sections:

A Sequel to College Algebra

by Tony Berard

Table of Contents

Introduction...4
Chapter 0: Review: What the Student Should Already Know..14
    Theorem Highlight A: The Pappus Line................32
Chapter 1: Fractions, Rational Points, and Denesting
       Radicals...37
Chapter 2: The Deluxe Toolkit............................61
    Theorem Highlight B: Desargues' Theorem............73
Chapter 3: Pythagorean Triples...........................79
Chapter 4: The Circle....................................92
    Theorem Highlight C: The Wallace-Simson Line.......105
Chapter 5: Lines and Parabolas: Part 1*..................114
Chapter 6: Lines and Parabolas: Part 2*..................129
    Theorem Highlight D: Some Square Erecting Theorems.145
    Theorem Highlight E: Two Special Quadrilaterals....152
Chapter 7: Linear and Parabolic Asymptotes...............161
Chapter 8: Some Tangency Problems........................169
    Theorem Highlight F: Frégier's Theorem.............191
    Theorem Highlight G: The Polar Line and Conic Section
       Diameters....................................203
Chapter 9: Triangle Geometry.............................211
    Theorem Highlight H: The Symmedian Point...........277
Chapter 10: Concyclic Points.............................237
Chapter 11: Lines and Conic Centers*.....................253
    Theorem Highlight I: The Eyeball Theorem...........260
Chapter 12: Lines and Ellipses: Part 1*..................265
Chapter 13: Lines and Hyperbolas: Part 1*................279
    Theorem Highlight J: Johnson's Circle..............285
Chapter 14: Lines and Ellipses: Part 2*..................290
Chapter 15: Lines and Hyperbolas: Part 2*................299
    Theorem Highlight K: The Pascal Line...............308
Appendix A: Finding More Terms of a Sequence.............313
Appendix B: Source Code for PythTrip.exe Program.......318
Appendix C: Help Me Write the Solutions Manual
       (or: I Need a Coauthor)......................325
Index..327

Note: The seven chapters designated with an * are chapters representing original contributions to mathematics as far as the author knows.

Introduction

My opening remark for this textbook is to say that I hope my math textbook is a good textbook. Typically, a good math textbook does not present any new information, but it does explain the many things that it contains well so that a student can comprehend the material. Having said this, I have seen from my years of experience tutoring math that students have trouble reading math textbooks. If I forget the details of a theorem while helping a student, I am the one that reads the student's textbook for insight. A careful author puts in the relevant details so that an expert, such as myself or an instructor, can glean from the textbook its insights. Thus, I understand that students will have trouble reading my math textbook because that is the way of things. Where this book will be judged is how well the experts—math tutors and math instructors get the insights that this book describes. Yes, well over half of this book is not new information, and this part I hoped to have explained well. However, a significant part of this book is new information, and it is on these pages that I hope to have really shined as an author because this new material is my own contribution to math. I want people to understand this part of the book more than the other parts of the book because this part is my part, and I have pride in my part. Of course, the rest of the textbook that is not my part is important, too. If it was not important, it would not even show up in this textbook.

This textbook grew out of my senior project at Lawrence Technological University. I did not finish the project during the term, but I got quite far. I have now completed the project and worked it into a college level math text, which includes many chapters that were not part of my original project. Regarding my project, it is an extended look into the conic sections of precalculus. In most situations, we seek to find the equation of the line passing through some specified point in the conic and intersecting it again some specified distance away although this is not always the case. With each new situation, the necessary formula will be derived using precalculus techniques. However, the conceptualization and resulting algebra will be quite intense. The many chapters included in this book that were not part of my project complement the material of my project nicely, and I had to do a great deal of research to write these chapters because the book for this course never existed before. I had to decide what to include and what to leave out. The chapters of original authorship, which represents my completed senior project, are flagged in the Table of Contents with an asterisk.

I was a student myself when I had the first insights for my new equations. Perhaps, other students will be inspired to think of new equations that contribute to the body of knowledge that is mathematics as I have done. I have been a tutor of mathematics for a number of years now, so some of my insights regarding educating students come from here as well.

We will discuss what the student needs to know before beginning this course. Some beginning algebra books will touch on the parabola, yet the treatment there is of a minimal nature. The circle, ellipse, and hyperbola are almost never mentioned in the beginning algebra textbook. Thus, it is not recommended that a beginning algebra student attempt this course. The treatment of conic sections in intermediate algebra is of an introductory nature, yet it is usually sufficiently thorough to serve as adequate prerequisite material for this course. However, it is not recommended in general that an intermediate algebra graduate take this course. Many times, the material on conic sections is taught at the very end of the college algebra course. Thus, it is rushed through or just plain omitted. If the student received a quality treatment of conics in the college algebra course, then that is the correct prerequisite material for this course. If the student received an incomplete treatment of conics in college algebra, the student usually has to wait until the second term of calculus to see the conics again. However, the subject is covered much more deeply in calculus, which is more than adequate prerequisite material for this course. Regardless of how thoroughly the student is prepared, I have included a review chapter to put all of the students in the class on the same page.

The instructor is not expected to teach the review chapter, so the responsibility rests with the student to ensure that he or she thoroughly understands this review chapter. Teachers provide office hours, and the student should make use of them to get up to speed on the review chapter if that is necessary. In addition, the review chapter serves as a ready reference to be able to look up prerequisite material easily to facilitate the working of problems in this book. Also, many colleges and universities have a tutoring center that students may get help from a qualified professional who is not an instructor but has the requisite knowledge to help students.

This course can further prepare students for the calculus series by giving them more grounding in algebra concepts and methodology. This course can further clean up lingering issues with students writing in mathematics as well since a number of the exercises ask the student to write. The other prerequisite course for calculus is trigonometry, and many students have an inadequate working knowledge of trigonometry that hampers their success in calculus. However, this textbook does not address the trigonometry issue because it focuses on the algebra issue. Perhaps, a later edition of this book will have an optional chapter or two on lines and conic sections using trigonometric and parametric equations since those mathematical objects can describe this new material as well; however, it is sufficient for this edition to stick with just the algebra.

The Theorem Highlight feature of this textbook portrays some famous theorems discovered by mathematicians. These theorem highlight discussions are sandwiched between the main chapters. The instructor may or may not teach the material in these theorem highlight discussions. The instructor may even decide to use some of this material this material for extra credit. The material on these theorem highlights is included mainly to expose the student to quality material that would not otherwise be seen by the student. In addition, since the title of this book is *Points, Lines, and Conic Sections: A Sequel to College Algebra*, it would be prudent to discuss some famous points, lines, and conic section theorems in these theorem highlight chapters so that the book can live up to its title. The theorem highlight on Desargues' theorem, however, is a steppingstone to ease the pain of the triangle geometry chapter. Other parts of this book have been designed to ease the pain of the triangle geometry chapter as well because that chapter is a most formidable one.

I have briefly stated the layout of my book—i.e. the chapters reflecting my completed senior project and the other chapters to complement my material to achieve a full textbook. I will dwell on this in more detail now.

Why did I write this book? The primary reason is that I wanted to showcase my contributions to mathematics. I have never seen my equations in any reference work on mathematics, and believe me, I have looked. I have also asked numerous college and university instructors and professors, and none of them have seen my equations before. Thus, I have pretty solid evidence that I have made a genuine contribution to mathematics. However, looking at the amount of material I had proved to me that it was not sufficient for a full semester course in college mathematics.

In the face of that disheartening revelation, I set out to research material of a comparable level of difficulty to my material. I thought about where my material would fit into the existing standard offering of mathematical courses. I decided that a student passing college algebra would have the skill set necessary to understand my material. Hence, I needed to find material to complement that. I found quite a bit of stuff on the Internet. A perusal of the Table of Contents shows much of what I found. I also had some books on synthetic geometry (proofs using axioms) and a few dictionaries on mathematics. I also found a number of papers that were published on the Internet. However, I couldn't use this material as I found it. No, it was presented for graduate level students. I had to bring the material down to a lower level of presentation, and I had a difficult time organizing the growing amount of material as well. As an embarrassing side note, in many cases I first had to understand the material myself because I had never seen almost all of it before, and I have a math degree. When I understood it, I was able to write about it.

At some point, I realized that I now had too much material. Thus, I had to select the best stuff from this mass. Of course, my own work would be the centerpiece of the book. What guided me in the selection of the rest of the book was the question, "Could a college algebra graduate handle this?" If not, "Could an appropriate build-up occur in the textbook to make it so?" Thus, after deciding what to keep and what to toss, I had to order the material by figuring out what needed to come before something else. Then, I had a roughly linear progression of the material building from simpler to more complex in sophistication.

A sad thing for new results in mathematics is that they are often graduate level and get published in relevant magazines and journals of mathematics and related sciences. Thus, professionals see and understand these new results, but people on a lower level of mathematical ability will never see or understand these results. Only a really famous result captures the public fascination, such as when Andrew Wiles proved Fermat's Last Theorem. I had read somewhere that only one-tenth of one percent of the mathematical community was qualified to evaluate the proof of Fermat's Last Theorem (and no, I am not in that select group). I am happy to say that my contribution to mathematics is not confined only to the mathematics professionals. Nay, anyone who has successfully completed college algebra has the capacity to understand my contribution. Also, I have incorporated a number of other recent finds in mathematics that are within reach of the college algebra graduate. Thus, the student gets to see firsthand the evolution of mathematics.

A wonderful thing about mathematics is that it is a multicultural enterprise. Where possible, I have noted who came up with a particular theorem I present. I note the birth date and death along with that mathematician's country of origin. In this fashion, the student gets to see that mathematics has become enriched by the efforts of many people from many cultures. I will endeavor to complete any unfinished documentation of the theorems presented by the time the second edition of this book rolls around. Unfortunately, the originators of many theorems just get lost in the mists of time.

A phenomenon that occurred to me while I was a student was that a lot of new concepts would get introduced in the higher-level math classes. We were learning more sophisticated methods of proof, but this was complicated with the newness of the concepts we were learning too. I had not seen these new concepts before, and I did a little bit of experimenting with them computationally to attempt to understand them. However, I couldn't dawdle at it too long because my homework was to *prove* things with these new concepts, not compute anything with them. I struggled in these proof classes as a result. The new methodology of proof and the new concepts together proved to be too much for me.

Happily, this new textbook of mine provides a rich assortment of new concepts for the student. Proofs are not the order of the day here. However, I do ask for a demonstration from time to time which is basically a proof. These times are rare, though, and I have put them in here mainly to stimulate the students with the gift and love of mathematics to do more. Thus, I'm happy if the student can do the arithmetic or algebraic manipulation asked for, not necessarily the deeper demonstrations.

If the student goes on to take graduate level courses in geometry, some of the material in these pages will bring an old familiarity to them when they get to those higher-level courses. Thus, these students can learn how to do the proofs that will be asked of them with at least a passing familiarity with the not so new concepts.

Because of all the topics presented in this book, it has become a valuable reference source. This is in stark contrast to the many math books that come out nowadays—at least at the lower levels of mathematics in which I tutor students. I have seen that when the community college I work at adopts a new textbook for a class, it isn't any different or better than the old one they discarded. When I discuss this matter with the other tutors, I hear something like, "Sure, some minor variations of topics are present, or the order in which the topics are presented is slightly altered; however, it is essentially the same textbook." Inevitably, we conclude that it made no difference to us.

I didn't always feel this way. I remember one time when one of my professors at Oakland University told me, "All calculus books are the same." I looked at him incredulously, and he added tapping his chest, "To me." He had a Ph.D. in mathematics, so freshman calculus was so far down the math ladder for him that all the calculus textbooks must have really looked the same to him. Now that I am a little higher up the math ladder than I was at the time, I understand his sentiment. He was just sharing with me what I was later to conclude with my fellow tutors. The new textbook for a subject is not any better than the old. Having said this, my new textbook is not like the old because I created this one from scratch and put many modern results in it, including my own results. Thus, this textbook *is* different. Another main feature of this book is that I have interwoven an introduction to triangle geometry without proofs into the text. Triangle geometry is a deep subject that has been developed over the course of the last few centuries. I have attempted to integrate this subject into my book (without the graduate level proofs, of course). Thus, this brings the subject down to the level of the college algebra graduate. A lot of exciting things in triangle geometry are now within reach of the student as long as the build-up is correct. I think I have done a good job with this build up, but any suggestions for improvement are welcome, of course. And of course, if any errors are found, please bring them to my attention so that I can fix it for the next edition! As a result of the build-up in which I have formatted this textbook, the book is best learned by going straight through without skipping around. We next probe the boundaries of where this book fits into the current curriculum of mathematics.

We have the traditional mathematical subjects taught in school comprising arithmetic, beginning algebra through precalculus, trigonometry, and geometry. These classes for the most part require the student to find the solution to a problem, which requires the use of already established theorems in mathematics. Sure, trigonometry and geometry have their proof portions, but these are minor compared to the rigor and quantity of proofs required by the higher abstraction mathematical courses. These traditional classes prepare the student for calculus, which is taught over several semesters. We can also include differential equations and linear algebra along the same lines as calculus in that these courses lay the foundation for higher-level work—i.e. proofs in mathematics.

Next, we get into the junior and senior level classes, which some are computational in nature while others are theoretical mathematics. With theoretical mathematics, the emphasis is on proving this or that from other things known. A proof in mathematics establishes that a theorem is true. For example, how do we know that multiplication is associative? We have known that $(ab)c = a(bc)$ for any and all values of a, b, and c. It is impossible to test every case for every possible value of a, b, and c. Mathematicians have *proven* that this is true, and because they have done so, we can switch around the order of multiplications in a series of numbers to be multiplied. As another example, how do we know that the Pythagorean theorem is true? We have used it when we deal with right triangles, but that is because we were told to use it in those situations. For the interested student, the math theorem that has the most proofs for it is the Pythagorean theorem. Just Google (or insert your favorite search engine here) it on the Internet. To give a taste of a proof that the student can understand, the following is a proof that establishes that there are an infinite number of prime numbers. I learned this proof in Discrete Mathematics, a junior level math class.

What to prove: There are an infinite number of prime numbers.
Method: Proof by contradiction.

Assume to the contrary that there are only a finite number, say $n \in \aleph$, of primes.

This method proceeds along correct lines of reasoning until we arrive at a contradiction. Hence, this contradiction only comes about because of the incorrect initial assumption, namely, that there are only a finite number of primes. Because of the contradiction, we are thus forced to conclude that the opposite (technically called the negation) of the assumption made incorrectly must be true. Thus, we proceed along correct lines of logic and continue until we get a contradiction.

Now, let p_n be the largest prime number. This is our initial (false) assumption.

Construct a new number Ω that is the product of the largest prime number and of all of the previous prime numbers, and then add one. Thus, we have the following:

$$\Omega = 2 \cdot 3 \cdot 5 \cdot 7 \cdot 11 \cdot 13 \cdot 17 \cdot ... \cdot p_{n-3} \cdot p_{n-2} \cdot p_{n-1} \cdot p_n + 1$$

What happens when we divide Ω by 2? We get a remainder of 1.
What happens when we divide Ω by 3? We get a remainder of 1.
What happens when we divide Ω by 5? We get a remainder of 1.

This pattern continues all the way to the end. The last few statements are presented.

What happens when we divide Ω by p_{n-3}? We get a remainder of 1.
What happens when we divide Ω by p_{n-2}? We get a remainder of 1.
What happens when we divide Ω by p_{n-1}? We get a remainder of 1.
What happens when we divide Ω by p_n? We get a remainder of 1.

This line of questioning has led us to discover that Ω is a prime number because it has no prime divisors (i.e. no remainders of zero). It is clear that $\Omega > p_n$ because of the way we constructed it. Thus, we have our contradiction in that we assumed that p_n was the largest prime number, and we found that we cannot assume a largest prime number such as p_n because we can always construct another prime number Ω that is larger than that. Hence, we are unable to construct a largest prime number. Hence, we conclude that there are an infinite amount of prime numbers. Q.E.D.

I will offer here another proof. Recall the theorem that states that two lines are perpendicular if the relationship $m_1 m_2 = -1$ holds. We say that if the slopes are negative reciprocals, then the lines are perpendicular. I was tutoring a student in college algebra, and we saw that property of lines in a highlight box in her textbook. She asked me, "Can you prove that?" I tried with trigonometry to do so and failed. I told her, "I guess not." She said, "That's OK." I would have tried again, but other students needed help. I though about it again the next day, and I came up with a proof. Here it is:

Let $m_1 = a/b$ where a is the rise and b is the run. We stipulate that both a and b are real and b is not zero. Next, using $m_1 m_2 = -1$, we solve for m_2 to find that $m_2 = -b/a$. We would like to graph these two lines to see the relationships of the slopes. Without loss of generality, we can let the intersection point of the two lines be the origin. Hence, the lines $y = m_1 x$ and $y = m_2 x$ become the lines $y = (a/b)x$ and $y = (-b/a)x$. Again, without loss of generality, assume $m_1 > 0$.

One line 1, we would like to plot some point on it. Choose (b, a), and we will call this P_1. One line 2, we choose point $(-a, b)$. This sets up two right triangles with sides a, b, c_1 and a, b, c_2. Because these are both right triangles with the same lengths for the legs of the triangle, the hypotenuses are equal to $\sqrt{a^2 + b^2}$. Thus, triangle OP_1P_2 is at least an isosceles triangle. If triangle OP_1P_2 satisfied the Pythagorean theorem, then angle P_1OP_2 would be a right angle. We set up the Pythagorean theorem to check.

$$c_1^2 + c_2^2 = d^2$$

$$(\sqrt{a^2+b^2})^2 + (\sqrt{a^2+b^2})^2 = (\sqrt{(b-(-a))^2 + (a-b)^2})^2 \quad \text{Note: RHS uses the distance}$$

formula.

$$a^2 + b^2 + a^2 + b^2 = (\sqrt{b^2 + 2ab + a^2 + a^2 - 2ab + b^2})^2$$

$$2a^2 + 2b^2 = (\sqrt{2a^2 + 2b^2})^2 = 2a^2 + 2b^2 \quad \text{We obtained an identity.}$$

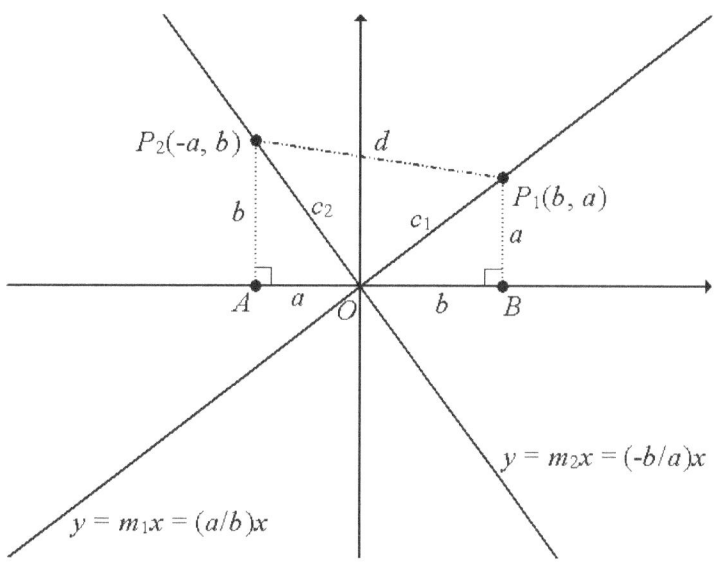

Figure 1: Diagram for Proving Two Lines Perpendicular

Thus, triangle OP_1P_2 satisfies the Pythagorean theorem. Thus, it is a right triangle. Hence, lines 1 and 2 cross at a right angle, which means that they are perpendicular. Q.E.D.

I feel bad that I was not able to answer her question on the spot, but perhaps this proof will redeem me. I saw her a few weeks later, and I was able to give her a copy of this proof. A belated answer is better than no answer, I suppose. We proceed now with our discussion about mathematics classes.

One class at the junior level is an introductory class to proofs in general, which prepares the student for graduate work (and senior level proof classes). These classes include abstract algebra, real and complex analysis, topology, number theory, numerical approximation methods, set theory, etc. These classes, while they may have their computational aspects, are heavily slanted towards proof. However, other graduate classes are heavily involved in computation, and these classes typically prepare the student for the industrial applications of mathematics. Many problems arise in industrial applications that require sophisticated methodologies and/or number crunching software power to solve. Linear programming and mathematical modeling in general requires powerful software implementing heavy-duty methodologies.

Now that we have seen an overview of the math curriculum, we ask what type of class is this, then? It is a brand new class, so we should have some idea of where it fits into our hierarchy. Mostly, it is a new class of the traditional sort. Thus, the deep proofs of the higher abstraction classes are not a part of this course. The computational aspects are present in this course (but not of the intense industrial sort), and the student will learn more about computational sophistication than previously. A little bit of proof is present in this course, but it is only a sideshow for the reason previously stated. However, another reason for *some* proof in this course is because I have found new equations relating to conic sections, and I have had to prove my results. The proofs presented are mostly sketches with the students filling in the details that need algebraic legwork. However, I have lighted the signposts along the road, so the students should not lose their way.

It is an approach of instructors at the higher levels of math that a student may not use a result until he or she has proved it. This makes sense at the higher levels because the student is making the trek up to the top of the mountain. Being able to use a result that the student has not *earned* is an affront to such an instructor. I understand this approach, but the students taking this course are not on that journey. We do not tell a student that he or she cannot use the Pythagorean theorem until he or she has proven it. If we did, then said student would not be able to find the length of the third side. To take this idea to its extreme, I have seen in set theory how to construct the sets "1" and "2" from the initial assumption of the empty set existing. From there, the binary operation addition is constructed. Then, comparing the cardinalities of the set "1 + 1" and the set "2," we proved that $1 + 1 = 2$ in class. Do we ask children to do this sort of thing to make sure that they have all of their crayons? No, we just teach them how to count and add.

This book, while it is a step above just counting and adding, is still about showing how to do the math. Most of the students in this course will not make it up the mathematical mountain of proofs, so this course allows them to peak at the treasures and get some benefit without truly owning it by way of being able to prove it. This is my answer for why I brought this stuff down from the mountain on the clay tablets to give it to the people. This material deserves to be seen by a wider audience. Even high schools that teach AP Calculus can teach this course because AP Calculus has precalculus (incorporating both college algebra and trigonometry) as its prerequisite; however, the students taking this course only need college algebra as the sole prerequisite. We might say, then, that this course should be numbered with a smaller number than freshman calculus because it has less prerequisites to get into it.

I have read in several places that the parabola is one of the most studied curves in recorded history. This is not too difficult to see considering that projectiles follow a parabolic path, and a parabola also has well known optical properties. Further, we have taught about the slope of a line, the distance between two points, and what the leading coefficient does to a parabola. Even back when Euclid was teaching, he had a book on Conics that inspired Apollonius. In light of this, it seems that it would be unlikely to find anything new about a parabola with these concepts. However, I am proud to say that I have seen something missed by so many. The math techniques to handle what I have seen are slight, to be sure. However, the spark I had that one day to blend the three concepts together to produce this work was something I treasure. If other mathematicians had that same spark, they did not follow up on it perhaps thinking it wouldn't lead anywhere.

It is my hope that this course broadens the mathematical horizons of the students who take it. Many classes that are higher than this one do not have a prerequisite per say, except for a certain amount of mathematical sophistication. This class should raise the student's mathematical sophistication. It is also my hope that this course brings students some joy in mathematics—the process of discovery and creativity that is the hallmark of all of the greatest pursuits of mankind, and mathematics may be counted among them.

Review: What the Student Should Already Know

This review chapter is provided so that students in this class can be prepared for the heavier material to follow in this course. The students may not have seen some of the material in this review chapter before now, but we will have occasion to use these types of concepts later. Thus, it would be wise for students to go through this chapter. If the student runs into trouble with this chapter, the student should seek the help of a tutor or use the instructor's office hours. I have also seen that study groups can be helpful as well. Some students prefer to be the lone wolf, so to speak, and do not profit by study groups. Use the resources available at your learning institution to help you succeed.

Arithmetic

The student should know how to compute (i.e. do arithmetic) with various types of numbers including decimals, fractions, radicals, integer and rational exponents, logarithms, and the transcendental numbers π and e. Of course, $\pi \approx 3.1415926535\ldots$ is the ratio of the circumference of the circle to its diameter, and $e \approx 2.7182818\ldots$ is the base of the natural logarithm. Converting fractions to decimals and vice versa may be necessary to aid in the computations and simplifications. Use of the change of base formula for logarithms may be necessary as well so that the result can be approximated with a calculator. The order of operations must be followed—recall PEMDAS, which stands for Parentheses, Exponents, Multiplication, Division, Addition, and Subtraction. Some examples of arithmetic problems follow.

1. Convert $7.3\overline{273}$ to a fraction.

2. Simplify $\dfrac{2 + \dfrac{3}{5.7}(1 - 0.6)^3}{\dfrac{1}{7} - \dfrac{8}{13}}$.

3. Compute $\dfrac{(5+\sqrt{7})^2 + (7-\sqrt{5})^2}{(2+\sqrt{8})^2 - (8-\sqrt{2})^2}$. Be sure to rationalize the denominator.

4. Simplify $2\log_5 7 - 3\log_5 2$. Approximate answer to 4 decimal places with the aid of a calculator.

5. Approximate $3\ln(3^{4.2} - 2^{1.7}) + \sqrt[11]{5e - 4\pi}$ to 4 decimal places with the aid of a calculator.

Formulas

The student should be able to solve for the other variables in a given formula. Thus, a formula is really several formulas—it depends on the number of variables in the formula. Given all but one of the variables in a formula, the student should be able to solve the formula for the unknown variable, and then plug in the given numbers to compute the result. The student should know about the quadratic formula and be able to use it to solve for variables that are quadratic in nature. Some examples of these types of problem follow.

1. Solve the formula $E = \pm m \cdot c^2$ for c. What do you think the minus sign represents?

2. Solve the formula $I = P \cdot R \cdot T$ for P.

3. Solve the formula $A = \dfrac{1}{2} h(B + b)$ for B.

4. Solve the formula $A = 2\pi r h + 2\pi r^2$ for r. Note that this equation is quadratic in r.

5. Solve the equation $21 \log_3^2 (2x) + 8 \log_3 (4x^2) - 41.2 = 0$ for x analytically. Approximate any answers found to four decimal places of accuracy.

6. If x_1 and x_2 are the real roots of the equation $ax^2 + bx + c = 0$, then show that the sum of the roots is $-b/a$ and the product of the roots is c/a. In addition, show that the equation can be written as $x^2 - (x_1 + x_2)x + x_1 x_2 = 0$. Finally, show that the roots of this last equation are x_1 and x_2 by using the quadratic formula on it.

Inequalities

The student should be able to use inequalities and convert them to interval notation if necessary. Getting zero on one side and factoring is useful because it allows the factors to determine the cut points. Usage of the sign test can determine the sign of each region. Then, it is a simple matter to construct the intervals and union them to solve the inequality. The student should recognize some basic inequality facts as well, such as adding a positive quantity to some value increases it. In symbols this says $a + b > a$ when $b > 0$. Another inequality falling into this category is the famous triangle inequality that states the lengths of any two sides of a triangle are greater than the length of the third side. In symbols this says that if a, b, and c are the lengths of the sides of a triangle, then $a + b > c$. In addition, the inequality $a + c > b$ is also true as is the inequality $b + c > a$.

1. Use the sign test to determine the solution to $\dfrac{4x^2 - 3x + 5}{x - 2} \le \dfrac{13}{3}(x - 8) + 40$.

2. Use the calculator to approximate the solution to $8 - \sqrt{3 - 5(x - 9)} < \sqrt[3]{3x^2 - 2}$. Use three decimal places of accuracy in the interval notation. Treat each side as a function of y, and graph both in the same viewing window. This is called the intersection method. Label the intersection points with three decimal places of accuracy. Determine whether or not the exact solution interval is obtainable, and explain why or why not.

Factoring and Expanding

The student should be able to factor integers and algebraic expressions. In addition, the student should be able to expand out algebraic expressions in various states of messiness. We will see time and again in this book that expanded out expressions usually have terms that cancel out, and the resulting expression after the dust clears is much simpler.

1. Expand $(1 + x)^3 - (1 - x)^2 + (x - 1)$.

2. Factor the integers 21, 65535, and 122221. Research to find the divisibility shortcuts for 7 and
    11 as well as the usual ones for 2, 3, 4, 5, 6, and 9.

3. Writing in mathematics: Factor $(7 + 27x^3)^4 - 1$. Explain in words as much as you can about each of the five factors. Include in your explanation a consideration of this expression equal to zero, and how these factors could help in finding the solutions for x.

4. Factor $2x^4 - x^3 - 11x^2 + 4x + 12$. The rational roots theorem will help.

Solving and Graphing Equations and Systems of Equations

The student should be able to solve and graph various one-variable and two-variable algebraic equations, including radical, rational, and systems of equations. The methods available to solve these equations and systems are numerous. Where possible, confirm with a calculator.

1. Solve the radical equation $\sqrt{4x + 1} - \sqrt{3x - 2} = 5$.

2. Solve the following rational equation $\dfrac{21}{x + 2} - \dfrac{1}{x - 4} = 2$.

3. Solve the following system of linear equations. Show all work.

$$\begin{cases} 2x - 3y + 7z = 4 \\ 7.3x - 4y + 5z = 101 \\ 2x/3 - 11.3\overline{17}y - 14z = 21 \end{cases}$$

4. Solve the following systems involving at least one quadratic equation. Show all work.

a) $$\begin{cases} \dfrac{x^2}{24} + \dfrac{y^2}{26} = 1 \\ 5x + y = 7 \end{cases}$$

b) $$\begin{cases} y^2 - 4y - 3x + 1 = 0 \\ 3y - 4x = 6 \end{cases}$$

c) $$\begin{cases} \dfrac{5}{x^2} + \dfrac{3}{y^2} = 32 \\ \dfrac{1}{xy} = 4 \end{cases}$$

d) $$\begin{cases} y^2 - 3x^2 = -27 \\ x^2 + 45 = y^2 \end{cases}$$

Slope

The student should know and be able to use the equation for slope.

$$m = \frac{y_2 - y_1}{x_2 - x_1}$$

The student should know and be able to use the perpendicular slope formula.

$$m_\perp = \frac{-1}{m} = -m^{-1}$$

The student should know and be able to use the slope of the inverse.

$$m_{inv} = \frac{1}{m} = m^{-1}$$

1. What is the equation of the line ℓ_1 through the points (9,-3) and (-11, 5). What is the equation of the line ℓ_2 perpendicular to ℓ_1 through the point (2, 7)? What is the equation of the inverse line of ℓ_1? What is the equation of the inverse line of ℓ_2? Graph all four of these lines together.

2. Solve the slope equation for y_1.

3. What is $m_\perp$ in terms of the variables in the slope equation?

4. Two perpendicular lines intersect at the point (7, 11). One of them has a slope of a million. Find the equations of the two lines.

5. A line has a slope of $\dfrac{5-\sqrt{2}}{\sqrt{11}-12}$. Find the equation of the line perpendicular to this

going through the point $(3,3.\overline{3})$.

Lines

The student should know and be able to use the various forms of equations for a line.

$ax + by + c = 0$	Equation of line in standard form.
$y = mx + b$	Slope-intercept form of a line.
$y - y_1 = m(x - x_1)$	Point-slope form of a line.
$y - y_1 = \dfrac{y_2 - y_1}{x_2 - x_1}(x - x_1)$	Two-point form of a line.
$\dfrac{x}{a} + \dfrac{y}{b} = 1, \ a \cdot b \neq 0$	Intercept form of a line.
$x = a$	Vertical line through the point $(a,0)$.
$y = b$	Horizontal line through the point $(0,b)$.

The student should be able to sketch the graphs of the basic types of line including any given points and the x and y-intercepts.

1. A line of 6% slope goes through the point (4, 5). Find the equation of the line.

2. Find the equation of the line parallel to $7x + 3y = 19$ and goes through the point (5/7, -61.3).

3. Two perpendicular lines intersect at the point (-14, 44). One of them is vertical. Find the equations of both lines.

4. Two points on a line are (5,6) and (-1,4). Use the two-point form equation of a line to find the equation of this line.

5. Represent the equation of the line in the previous question in the various possible forms.

6. Why is the line $y = 4x$ not able to be represented in the intercept form of the equation of a line?

Absolute Value

The student should understand what absolute value means and be able to work with it in algebraic equations and inequalities.

1. Solve the absolute value equation $|2x^2 - 11| - |1 - 5x| = 2x + 7$. Treat each side of the equation as a function, and graph the two functions labeling the intersection points. This is called the intersection method. Also, rearrange the equation to get zero on one side, and graph that function. Label the roots. This is called the root method.

2. Solve the absolute value inequality $|1 - x| - 2 > \dfrac{6 \cdot |2 - x^2|}{3 - 5x}$.

3. Solve the absolute value equation $|1 - y| = |2x + 5|$ for x and for y. Graph the solutions.

4. Solve the absolute value equation $|1 - 2xy| = |4 - x + 3y|$ for x and y. Graph the solutions.

Distance

Given any two points, the student should be able to compute the distance between them with the equation $d = \sqrt{(x_1 - x_2)^2 + (y_1 - y_2)^2}$. The distance d between the line $ax + by + c = 0$ and the point $P(x_0, y_0)$ not on the line is $d = \dfrac{|ax_0 + by_0 + c|}{\sqrt{a^2 + b^2}}$.

1. Find the distances of the sides of the triangle with points at (1,2), (-1,-7), and (-3, 5).

2. Find the distances of the sides of the triangle with points at (1.2, 3.4), (3/7, -4), and $(0.00\overline{1}, -\sqrt{5000})$.

3. Solve the distance equation for x_2.

4. Compute the distance from the point (1,2) to the line $y = \dfrac{-3}{4}x + \dfrac{22}{7}$.

5. Compute the distance from the point $(-2.8\overline{65}, 5\sqrt{3})$ to the line $21x - 5y - 29 = 0$.

Circles

$x^2 + y^2 = r^2$ Equation of circle centered at origin with radius r.

$(x - h)^2 + (y - k)^2 = r^2$ Equation of circle centered at (h, k) with radius r.

$x^2 + y^2 + dx + ey + f = 0$ Reduced general second-degree equation of a circle.

A tangent line to a circle is a line that "skims" the edge of the circle at just one point. A tangent line to a circle does not enter the interior of a circle—it intersects the boundary of the circle at one point. A tangent line to a circle at a point on the circle and the line corresponding to the radius of the circle at that point are perpendicular. Thus, the perpendicular slope equation stated above is useful here. Given three points in the plane, the student should be able to find the equation of the circle through them.

1. Find the equation of the circle going through the points (0, 0), (1, 5), and (-5, 9). Sketch the circle and label the diagram including the center.

2. The center of a circle lies midway between (3.2, 7.9) and (-40, -4.1). The radius of the circle is twice the distance between these points. Find the equation of the circle.

3. Find the equation of the tangent line to the circle $(x-5)^2 + (y+9)^2 = 42$ at the point $(3, -9+\sqrt{38})$.

Parabolas

Given three non-collinear points, the student should be able to find the equation of the left-right or up-down parabola through them. Given any parabola, the student should be able to find or identify the vertex, focus, latus rectum, axis of symmetry, and directrix. The student should know the geometric significance of the directrix and how this relates to the eccentricity *e*. The student should be able to sketch a graph of any parabola with these components included. The student should be able to compute the length of the latus rectum. A line segment through a focus of the parabola that touches the parabola in two places is called a focal chord. The latus rectum is a special case of a focal chord, and its length is sometimes called the focal diameter. We mention in passing that the distance between the focus and the directrix is called the focal parameter, and it is half the length of the focal diameter. We won't discuss focal parameters again in this book.

$y = ax^2$, $x^2 = 4py$	Equations of a parabola with vertex at the origin.
$y = a(x-h)^2 + k$, $4p(y-k) = (x-h)^2$	Equations of parabola with vertex at (h, k).
$(-b/(2a), f(-b/(2a)))$	Coordinates of vertex of parabola $y = f(x) = ax^2 + bx + c$.
$x = ay^2$, $y^2 = 4px$	Equations of a left-right parabola with vertex at the origin.
$x = a(y-k)^2 + h$, $4p(x-h) = (y-k)^2$	Equations of left-right parabola with vertex at (h, k).

$(g(-b/(2a)), -b/(2a))$ Coordinates of vertex of parabola $x = g(y) = ay^2 + by + c$.

$\ell = |4p| = \left|\dfrac{1}{a}\right| = \dfrac{1}{|a|}$ Length ℓ of the latus rectum called the focal diameter.

1. Sketch the graph $y = -x^2 + 6x - 5$ by hand. Label all relevant parts listed above.

2. Sketch the graph $y = \left(x - \dfrac{1}{2}\right)^2 + 7$ by hand. Label all relevant parts listed above.

3. Find the equation of the up-down parabola going through the points (0,0), (1,5), and (3,-11).

4. Find the equation of the left-right parabola going through the points (1,1), (4,4), and (-1, 5).

5. Choose three points on the parabola $y = -3(x-4)^2 + 6$. For each point in turn, compute the distance from the selected point to the focus and from the selected point to the directrix. For each point in turn, place these two computed distances in a ratio with the distance to the focus in the numerator and the distance to the directrix in the denominator. Make any relevant comments.

6. Compute the length of the latus rectum for the parabola $x = 4(-2)y^2$.

Ellipses

$a > b$ Always true for any ellipse.

$c^2 = a^2 - b^2, e = c/a$ Equations for finding foci and eccentricity.

$\ell = \dfrac{2b^2}{a}$ Length ℓ of a latus rectum.

$\dfrac{x^2}{a^2} + \dfrac{y^2}{b^2} = 1, \; a \cdot b \neq 0$ Equation of wide ellipse centered at the origin.

$x = \pm\dfrac{a^2}{c}$ Equations of directrices of wide ellipse centered at origin.

$\dfrac{(x-h)^2}{a^2} + \dfrac{(y-k)^2}{b^2} = 1, \; a \cdot b \neq 0$ Equation of wide ellipse centered at (h, k).

$x = \pm\dfrac{a^2}{c} + h$ Equations of directrices of wide ellipse centered at (h, k).

$\dfrac{x^2}{b^2} + \dfrac{y^2}{a^2} = 1, \; a \cdot b \neq 0$ Equation of tall ellipse centered at the origin.

$$y = \pm \frac{a^2}{c}$$

Equations of directrices of tall ellipse centered at origin.

$$\frac{(x-h)^2}{b^2} + \frac{(y-k)^2}{a^2} = 1$$

Equation of tall ellipse centered at (h, k).

$$y = \pm \frac{a^2}{c} + k$$

Equations of directrices of tall ellipse centered at (h, k).

The student should be able to sketch the graph of any tall or wide ellipse and include the vertices, foci, major and minor axes, directrices, latera recta, and center. The student should be able to compute the eccentricity and the length of a latus rectum. The student should know the geometric significance of the directrices and how they relate to the eccentricity e. A line segment through a focus of the ellipse that touches the ellipse in two places is called a focal chord. A latus rectum is a special case of a focal chord. A line segment that goes through the center of the ellipse with endpoints on the ellipse is called a diameter of the ellipse. Conjugate diameters are such that a system of chords parallel to one conjugate diameter are bisected by the other conjugate diameter. Call one semi-diameter p and the other q. The first theorem of Apollonius (an ancient Greek mathematician) is that $p^2 + q^2 = a^2 + b^2$. The second theorem of Apollonius is that the area of the parallelogram of the tangent lines to the ellipse at the endpoints of two conjugate diameters is the constant ab.

1. Sketch the graph of the ellipse $\frac{x^2}{25} + \frac{y^2}{36} = 1$ by hand. Label all the relevant parts.

2. Sketch the graph of the ellipse $\frac{x^2}{49} + \frac{y^2}{36} = 1$ by hand. Label all the relevant parts.

3. Sketch the graph of the ellipse $\frac{(x-2)^2}{4} + \frac{(y+3)^2}{9} = 1$ by hand. Label all the relevant parts.

4. Sketch the graph of the ellipse $\frac{(x+4)^2}{16} + \frac{(y-1)^2}{4} = 1$ by hand. Label all the relevant parts.

5. Compute the eccentricities of each of the above ellipses by using the formula given above as well as by choosing a point on the ellipse and computing the ratio of the distance from that point to its focus to the distance from that point to its associated directrix.

6. Compute the length of a latus rectum for each of the above four ellipses.

7. Let the slope of a conjugate diameter be 1. Find the equation of the other conjugate diameter for each of the above four ellipses. Show that the two theorems of Apollonius are true for each pair of conjugate diameters in the above four ellipses.

Hyperbolas

$a, b \in \Re$
No magnitude relationship between a and b exists for the hyperbola.

$c^2 = a^2 + b^2$, $e = c/a$
Equations for finding foci and eccentricity.

$\ell = \dfrac{2b^2}{a}$
Length ℓ of a latus rectum.

$\dfrac{x^2}{a^2} - \dfrac{y^2}{b^2} = 1$
Equation of left-right hyperbola centered at origin.

$y = \pm\dfrac{b}{a}x$
Asymptote lines of left-right hyperbola centered at origin.

$x = \pm\dfrac{a^2}{c}$
Equations of directrices of left-right hyperbola centered at origin.

$\dfrac{(x-h)^2}{a^2} - \dfrac{(y-k)^2}{b^2} = 1$
Equation of left-right hyperbola centered at (h, k).

$y = \pm\dfrac{b}{a}(x-h) + k$
Asymptote lines of left-right hyperbola centered at (h, k).

$x = \pm\dfrac{a^2}{c} + h$
Equations of directrices of left-right hyperbola centered at (h, k).

$\dfrac{y^2}{a^2} - \dfrac{x^2}{b^2} = 1$
Equation of up-down hyperbola centered at origin.

$y = \pm\dfrac{a}{b}x$
Asymptote lines of up-down hyperbola centered at origin.

$y = \pm\dfrac{a^2}{c}$
Equations of directrices of up-down hyperbola centered at origin.

$\dfrac{(y-k)^2}{a^2} - \dfrac{(x-h)^2}{b^2} = 1$
Equation of up-down hyperbola centered at (h, k).

$$y = \pm \frac{a}{b}(x-h) + k \qquad \text{Asymptote lines of up-down hyperbola centered at } (h, k).$$

$$y = \pm \frac{a^2}{c} + k \qquad \text{Equations of directrices of up-down hyperbola centered at}$$
$$(h, k).$$

In the hyperbola equations, it is customary to place the a^2 with the positive term at the left and the b^2 with the negative term on the right although we understand that $a \leq b$ may hold as well as the mandatory relationship $a > b$ for ellipses. This is because the formula for c makes no magnitude demands on a and b in the hyperbola. The student should be able to sketch the graph of any up-down or left-right hyperbola by first drawing the representative rectangle and asymptote lines. The drawing should also include the vertices, foci, major and minor axes, directrices, latera recta, and center. A useful mnemonic I have found for students is the word "yud." It reminds the student that if the positive first term is y, then it is an up-down hyperbola. The student should be able to compute the eccentricity and the length of a latus rectum. The student should know the geometric significance of the directrices and how they relate to the eccentricity e. The student should be able to sketch in the conjugate hyperbola for a given hyperbola and be able to write its equation. The four foci of the pair of conjugate hyperbolas lie on a circle with the same center as the hyperbolas with radius c^2. Thus, if the center of the pair of conjugate hyperbolas is (h, k), then the equation of the circle is $(x - h)^2 + (y - k)^2 = c^2$, where $c = \sqrt{a^2 + b^2}$. If $a^2 = b^2$, then both the hyperbola and its conjugate are rectangular. A line segment going through a focus and touching the hyperbola in two places on the same branch is called a focal chord. The latus rectum is a special case of a focal chord.

The equation of the infinite family of left-right hyperbolas sharing the asymptote lines $y = \pm \frac{b}{a}(x-h) + k$ is $\dfrac{(x-h)^2}{\lambda^2 a^2} - \dfrac{(y-k)^2}{\lambda^2 b^2} = 1$, where $\lambda \in \Re, \lambda > 0$. The equation of the infinite family of up-down hyperbolas sharing the asymptote lines $y = \pm \frac{a}{b}(x-h) + k$ is

$\dfrac{(y-k)^2}{\lambda^2 a^2} - \dfrac{(x-h)^2}{\lambda^2 b^2} = 1$, where $\lambda \in \Re, \lambda > 0$. If we allow a^2 and b^2 to vary such that

$a^2 + b^2 = c^2$, while holding c^2 constant, then we get an infinite family of hyperbolas, along with their conjugates, that all have their foci on the circle $(x - h)^2 + (y - k)^2 = c^2$.

1. Sketch the graph of the hyperbola $\dfrac{x^2}{9} - \dfrac{y^2}{4} = 1$. Include all relevant parts.

2. Sketch the graph of the hyperbola $\dfrac{y^2}{9} - \dfrac{x^2}{25} = 1$. Include all relevant parts.

3. Sketch the graph of the hyperbola $\dfrac{(x-2)^2}{4} - \dfrac{y^2}{25} = 1$. Include all relevant parts.

4. Sketch the graph of the hyperbola $\dfrac{(y+6)^2}{49} - \dfrac{(x-3)^2}{25} = 1$. Include all relevant parts.

5. Compute the lengths of the latera recta in the above four hyperbolas.

6. Compute the eccentricity of the above four hyperbolas using the appropriate formula and by constructing a ratio of the distance from a point on the hyperbola to the focus to the distance from that point on the hyperbola to the associated directrix.

7. Write the conjugates of the above hyperbolas, and sketch the graph for one of the conjugates. Include all relevant parts.

8. Find an equation of a different hyperbola sharing the asymptote lines for the hyperbola in the third problem in this mini problem set. Explain how this hyperbola was obtained.

9. Find an equation of a different hyperbola sharing the same set of focus points as the hyperbola in the fourth problem in this mini set.

Sequences and Series

A sequence is a function having the natural numbers as its domain. A series is a summation of some or all of the terms of a sequence, which is specified by the notation. Oftentimes, sequence and series topics are included in a college algebra textbook, but they are usually omitted or treated superficially in the class because of time constraints. Then, the student has to wait until the second term of calculus to come back to the topic; however, the topic is treated much deeper in the calculus course. Beyond the calculus courses is a class called Discrete Mathematics, which delves deeply into integer sequences. Our review here will not be that deep, of course. The interested student is referred to Neil Sloane's Online Encyclopedia of Integer Sequences website. The web address is www.research.att.com/~njas/sequences. Incidentally, I met Neil Sloan when he gave a guest lecture at Oakland University when I was a student there. I submitted an integer sequence to his website to see if it existed. It did not, so he added it to his collection.

To denote a sequence, we use set notation because a sequence is a set of terms. Thus, $\{a_n\}$ is a set of terms with a general term a_n. The letter a identifies the sequence a_n like f identifies the function $f(x)$. The subscript n identifies the specific term under consideration. Thus, a_1 is the first term of the sequence. The sequence $\{a_n\}$ may be written out in open form as $a_1, a_2, a_3, a_4, ...$ when we specify the first few terms of the sequence and show it continues with the ellipsis (the three dots). The number of terms that we write down is arbitrary, but we will generally write only enough terms to be able to discern the sequence. We may also specify a sequence with an expression for the general term. Thus, $a_n = 2n + 5$ is an example with a general term of $2n + 5$. Another example is $a_n = \dfrac{5 \cdot \sqrt{n+1}}{n^2 - 2}$ with a general term of $\dfrac{5 \cdot \sqrt{n+1}}{n^2 - 2}$. These two examples are in closed form.

Adding a set amount to an initial quantity each time forms an arithmetic sequence or progression. Thus, the sequence 5, 9, 13, 17, 21, ... has an initial quantity a_1 (or first term) of 5; and it adds 4 to each term to produce the next. The set amount added each time is called the common difference, d. Thus, we write the formula for an arithmetic sequence as $a_n = a_1 + (n - 1) \cdot d$. Inserting a called for number of arithmetic means is an important concept, but we won't get into it in this book. We obtain d by subtracting any term from its successor. Thus, $d = a_{n+1} - a_n$.

The sum S_n of the first n terms of an arithmetic sequence is $S_n = \dfrac{n}{2}[2a_1 + (n-1) \cdot d] = \dfrac{n}{2}(a_1 + a_n)$. Summing a sequence is called a series, so summing an arithmetic sequence or progression yields an arithmetic series. We may wish to sum a certain number of consecutive terms in an arithmetic progression starting at some term other than the first. The notation $\displaystyle\sum_{i=n_1}^{n_2}(mi + b)$ allows us to do this. With this notation, we sum from the terms n_1 to n_2. When working problems, we just compute the series from term 1 to term n_2 and subtract the series from term 1 to term $n_1 - 1$. The common difference between the terms is m, which acts like the slope in the slope-intercept equation. The b is just like the b in $y = mx + b$.

Multiplying a term by a set amount each time to produce the next term forms a geometric sequence or progression. Thus, the sequence 1, 2, 4, 8, ... has an initial quantity a_1 (or first term) of 1; and it multiplies each term by 2 to produce the next. The set amount that we multiply by each time is called the common ratio, r. We thus write a geometric sequence as $a_n = a_1 \cdot r^{n-1}$. Inserting a called for number of geometric means is an important concept, but we won't get into it in this book. We obtain r by dividing any term by its predecessor. Thus, $r = a_n \div a_{n-1}$.

The sum S_n of the first n terms of a geometric sequence is $S_n = \dfrac{a_1(1-r^n)}{1-r}, r \neq 1$.

Summing a geometric sequence or progression yields a geometric series. We may wish to sum a certain number of consecutive terms in a geometric progression starting at some term other than the first. The notation $\sum_{k=n_1}^{n_2} a_1 \cdot r^{k-1}$ allows us to do this. With this notation, we sum from the terms n_1 to n_2. When working problems, we just compute the series from the terms 1 to n_2 and subtract from this the series from the terms 1 to $n_1 - 1$. The common ratio between the terms is r. Often, we want the geometric series to have infinitely many terms. Then, the formula for its sum is $S_\infty = \dfrac{a_1}{1-r}, |r| < 1$. If $|r| > 1$, then the series becomes unbounded.

Sometimes, we will have the first few terms of a sequence, and we must deduce the next term or the nth term. We may be able to deduce a pattern in the terms, which would enable us to deduce the general term with that pattern. However, we could just as well think of the sequence in a different way and arrive at a different general term that yields different subsequent terms. For example, we may use a system of equations to find a general term. Alternatively, we may observe that the sequence follows a known law from physics, chemistry, or economics. The use of different techniques will generate different general terms, but the computed initial terms will be identical to the given initial terms. Thus, how do we decide which next term is correct, and which general term is correct? In the abstract way we ask the question, we do not have a way to tell which is correct (if any). We must use our judgment to select the best answer from those we have at hand. If we do not think that some of the sequences at hand are correct from an experiential point of view, we may cast them from the list. Then, we can use our judgment to select from those that remain. See *Appendix A* for some sequence finding techniques.

The open and closed forms of sequences and series discussed thus far are explicit in nature. That is, they define precisely the terms in the sequence. An implicit form of defining sequences and series exists, and it is called the recursive form. We specify how to find the new term in terms of previous terms. Thus, the recursive equation $a_{n+1} = a_n + 2$ says that to find the new term a_{n+1}, we take the most recently computed term a_n and add 2 to it. In order to anchor a recursive formula to a single sequence or series, we must specify the necessary previous terms. Thus, the recursive equation $a_{n+1} = a_n + 2$, $a_1 = 5$ will generate the sequence 5, 7, 9, 11, ... A recursive formula such as $a_n = 3a_{n-1} + 5a_{n-3} - n^2 - 6n + 16$ requires that two terms having one intervening member be all specified. Thus, $a_n = 3a_{n-1} + 5a_{n-3} - n^2 - 6n + 16$, $a_1 = 2$, $a_2 = 11$, $a_3 = 5$ will generate the sequence 2, 11, 5, 1, 19, 26, 8, 23, 80, 136, 352, ... We can then ask such questions as what is a_{100}? However, in many instances of this sort, we may require a computer program to do the computations for us. It may be possible to convert a recursive equation into a general term equation to be able to compute specific terms directly, but that topic is taught in the more advanced Discrete Mathematics course. As one of my former instructors used to say, "Stay tuned for coming attractions!"

Recursive equations need not be restricted to a sequence of single numbers. We can have a sequence of ordered pairs as well. We will now provide a couple of examples. The first example is for finding consecutive powerful numbers, and the second example is for finding when a triangular number equals a square number. Later in the book, we will have need for a sequence of ordered triples called Pythagorean triples. Thus, a couple of simpler sequences of ordered pairs is necessary to include in the review chapter. Both of these are from number theory.

A powerful number is one in which each prime factor in its prime factorization appears at least as a square. Thus, every perfect square is a powerful number. In addition, every cube is a powerful number. Every higher order exponent of a single factor is likewise a powerful number. Finally, every multiplicative combination of these powerful numbers is also a powerful number.

Do consecutive powerful numbers exist? Yes, 8 and 9 is the first such consecutive pair (excluding 0 and 1). An infinite number of consecutive powerful numbers exists. It is known that if the ordered pair (x, y) satisfies Pell's equation $x^2 - 2y^2 = \pm 1$ with $x, y \in Z$, then $8x^2y^2$ and $(x^2 + 2y^2)^2$ are consecutive powerful numbers such that the larger is always a perfect square. Note that consecutive powerful numbers exist that do not follow the pattern in Pell's equation. Let (x, y) be $(1, 1)$, which satisfies Pell's equation. Then, $8x^2y^2$ and $(x^2 + 2y^2)^2$ compute to 8 and 9 respectively. Let such an ordered pair (x, y) satisfying Pell's equation be the first ordered pair of a sequence of ordered pairs satisfying Pell's equation. Thus, $(x, y) = (1, 1)$ becomes $(x_1, y_1) = (1,1)$. Then, a sequence of ordered pairs (x_n, y_n) satisfying Pell's equation is found with the following system of recursive equations:

$$\left.\begin{cases} x_{n+1} = x_n + 2y_n \\ y_{n+1} = x_n + y_n \end{cases}\right\} \text{ where } x_1 = 1 \text{ and } y_1 = 1.$$

Another set of recursive equations will be presented next. These equations describe the times when a square number is equal to a triangular number. The square numbers are far more famous than the triangular numbers. However, the triangular numbers can be thought of as the arrangement of bowling pins. We have one pin in the front. Thus, 1 is the first triangular number. We have two pins behind the first. The first pin, together with the two directly behind, forms a triangle. Hence, the second triangular number is 3. The next triangular number is the sum of the previous triangular number and the next row of pins. Hence, the third triangular number is 3 + 3 = 6. The next one is 6 + 4 = 10. The next one is 10 + 5 = 21. Yeah, I know; the bowling pins stop at the fourth triangular number. But, you get the idea. The equation $T_n = T_{n-1} + n$ where $n = 2, 3, 4, \ldots$ and $T_1 = 1$ describes the triangular numbers. We also have it that the nth triangular number is able to be computed with the equation $T_n = \dfrac{n(n+1)}{2}$. Of course, the square numbers are $S_n = n^2$. The recursive equation for the square numbers is $S_n = S_{n-1} + 2n - 1$, $n = 2, 3, 4 \ldots$ and $S_1 = 1$.

We can now present the recursive equations for when a triangular number is equal to a square number. These equations were derived in a recent year as of this writing by Armando Guarnaschelli of Argentina. In his equations, the x's are the lengths of the side of the square, and the y's are the lengths of the side of the triangle. Thus, these equations tell us that the xth square number and the yth triangular number are equal.

$$\begin{cases} x_n = 3x_{n-1} + 4y_{n-1} + 1 \\ y_n = 2x_{n-1} + 3y_{n-1} + 1 \end{cases} \text{ where } x_1 = 1 \text{ and } y_1 = 1.$$

1. Write the first five terms of the sequence $a_n = 3n - 11$.

2. Write the first four terms of the sequence $a_n = \dfrac{n}{3-n}$. What can we say about this sequence?

3. Find the general term for the sequence 14, 17, 20, 23, …

4. Find the general term for the sequence 3, 12, 27, 48, …

5. Write the first five terms of the sequence $a_n = -2 \cdot a_{n-1} - 7n^2, a_1 = 9$.

6. Find the general term for the sequence -10, 0, 16, 38, …

7. Find the general term for the sequence $\dfrac{3}{4}, \dfrac{4}{10}, \dfrac{5}{28}, \dfrac{6}{82}, \ldots$

8. Find the general term for the sequence 9, 30, 66, 123, …

9. Write the first four terms of the sequence $a_n = 2a_{n-1} + 3a_{n-2} - 5n, a_1 = 7, a_2 = 4$.

10. Write the first four terms of the sequence $a_n = n + \dfrac{1}{a_{n-1}}, a_1 = \dfrac{4}{7}$.

11. Sum the first ten terms of the sequence 1, 4, 7, 10, …

12. Sum the first fifteen terms of the sequence 6, 3, $\dfrac{3}{2}, \dfrac{3}{4}, \ldots$

13. What is a_{100} for the sequence 6, 13, 20, 27, …

14. What is a_{13} for the sequence 3, 6, 12, 24, …

15. Sum the sequence 7, 14, 21, 28, ... from a_{12} to a_{17}.

16. Sum the sequence $a_n = 4 \cdot \left(\dfrac{3}{11}\right)^{n-1}$ from the eleventh term to the twentieth term. What is the sum of all of the terms in this sequence?

17. Compute the ordered pairs (x_2, y_2) through (x_5, y_5) that satisfy Pell's equation $x^2 - 2y^2 = \pm 1$ when (x_1, y_1) is (1, 1) using the following system of recursive equations:

$$\begin{cases} x_{n+1} = x_n + 2y_n \\ y_{n+1} = x_n + y_n \end{cases}$$

18. Using the ordered pairs generated in problem 17, compute the corresponding pairs of consecutive powerful numbers with the expressions $8x^2y^2$ and $(x^2 + 2y^2)^2$.

19. Write out the first five (x_n, y_n) ordered pairs of numbers that yield when a square number is equal to a triangular number. Compute these five pairs of values using the direct computation formulas to confirm that they are equal.

Combining These Ideas

The student should be able to combine any of these concepts en route to completing any requested task within the scope of these ideas. However, combining some of these concepts could result in a fourth degree (or even higher) polynomial function, which may be too difficult for the student to handle at this point. Sometimes, we have ways to avoid the fourth power. Some of these methods will be shown at the appropriate places in the text.

Exercises

Even though the instructor is not expected to teach this chapter, any exam the instructor proctors may contain questions like those found in this review chapter. In addition, some of the concepts in this review chapter were not in the college algebra course but were placed here to prepare the student for later topics in this book. Thus, make sure you know how to do these problems.

1. Find the intersection point of the line $4x - 3y + 11 = 0$ and $y - 5 = 17(x - 5/11)$.

2. Find the equation of the line perpendicular to $y = 3x - 2$ and goes through the focus of the parabola $7y - 5\sqrt{7} \cdot x^2 + 4.8x - 6\frac{16}{19} = 0$.

3. Find the intersection points between the ellipse $2x^2 - 5x + 3y^2 + 11y - 40 = 0$ and the line $y = 2.8x + 2$ analytically. Sketch a nice graph including the vertices, foci, and center of the ellipse.

4. Given the points (5,-2), (9, 11), and (-4, 9), find the equations for the circle, up-down parabola, and left-right parabola that goes through them. Graph each one separately. For each parabola, include the vertex, axis of symmetry, directrix, focus, and latus rectum. Compute the length of the latus rectum in each case.

5. To the three points in question 4, add the point (3, 7), and find the equations of the wide ellipse and left-right hyperbola that go through all four of the points. Sketch each one on a separate graph; and identify the vertices, foci, directrices, representative rectangle and asymptote lines (for the hyperbola only, of course), latera recta, and center.

6. Solve $\dfrac{2x-5}{x^2-4} + \dfrac{11}{2-x} = \dfrac{x^2+4}{x^2-3x-10}$ analytically.

7. Solve $\dfrac{3}{4x-5} + \sqrt{3x} = 5.7$ analytically.

8. Given a circle and a chord, a triangle is to be constructed such that two of its vertices are the endpoints A and B of the chord. The third vertex of the triangle is some other point C on the circle. The larger part of the circle that is cut by chord we call the major arc, and the smaller part of the circle cut by the chord we call the minor arc. It is a fact that the isosceles triangle is the largest possible triangle in terms of both area and perimeter that can be so constructed. Let the circle be $x^2 + y^2 = 25$ and the chord be $y = x + 1$. Find the point C on the major arc that makes the maximum triangle. Find the point C' on the minor arc that makes the maximum triangle. Compute the area and perimeter of both of these maximum triangles.

Theorem Highlight: The Pappus Line

The theorem of Pappus, named after Pappus of Alexandria (fl. 300 – 350), an ancient Greek mathematician, considered in this highlight states that given any two lines and any three points on each, the meets of the cross-joins lie on a straight line.

We first need to understand what joins and cross-joins are. If we rank the three given points on each line by when they occur from left to right according to their x-coordinates, then we do not join pairs of points with the same rank. All other joins of point pairs from the two lines are joins. Thus, the point on ℓ_1 with a rank of 1 will join with the points on ℓ_2 with ranks of 2 and 3. The point on ℓ_1 with a rank of 2 will join with the points on ℓ_2 with ranks of 1 and 3. The point on ℓ_1 with a rank of 3 will join with the points on ℓ_2 with ranks of 1 and 2. This makes for a total of six joins. To have a consistent way of referring to these points, we will agree to call the points on ℓ_1 with a P and the points on ℓ_2 with a Q. Combining the ranks with these letters, we have the points P_1, P_2, P_3, Q_1, Q_2, and Q_3. We can label these joins with a notation such as join(P_1,Q_2). As an additional agreement, we agree to put the point from line one in the first position and the point from line two in the second position. The notation join(P_1,Q_2) denotes the join from line one's first point to line two's second point. A cross-join can be labeled with a notation such as join(P_1,Q_2)×join(P_2,Q_1). The other two cross-joins are join(P_1,Q_3)×join(P_3,Q_1) and join(P_2,Q_3)×join(P_3,Q_2).

Since two points determine a line, a join is a line. So, finding a join means we find the equation of the line going through the two points denoted in the join notation we have just developed. A cross-join is a pair of lines, which is equivalent to a system of linear equations in two variables. The meet of a cross-join is equivalent to an intersection of two lines. Thus, finding the meet of a cross-join means we find the coordinates of the intersection point of the two lines denoted in the cross-join notation. With three meets of the cross-joins, we have three ordered pairs to denote these meets. According to this theorem by Pappus, the meets of the three cross-joins lie on a straight line.

Let us make up an example and verify that the meets of the cross-joins lie on a straight line, which is, of course, the Pappus Line. Let the first line be $\ell_1 : y = \frac{1}{4}x - 3$, and let line two be $\ell_2 : y = \frac{-2}{11}x + 5$. Let the x-coordinates on $\ell_1 \in \{1,3,6\}$ and $\ell_2 \in \{2,4,9\}$. Our strategy will be to use two of the meets of the cross-joins to establish the Pappus line, and use the third meet of the cross-joins as a verification point. If the three points are collinear, then we will have verified the Pappus line for this situation. The first thing we should do is to find the y-coordinates of the six given points.

$$P_1(x, y) = \left(1, \frac{1}{4} - 3\right) = \left(1, \frac{-11}{4}\right) \qquad Q_1(x, y) = \left(2, \frac{-4}{11} + 5\right) = \left(2, \frac{51}{11}\right)$$

$$P_2(x, y) = \left(3, \frac{3}{4} - 3\right) = \left(3, \frac{-9}{4}\right) \qquad Q_2(x, y) = \left(4, \frac{-8}{11} + 5\right) = \left(4, \frac{47}{11}\right)$$

$$P_3(x, y) = \left(6, \frac{3}{2} - 3\right) = \left(6, \frac{-3}{2}\right) \qquad Q_3(x, y) = \left(9, \frac{-18}{11} + 5\right) = \left(9, \frac{37}{11}\right)$$

Next, we can find the equations of the six joins.

$$\text{join}(P_1, Q_2) = \text{join}\left(\left(1, \frac{-11}{4}\right), \left(4, \frac{47}{11}\right)\right)$$

$$m = \frac{\frac{47}{11} - \frac{-11}{4}}{4 - 1} = \frac{103}{44}$$

$$y - y_1 = m(x - x_1)$$

$$y - \frac{47}{11} = \frac{103}{44}(x - 4) \quad \text{We used } Q_2 \text{ here, but } P_1 \text{ will yield the same line.}$$

$$y = \frac{103}{44}x - \frac{56}{11}$$

$$\text{join}(P_1, Q_3) = \text{join}\left(\left(1, \frac{-11}{4}\right), \left(9, \frac{37}{11}\right)\right) \Rightarrow y = \frac{269}{352}x - \frac{1237}{352}$$

$$\text{join}(P_2, Q_1) = \text{join}\left(\left(3, \frac{-9}{4}\right), \left(2, \frac{51}{11}\right)\right) \Rightarrow y = \frac{-303}{44}x + \frac{405}{22}$$

$$\text{join}(P_2, Q_3) = \text{join}\left(\left(3, \frac{-9}{4}\right), \left(9, \frac{37}{11}\right)\right) \Rightarrow y = \frac{247}{264}x - \frac{445}{88}$$

$$\text{join}(P_3, Q_1) = \text{join}\left(\left(6, \frac{-3}{2}\right), \left(2, \frac{51}{11}\right)\right) \Rightarrow y = \frac{-135}{88}x + \frac{339}{44}$$

$$\text{join}(P_3, Q_2) = \text{join}\left(\left(6, \frac{-3}{2}\right), \left(4, \frac{47}{11}\right)\right) \Rightarrow y = \frac{-127}{44}x + \frac{174}{11}$$

The next thing we need to do is to find the meets of the three cross-joins. When we have these three intersection points, we can use two of them to find the Pappus line. The third point will then be placed into the Pappus line equation, and we should get an identity if we did our work correctly. If we get a contradiction, then something in our work is wrong because the theorem states that these three points all lie on a straight line.

$$\text{join}(P_1, Q_2) \times \text{join}(P_2, Q_1) \Rightarrow$$

$$\left\{ \begin{array}{l} y = \dfrac{103}{44}x - \dfrac{56}{11} \\[2mm] y = \dfrac{-303}{44}x + \dfrac{405}{22} \end{array} \right\}$$

$$\therefore (x, y) = \left(\frac{517}{203}, \frac{7779}{8932} \right) \quad \text{This is one of the meets of the cross-joins.}$$

$$\text{join}(P_1, Q_3) \times \text{join}(P_3, Q_1) \Rightarrow (x, y) = \left(\frac{3949}{809}, \frac{15387}{71192} \right)$$

$$\text{join}(P_2, Q_3) \times \text{join}(P_3, Q_2) \Rightarrow (x, y) = \left(\frac{5511}{1009}, \frac{2367}{44396} \right)$$

These are the other two cross-join meets.

Now, we can use two of these cross-join meets to find the Pappus line, and we can check our work with the third point. Let us use the first two points to find the Pappus line.

$$m = \frac{y_2 - y_1}{x_2 - x_1} = \frac{\dfrac{15387}{71192} - \dfrac{7779}{8932}}{\dfrac{3949}{809} - \dfrac{517}{203}} = \frac{-543}{1936}$$

$$y - y_1 = m(x - x_1) \Rightarrow y - \frac{7779}{8932} = \frac{-543}{1936}\left(x - \frac{517}{203} \right)$$

$$\therefore y = \frac{-543}{1936}\left(x - \frac{517}{203} \right) + \frac{7779}{8932} \quad \text{This is the equation of the Pappus line.}$$

Checking the third point should yield an identity, which shows that the three points are collinear. We just punch this verification into a graphing calculator to check for now.

$$\frac{2367}{44396} \equiv \frac{-543}{1936}\left(\frac{5511}{1009} - \frac{517}{203}\right) + \frac{7779}{8932} = \frac{2367}{44396} \checkmark$$

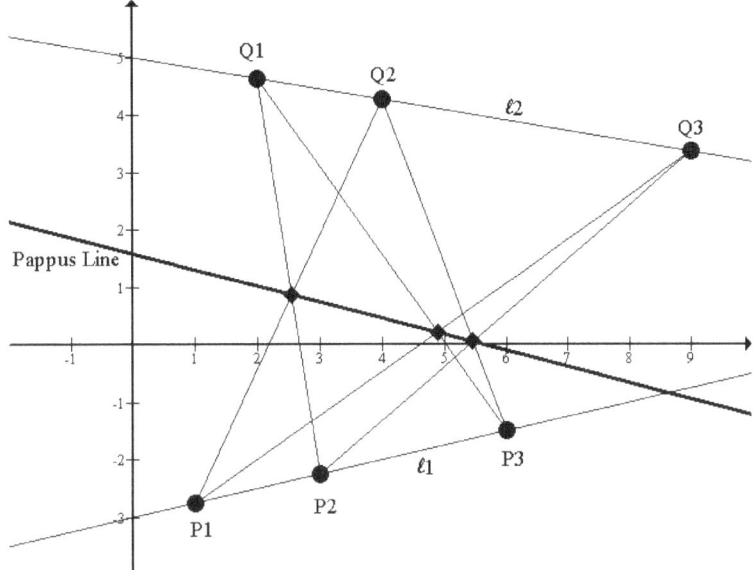

Figure 1: The Pappus Line

Our situation appears in the diagram labeled Figure 1. Note that the meets of the cross-joins are shown with a diamond shape while the given points are shown with a circle shape. The Pappus Line is labeled as are the given points and lines. However, the diagram would be terribly cluttered with coordinates and equations, so aesthetics dictates they be left out.

Exercises

1. Find the equation of the Pappus line for the following circumstances:

$$\ell_1 : y = 2x + 3, x \in \{-5, 0, 5\}$$
$$\ell_2 : y = -3x - 11, y \in \{-5, 0, 3\}$$

2. Find the equation of the Pappus line for the following circumstances:

$$\ell_1 : y = \frac{1}{3}x - \frac{3}{7}, y \in \{1.1, 1.2, 1.3\}$$

$$\ell_2 : y = 19.11x - 4.5, y \in \left\{-5.1, 0.002, 6\frac{4}{9}\right\}$$

35

3. Find the equation of the Pappus line for the following circumstances:

$$\ell_1 : y = \frac{2}{\sqrt{3}} x + \sqrt{\frac{3}{11}}, x \in \{\sqrt{3}, 2\sqrt{5}, 3\sqrt{11}\}$$

$$\ell_2 : y = \sqrt{5}x + \sqrt{6}, x \in \{-3, -2.07, 100\sqrt{7}\}$$

4. Writing in mathematics: Construct and analyze a problem given ℓ_1 (which the student makes up) and the Pappus line (also which the student makes up) and what is required in terms of the points on the lines to find ℓ_2.

Fractions, Rational Points, and Denesting Radicals

The title of this chapter is *Fractions, RationalPoints, and Denesting Radicals*. Thus, the first item on our agenda is fractions. What is a fraction? A fraction is also known as a rational number. A rational number is defined as the quotient $\dfrac{p}{q}$ such that both p and q are integers with $q \neq 0$ and $gcf(p, q) = 1$. Note that any integer is a rational number because $q = 1$. Also note that a quotient of integers $\dfrac{m}{n}$ where $gcf(m, n) = w$ can be reduced to the form $\dfrac{p}{q}$ by dividing both m and n by w. Thus, $\dfrac{m \div w}{n \div w} = \dfrac{p}{q}$.

We need to have a way to handle tougher fractions than we may have been previously accustomed to dealing with. One such example occurred in the Pappus line discussion. We needed to compute the expression $\dfrac{-543}{1936}\left(\dfrac{5511}{1009} - \dfrac{517}{203}\right) + \dfrac{7779}{8932}$ to see if it was equal to $\dfrac{2367}{44396}$. The note there said to just punch it into a graphing calculator for now, but it would be nice to be able to crunch this out by hand. Thus, this will be our first example problem in this chapter.

The first thing we need to do is to perform what is inside the parentheses, which is just following PEMDAS. We must assume that the reader knows how to do basic arithmetic, which includes adding, subtracting, multiplying, and dividing integers of a reasonable size. Thus, these computations will be left for the graphing calculator (also called the grapher) or a CAS (Computer Algebra System) to do. We also must assume that the reader can find the prime factorization of a given integer of a reasonable size by dividing by the successive primes 2, 3, 5, 7, 11, 13, 17, 19, 23, 29, … while also checking for repeated factors. When a factor is found, we take the integer quotient as the new dividend. Of course, when the quotient < dividend, we stop the process and conclude that the last dividend is a factor. We compute inside the parentheses.

$$\dfrac{5511}{1009} - \dfrac{517}{203} \qquad \text{Note that 1009 is prime, and } 203 = 7 \cdot 29$$

$$\dfrac{5511 \cdot 203 - 517 \cdot 1009}{1009 \cdot 203} = \dfrac{597080}{203 \cdot 1009}$$

We know the factors of the denominator, so it is enough to check divisibility of the numerator by these factors (i.e. we do not need to factor 597080). We find that 597080 does not divide evenly by 7, 29, or 1009. We can now move to the rest of the expression.

$$\frac{-543}{1936}\left(\frac{5511}{1009}-\frac{517}{203}\right)+\frac{7779}{8932}=\frac{-543}{1936}\left(\frac{597080}{203\cdot1009}\right)+\frac{7779}{8932}$$

Note that $543=3\cdot181$. None of the factors of 543 cancels with any of the factors of $203\cdot1009$, but gcf(1936,597080) = 8. If the last three digits of an integer is a multiple of 8, then the integer is a multiple of 8. So, we can reduce by that factor.

$$\frac{-543}{1936}\left(\frac{597080}{203\cdot1009}\right)+\frac{7779}{8932}=\frac{-543}{242}\left(\frac{74635}{203\cdot1009}\right)+\frac{7779}{8932}$$

We now have two fractions added together, so we will factor the denominators to get a LCD.

$$242\cdot203\cdot1009=(2\cdot11\cdot11)(7\cdot29)(1009)$$
$$8932=(2\cdot2\cdot7\cdot11\cdot29)$$
$$LCM(242\cdot203\cdot1009,8932)=2\cdot2\cdot7\cdot7\cdot11\cdot11\cdot29\cdot1009=99{,}136{,}268$$

We see that the first denominator is missing a 2, and the second denominator is missing $11\cdot1009$. So we can convert the fractions and perform the arithmetic on the numerators over the LCD.

$$\frac{-543\cdot74635\cdot2+7779\cdot11\cdot1009}{2\cdot2\cdot7\cdot11\cdot11\cdot29\cdot1009}=\frac{5285511}{2\cdot2\cdot7\cdot11\cdot11\cdot29\cdot1009}=$$

$$\frac{7\cdot11\cdot29\cdot2367}{2\cdot2\cdot7\cdot11\cdot11\cdot29\cdot1009}=\frac{2367}{44396}$$

The problem of the last fraction in its bare integer form is most interesting. Suppose we were asked to reduce the fraction 5285511/99136268, which will be our next example. Without resorting to the cumbersome task of factoring, we can find the *gcf*. This can be handy for nasty integers in the fraction. To do this, we long divide the two integers putting the smaller number in the divisor spot. The *gcf* is also a factor of the remainder. Doing the long division yields a quotient of 18 with a remainder of 3997070—a seven-digit number. This is the same number of digits that the number 5285511 has. So we will long divide those two next, which yields the following computation 5285511/3997070 = 1 R 1288441. Repeating the process again, we compute 3997070/1288441 = 3 R 131747, and we are down to six digits. When it divides evenly, we will know that we have found the *gcf*. Continuing, we get 1288441/131747 = 9 R 102718. The next one is 131747/102718 = 1 R 29029, and we are down to five digits. Pressing on with the process yields 102718/29029 = 3 R 15631. Another iteration yields 29029/15631 = 1 R 13398. We also note that since the *gcf* is contained in both of the numbers of the fractions anywhere along the process, the *gcf* is contained in the difference of any two numbers obtained because it could just be factored out directly. Thus, 15631 – 13398 = 2233. Finally, we compute 13398/2233 = 6 R 0. Getting the remainder of zero means the last divisor is the *gcf*.

This process performed independently by two people will usually see two routes through to the *gcf* since a subtraction or long division may be performed at any step and also since the selection of what the next fraction is going to be is something of an art as well. Of course, the goal is to find the *gcf* in as few steps as possible, but any route that gets the job done is good enough if it is not too protracted. Another observation is that even a route that is a bit protracted is preferred over a monstrous factoring because a monstrous factoring could be prohibitive by hand; however, the grapher can factor larger integers than a person can rather quickly. For example, I asked mine (a TI-89) to factor 1587946254779, and it answered with $199 \cdot 9311 \cdot 857011$ in just a few seconds. Without technological assistance, it would take a person considerably longer than that to perform the feat. Completing the example, we divide the original numerator and denominator by the *gcf* to obtain the following:

$$\frac{5285511}{99136268} = \frac{5285511 \div 2233}{99136268 \div 2233} = \frac{2367}{44396}$$

Why does this procedure work? We try not to do things in math because someone says to do it this way or that way. We would like to know why we do what we do. To that end, the following is a rationale for why this procedure gets us the *gcf* without factoring the two numbers.

Suppose we want to reduce some improper fraction $\frac{a}{b}$, but the integers a and b may be too large or unwieldy to factor. We can assume that a *gcf* exists because if the two numbers are relatively prime, the *gcf* is 1. If the two numbers are not relatively prime, then the *gcf* > 1. Either way, the *gcf* exists and is an element of the positive integers. Having said this, we can now write the expressions $a = c \cdot gcf$ and $b = d \cdot gcf$. Then, the fraction becomes $\dfrac{c \cdot gcf}{d \cdot gcf}$. We can now perform long division on the fraction. Since $a > b$, we can claim that the quotient is some positive integer that we will call q.

$$
\begin{array}{r}
q \\
d \cdot gcf \overline{\smash{)}\, c \cdot gcf} \\
-q \cdot d \cdot gcf \\
\hline
c \cdot gcf - q \cdot d \cdot gcf
\end{array}
$$

We look at the bottom line for the remainder and observe that the *gcf* is contained in both terms, so it could be factored out. Since this is true, the *gcf* is always a factor of the remainder of a long division problem that is based on an improper fraction of two positive integers. Carrying this idea further, as we progress in the process of finding the *gcf*, the remainder becomes one of the integers in the next fraction. The other number already had the *gcf*. Thus, the next iteration finds a new remainder that also has this *gcf*. The terminal point for this argument is when we get a zero remainder. However, even zero has the *gcf* because $0 \cdot gcf = 0$. In the case of a proper fraction, we can invert the fraction, and then find the remainder. So it is immaterial whether the original fraction is proper or improper in terms of utilizing this procedure to find the *gcf*. This demonstration is not a rigorous proof, but it does provide a rationale for why this method of reducing fractions works. Next, we will look at Midy's theorem, named after E. Midy, a French mathematician.

Midy's theorem (1836) states that if a reduced fraction p/q where q is a prime number has an even number of digits in its repeating block, then the first half of the block added to the second half of the block yields all nines. A few examples illustrating Midy's theorem are the following:

- $4/11 = 0.\overline{36}$, then observe that there are two (an even number) digits in the repeating block. Hence, take the first digit, and add it to the second digit to get $3 + 6 = 9$. This string of nines is one member long.

- $5/7 = 0.\overline{714285}$, then observe that there are six (an even number) digits in the repeating block. Hence, take the first three digits, and add it to the second three digits to get $714 + 285 = 999$. This string of nines is three members long.

- $4/23 = 0.\overline{3043478260869565217391}$, then observe that there are 22 (an even number) digits in the repeating block. Hence, take the first 11 digits, and add it to the second 11 digits to get $30,434,782,608,695 + 69,565,217,391 = 99,999,999,999$. This string of nines is eleven members long.

Can we do this in reverse? In other words, can we take a string of nines (an even or an odd number of them), and split them into a sum. Then, can we append the two portions together to construct a repeating block to yield a fraction meeting the requirements of Midy's theorem? Let's give it a try. $9 = 8 + 1 \Rightarrow 0.\overline{81} = 81/99 = 9/11$. This worked because the denominator is prime. How about $99 = 23 + 76 \Rightarrow 0.\overline{2376} = \dfrac{24}{101}$. This also worked because the denominator is prime. Let us try a more elaborate example:

$$9999999 = 1234567 + 8765432 \Rightarrow 0.\overline{12345678765432} = \dfrac{1234568}{10000001} = \dfrac{2^3 \cdot 154321}{11 \cdot 909091}$$

We see that it doesn't work in reverse because the denominator in this last example is not prime. Hence, Midy's theorem is one-way only. We start with a fraction as specified in the theorem. Then, we will see the magic of the marching nines.

Midy's theorem has an extension. We start with a reduced fraction p/q where q is a prime number as before. Then, consider the length £ of the repeating block in the decimal expansion of this fraction. We factor £ into all possible cases of two factors $f_1 > 1$ and f_2 such that $f_1 \cdot f_2 = $ £. The extended Midy's theorem states that if the repeating block of the decimal expansion of p/q is divided into $f_1 > 1$ blocks of length f_2, then the sum of these f_1 blocks will be a multiple k of $10^{f_2} - 1$.

Our example for Midy's theorem extended will be to analyze fully the fraction $\dfrac{9}{19}$. Its decimal expansion is $0.\overline{473684210526315789}$. Thus, we have the overall block length £ equal to 18. We construct the table of possible factors f_1 and f_2 to yield the following:

f_1	f_2	$f_1 \cdot f_2$	Expression	Sum	$10^{f_2} - 1$	k
2	9	18	473684210+526315789	999999999	999999999	1
3	6	18	473684+210526+315789	999999	999999	1
6	3	18	473+684+210+526+315+789	2997	999	3
9	2	18	47+36+84+21+05+26+31+57+89	396	99	4
18	1	18	4+7+3+6+8+4+2+1+0+5+2+6+3+1+5+7+8+9	81	9	9

We think of the prime numbers as kind of funny, like they don't really fit in anywhere because they are so strange. Composite numbers are more pleasing to us because we can do more with them. For example, a fraction with two composite numbers will more probably reduce than one having a prime number—from our experience anyway.

Running into a prime number seems to cause trouble. But look at that chart! The prime numbers as denominators cause the most striking results in Midy's extended theorem. Any prime denominator is capable of such display as long as the length £ of the repeating block has a few factors f_1 and f_2 into which it can decompose. Midy's extended theorem reveals a remarkably beautiful pattern hidden behind the ugliness of the prime numbers.

Our next topic of discussion is called the dyadic rational numbers. These are fractions of the form $\dfrac{a}{2^b}$ where a is an element of the integers and b is an element of the whole numbers. Observe that the markings on a ruler are dyadic fractions rather than decimal fractions, which are of the form $\dfrac{a}{10^b}$ with the same restrictions as before. For example, three-sixteenths of an inch is a dyadic fraction. Also notice that a gallon is subdivided into dyadic fractions, except that a fifth of liquor is a fifth of a gallon. We do not have a word for fifth of a gallon other than a fifth. The measurements of $1/2^b$ of a gallon where b is 0, 1, 2, 3, 4, 7, and 8 are called gallons, half-gallons, quarts, pints, cups, ounces, and tablespoons, respectively.

We can add or subtract any two dyadic fractions, and the result is a dyadic fraction. We can also multiply any two dyadic fractions, and the result is a dyadic fraction. However, division is where this process ends. Dividing two dyadic fractions will not necessarily produce another dyadic fraction.

A property of dyadic fractions is that the decimal expansion of a dyadic fraction will always terminate. This is contrasted with a fraction that has a repeating decimal expansion.

Another base exists for denominators having a terminating decimal expansion. This is base 5. Hence, fractions of the form $\dfrac{a}{5^b}$ with the same restrictions as before will terminate. We may add, subtract, and multiply fractions of this type, and the result will be another fraction of this type. Hence, any result from these three operations on a pair of fractions on this type yields a fraction having a terminating decimal expansion.

Let us make a set of the fractions having terminating decimal expansions: $\left\{ \dfrac{a_1}{2^{b_1}}, \dfrac{a_2}{5^{b_2}} \right\}$ where the subscripted variables may not necessarily be equal. We may add, subtract, and multiply any two fractions of this form, and we will get a fraction having a terminating decimal expansion. However, the form of the fraction may no longer match these two forms. We may expand the set to the following: $\left\{\dfrac{a_1}{2^{b_1}}, \dfrac{a_2}{5^{b_2}}, \dfrac{a_3}{2^{b_3} \cdot 5^{b_4}}\right\}$ where the

subscripted variables may not necessarily be equal. Thus, we will get a new fraction that fits one of these three forms when we add, subtract, or multiply any two fractions from

this set. However, we may shorten the set membership to the following: $\left\{\dfrac{a}{2^{b_1} \cdot 5^{b_2}}\right\}$ with

the same restrictions as before. Then, if $b_1 > b_2 = 0$, we will have a dyadic fraction on our hands. But, if $0 = b_1 < b_2$, then we will have a fraction with a power of 5 in its denominator. Lastly, if neither b_1 nor b_2 are zero, then some product of powers of two

and five will be in the denominator. Thus, any fraction belonging to the set $\left\{\dfrac{a}{2^{b_1} \cdot 5^{b_2}}\right\}$

will have a terminating decimal expansion. If we add, subtract, or multiply any two members of this set, we will get another member in this set.

It is a fact that only members of this set of fractions have terminating decimal expansions. If a reduced fraction is not a member of this set, it will have a repeating decimal expansion. Thus, the way to tell if a reduced fraction has a repeating decimal expansion is to inspect the prime factorization of its denominator. If anything other than twos or fives are present, then the decimal expansion will repeat. If only twos and fives are present, then the decimal expansion will terminate.

We can long divide a fraction to get its decimal expansion, and if we did not know in advance whether it terminated or repeated, we would find out soon enough which type of decimal expansion it had. This is because the set of possible remainders is bounded by the size of the divisor. For example, the fraction 1/3 has at most two remainders (a remainder of 0 means that the decimal terminated, and a remainder of three means that we could have upped the quotient by one to get a remainder of 0). This means that we cannot expect to get a repeating block of more than two digits. It turns out that fractions with three as a denominator have just one digit in their repeating blocks. However, fractions with seven as a denominator have the maximum possible number of digits in their repeating blocks, which is six. Fractions with eleven have just two digits in their repeating blocks. So denominators of eleven under-perform in terms of having a potential of ten digits in their repeating blocks but only weighing in at two digits.

Suppose we want to convert a repeating decimal to a fraction. How do we do that? It is best to have just the decimal portion of the given number. For example, if we are given the number $12.0\overline{103}$, we just take the decimal portion of the number $0.0\overline{103}$ to convert that to a fraction. Armed with that, we can construct the mixed number 12 + that fraction and do whatever else we need to do from there. Thus, we are really just interested in the decimal portion of a number for conversion to a fraction. Here is how to do it.

Step 1: Set the decimal expansion equal to x.
Step 2: Above the equation in step one, construct another equation: multiply both sides of the equation in step 1 by 10 raised to the power of the number of digits in the repeating block.
Step 3: Subtract the bottom equation from the top equation. Observe that the repeating portions line up, and so they drop out leaving a terminating decimal.
Step 4: Solve the equation obtained in step 3 for x.
Step 5: By the transitive property of mathematics (if $a = b$ and $b = c$, then $a = c$), we have that the original repeating decimal expansion $= x$ (i.e. if $a = b$), and we have it that $x =$ some fraction in step 4 (i.e. and $b = c$), then we can conclude that the original decimal expansion is equal to the fraction in step 4 (i.e. then $a = c$).

We illustrate the method with an example. Convert $0.0\overline{103}$ to a fraction. We observe that there are two digits in the repeating block. Thus, we compute ten raised to the second power as $10^2 = 100$. Hence, when we get to step 2, this is the number we will need.

$$
\begin{array}{rll}
100x & = & 1.0\overline{303} \quad \text{Step 2} \\
\underline{-x} & = & -0.0\overline{103} \quad \text{Step 1 and Step 3} \\
99x & = & 1.02 \qquad \text{Step 4's equation to solve}
\end{array}
$$

We solve the equation in step 4 as follows:

$99x = 1.02$

$$x = \frac{1.02}{99} = \frac{1.02}{99} \cdot \frac{100}{100} = \frac{102}{9900} = \frac{102 \div gcf(102,9900)}{9900 \div gcf(102,9900)} = \frac{102 \div 6}{9900 \div 6} = \frac{17}{1650}$$

Thus, we may conclude by the transitive property that $0.0\overline{103} = \dfrac{17}{1650}$. As a check,

we can long divide 1650 into 17 to see that we get $0.0\overline{103}$ as a quotient.

The next topic of discussion is the mediant. Let the two fractions $\dfrac{a}{c} < \dfrac{b}{d}$ be two reduced fractions with both c and d positive. Then, the mediant of these two fractions is $\dfrac{a+b}{c+d}$. One property of the mediant is that it is strictly between the two originating fractions. Hence, $\dfrac{a}{c} < \dfrac{a+b}{c+d} < \dfrac{b}{d}$. Another property of the mediant is that it is the simplest fraction between the two originating fractions. The word simplest here means the fraction with the smallest denominator.

We will next discuss Farey sequences. Farey sequences are named after John Farey (1766 – 1826), an English civil engineer and mathematician. However, one of my sources lists him as a geologist. Farey sequences are labeled as F_n with n representing the order of the sequence. The first Farey sequence is $F_1 = \left\{\dfrac{0}{1}, \dfrac{1}{1}\right\}$. Each new Farey sequence is obtained by computing the mediant of each neighboring pair of fractions in the previous Farey sequence and inserting the results (maintaining the ranking from smallest to largest) into the previous Farey sequence. In addition, we eliminate from contention those mediants that have a denominator greater than the order n of the Farey sequence. Thus, to find F_2, we compute the mediant of the two fractions $\dfrac{0+1}{1+1} = \dfrac{1}{2}$. This is reduced with the denominator 2 not exceeding the order $n = 2$, so we insert this result into F_1 while maintaining the ranking from smallest to largest. Thus, $F_2 = \left\{\dfrac{0}{1}, \dfrac{1}{2}, \dfrac{1}{1}\right\}$. Next, we can find F_3 by finding the mediants of the first pair of neighbors and the second pair of neighbors, and we just insert these reduced fractions in the appropriate places in the growing list. Doing this yields $F_3 = \left\{\dfrac{0}{1}, \dfrac{1}{3}, \dfrac{1}{2}, \dfrac{2}{3}, \dfrac{1}{1}\right\}$. None of the computed mediants had to be thrown out.

The next order Farey sequence is F_4. This is the first list when we have to remove from contention at least one of our computed mediants. The mediant between $\dfrac{0}{1}$ and $\dfrac{1}{3}$ is $\dfrac{1}{4}$. This mediant is a keeper because the denominator is not larger than the order 4 we are building. However, the next mediant is between $\dfrac{1}{3}$ and $\dfrac{1}{2}$, which is $\dfrac{2}{5}$. Here, the denominator of 5 exceeds the order 4 of the Farey sequence we are building. Thus, we cast this one aside—it will not be included in F_4. One other candidate mediant will be cast aside for F_4. The other mediant will be a keeper. Our final result yields $F_4 = \left\{\dfrac{0}{1}, \dfrac{1}{4}, \dfrac{1}{3}, \dfrac{1}{2}, \dfrac{2}{3}, \dfrac{3}{4}, \dfrac{1}{1}\right\}$.

Another way to view the next Farey sequence is that the next larger denominator will appear. Thus, F_5 will see fifths appear, and F_6 will see sixths appear. If our next up order is a prime number p, then we will get p new terms from the previous Farey sequence. If our next up order q is not prime, then we will get fewer than q new terms. However, we will always get at least two new terms because $\dfrac{1}{q}$ and $\dfrac{q-1}{q}$ are both reduced fractions with denominators not exceeding q. We get a growing list length that is not uniform, then. We do have a way a computing the list length that we expect for F_n, written as $|F_n|$, as n becomes ever larger. It is $|F_n|$ approaches $\dfrac{3n^2}{\pi^2}$. We round off this value to the nearest natural number. F_7 has 19 members, and our formula tells us that it will have 14.89 = 15 members. It's still early in the game though, and this approximating formula gets better as n gets larger.

The curious reader may ask, "What's pi doing in this equation?" Ah, that is a good question. I remember reading somewhere once that some mathematician had computed some risk assessment tables for someone using some probability equations that had pi in them. The person responded something to the effect that the mathematician had made a mistake because people dying (or whatever category the risk assessment was measuring) had nothing whatsoever to do with a circle. At least this person knew enough about math to know that pi had something to do with a circle. However, pi shows up in an amazing amount of formulas in mathematics. In addition, oftentimes, infinite sequences of rational numbers add up to some rational number times pi (or some power of pi). Thus, do not think that pi's use is restricted to a circle. Pi and its cousin e get around to a lot of happening places in mathematics.

Let us return to discussing Farey sequences. Suppose $\dfrac{a}{c} < \dfrac{b}{d}$ is a neighboring pair in a Farey sequence. These neighbors are called a Farey pair. A Farey pair has the following property:

$$\frac{b}{d} - \frac{a}{c} = \frac{1}{cd}$$

We can also say that if the difference of two fractions is 1 divided by the product of their denominators (let's say c and d again), then they are a Farey pair in the Farey sequence of order $max(c, d)$. The max function returns the largest value input into it.

If $\dfrac{a}{c} < \dfrac{b}{d}$ is a Farey pair, then the first term that appears between this pair is their mediant $\dfrac{a+b}{c+d}$, and it will be a keeper in the Farey sequence of order $c + d$. Thus, for example, $\dfrac{1}{3}$ and $\dfrac{2}{5}$ is a Farey pair in F_5. Their mediant is $\dfrac{3}{8}$, but it won't be seen until the Farey sequence of order $3 + 5 = 8$, which is F_8.

Farey sequences play a number of roles in higher mathematics. We have only introduced them here along with a few properties. As the student progresses in mathematics, more details of Farey sequences will come.

In the next section of this chapter, we will provide the details for finding rational points on a line and the conic sections. We will begin with finding rational points on the line. But first, we will set down a few observations.

- A rational number $\pm$ a rational number is a rational number.
- A rational number $\pm$ an irrational number is an irrational number.
- An irrational number added to an irrational number can be either zero, a rational number, or an irrational number. The case zero happens only when the irrational numbers are opposites. A rational number occurs when the irrational parts of the irrational number are opposites. In general, however, aside from these two special cases, the result is irrational.
- Zero times an irrational number is zero.
- A rational number times a rational number is a rational number.
- A rational number times an irrational number can be either zero or an irrational number. The case zero happens only when the rational number is zero.
- An irrational number times an irrational number can be either a rational number or an irrational number. Typically, the rational case occurs with roots coming out of the radical by having just enough exponent to clear the radical between the two factors being multiplied. In addition, the irrational factors of the two numbers can cancel leaving a rational number.

Let $f(x) = mx + b$, and we desire to find rational points on this line. This is to say that we want to find ordered pairs on this line such that $(x, y) \in \boldsymbol{Q}^2$. We see that $y = mx + b$. The first case is that if we let b be rational, we can get a rational y-intercept regardless of the value of a non-infinite m. If we grant that x and y are both rational, then m and b must both be rational or irrational (excluding the $x = 0$, which is the y-intercept case). This is because a rational number multiplied by an irrational number is an irrational number. Thus, if m is irrational, then mx is irrational. If we want $mx + b$ to be rational, then b must also be irrational. However, not just any irrational number will do for b. It must be a special irrational number that creates the property that $mx + b$ is rational.

For example, suppose $m = \pi$. If x and y are rational, then we have it that b is a rational number minus an irrational number, which is an irrational number as we have already observed. However, we cannot get more specific without an example ordered pair (x, y). Let $(x, y) = (2, 4)$. What does b need to be in this instance? We have $y = mx + b$ with three of the four values given. We can solve for b. We get $b = y - mx = 4 - \pi \cdot 2 = 4 - 2\pi$. Thus, our equation of the line is $y = \pi x + 4 - 2\pi$, and the point $(2, 4)$ is a rational point on it.

Are there other rational points on this line? It is clear that the problem is with the π's. We can factor out a π in the equation to get the following: $y = \pi(x - 2) + 4$. The only way to knock the π out of this equation is to multiply π by zero. This can only happen when x is 2. Hence, the only rational point on this line is the point $(2, 4)$.

Starting fresh, let us suppose now that $b = e$, the base of the natural logarithm. We now have the line $y = mx + e$. If we insist that x and y are rational, what does that say for m? Let us choose a rational (x, y) pair and see what happens. Choose $(x, y) = (3, 7)$. Now, our equation becomes $7 = m \cdot 3 + e$ implying that $m = \dfrac{7 - e}{3}$. This makes the equation of our line $y = \dfrac{7 - e}{3} x + e$. Again, the problem here is with the e's. Let's rewrite the equation with that factored out. We get $y = e\left(1 - \dfrac{1}{3}x\right) + \dfrac{7}{3}x$. The only rational multiplicative way to knock the e out of this equation is to multiply it by zero, which only happens when x is 3. Hence, the only rational point on the line $y = \dfrac{7 - e}{3} x + e$ is the point $(3, 7)$.

What happens when we decide to choose some irrational numbers for m and b that have a common irrational factor? Thus, choose $m = \pi$ and $b = 2\pi$. The common irrational factor here is π. This yields the equation $y = \pi x + 2\pi$. Will we get a rational point? We factor the π out of the right hand side to get $y = \pi(x + 2)$. We can multiplicatively knock the π out if $x = -2$. Thus, the only rational point on the line is $(-2, 0)$, an x-intercept case.

Suppose we choose some irrational numbers for m and b that do not have a common irrational factor? For example, choose $m = \pi$ and $b = \pi \cdot \sqrt{7}$. The equation then becomes $y = \pi x + \pi \cdot \sqrt{7}$. Here, the π can factor out, but the $\sqrt{7}$ will not. Hence, the only way to get a rational y is for x to be $-\sqrt{7}$. Thus, we lose the ability to get a rational point because only one of the coordinates can be rational.

Of course, if both *m* and *b* are rational, then choosing any rational *x* will always yield a rational *y*. However, by choosing an irrational *m* or *b*, we can get at most just one rational point on the line. We can do better with conic sections. Let us look at the parabola next. The equation of a parabola is $y = ax^2 + bx + c$. We have more leeway here because we have more constants and a variable raised to two different powers. Let us choose a rational point $(x, y) = (5, 8)$. Let us choose an irrational $c = \sqrt{2}$. Our parabola now becomes $8 = a \cdot 5^2 + b \cdot 5 + \sqrt{2}$. This implies that $\sqrt{2} = 8 - 25a - 5b$. Because we have two variables left, we can choose one to be anything we like and solve for the other. Let *a* be 4. We solve for *b* to find that it is $b = \dfrac{-(92 + \sqrt{2})}{5}$. This makes the equation of

the parabola $y = 4x^2 - \dfrac{92 + \sqrt{2}}{5}x + \sqrt{2}$. We factor out the irrational problem, and set it

equal to the rest of the terms, which are all rational. We get $\sqrt{2}\left(1 - \dfrac{x}{5}\right) = y - 4x^2 + \dfrac{92}{5}x$.

The only way to knock the $\sqrt{2}$ out of the equation is when $x = 5$. Thus, this parabola only has the one rational point (5, 8).

What if we let $b = 4$ instead of *a* in the previous example? We then plug $b = 4$ in the

equation $\sqrt{2} = 8 - 25a - 5b$, and solve for *a*. When we do this, we get $a = \dfrac{-(12 + \sqrt{2})}{25}$.

Next, we plug in our values of *a*, *b*, and *c* into the parabolic equation to get the following:

$y = -\dfrac{12 + \sqrt{2}}{25}x^2 + 4x + \sqrt{2}$. Here, we factor out the irrational problem to see that

$\sqrt{2}\left(1 - \dfrac{x^2}{25}\right) = y - 4x + \dfrac{12}{25}x^2$. We see that we get two values of *x* that knock out the

offensive $\sqrt{2}$. These *x* values are ± 5, and they correspond to (-5, -32) and (5, 8).

We will look at one more parabola example that branches $y = ax^2 + bx + c$. Choose the rational point (-3, 4) on the parabola, and let $a = \sqrt{3}$. Plugging these in the equation yields $4 = 9\sqrt{3} - 3b + c$. From here, we will let $b = \sqrt{3}$, and we solve for $c = 4 - 6\sqrt{3}$. Then, we plug in the *a*, *b*, and *c* values into the parabolic equation to get $y = \sqrt{3} \cdot x^2 + \sqrt{3} \cdot x + 4 - 6\sqrt{3}$. Of course, the problem is the pervasive $\sqrt{3}$. We factor it out and isolate it as before to yield the following: $\sqrt{3}(x^2 + x - 6) = y - 4$. The way to knock out the $\sqrt{3}$ is when that trinomial is zero, and that happens when $x = -3$ and $x = 2$. These correspond to the rational points (-3, 4) and (2, 4).

Instead of letting $b = \sqrt{3}$ in $4 = 9\sqrt{3} - 3b + c$, suppose we let b be the rational number 11. What happens then? We get $c = 4 - 9\sqrt{3} + 33 = 37 - 9\sqrt{3}$. Then, the equation $y = \sqrt{3} \cdot x^2 + 11x + 37 - 9\sqrt{3}$ becomes $\sqrt{3}(x^2 - 9) = y - 11x - 37$. We just get the two rational points (-3, 4) and (3, 70).

In the above examples, we were making choices. A different choice would yield a different line or parabola. Upon having that new line or parabola, we proceed to find the rational points on it by isolating and factoring the irrational component of the equation. If we choose constants that do not have the entire irrational component as a common factor, then we will not be able to get any rational points on the parabola. If we choose constants that are all rational, then any rational x will yield a rational y, but not vice versa.

We move on to the ellipse and hyperbola. We will not toy with the irrationality of constants a, b, h, and k with the ellipse and hyperbola. Thus, a, b, h, and k in the four equations $\dfrac{(x-h)^2}{a^2} + \dfrac{(y-k)^2}{b^2} = 1$, $\dfrac{(x-h)^2}{a^2} - \dfrac{(y-k)^2}{b^2} = 1$, $\dfrac{(x-h)^2}{b^2} + \dfrac{(y-k)^2}{a^2} = 1$, and $\dfrac{(x-h)^2}{b^2} - \dfrac{(y-k)^2}{a^2} = 1$ are all rational. Hence, the squares of a and b are perfect squares—meaning that they can either be perfect square integers or perfect square fractions. Also, h and k are rational, which means that the center of the ellipse or hyperbola is a rational point. Classes that are more advanced than this one can plumb the depths that allow various combinations of the irrationality of a, b, h, and k. We also state, for the sake of clarity, that a and b do not have to have anything to do with a Pythagorean triple—they just need to be rational.

We will use the convention (A, B, C) for a Pythagorean triple to mean (short leg, long leg, hypotenuse). We use capital letters so as not to confuse the lowercase a and b in the ellipse and hyperbola constants with the uppercase A and B of a Pythagorean triple. The upcoming ordered pair expressions were derived by the author with the help of one of his friends, who is also a math tutor. These expressions may very well be an original contribution to mathematics as I have never seen them before.

We begin with a wide ellipse $\dfrac{(x-h)^2}{a^2} + \dfrac{(y-k)^2}{b^2} = 1$. We can find rational points on this ellipse first by selecting a Pythagorean triple (A, B, C). Then, the eight rational points

$\left(\dfrac{\pm B \cdot a}{C} + h, \dfrac{\pm A \cdot b}{C} + k\right)$ and $\left(\dfrac{\pm A \cdot a}{C} + h, \dfrac{\pm B \cdot b}{C} + k\right)$ are on the ellipse. Next, we detail

a tall ellipse's $\dfrac{(x-h)^2}{b^2} + \dfrac{(y-k)^2}{a^2} = 1$ rational points. Given a Pythagorean triple

(A, B, C), the eight rational points $\left(\dfrac{\pm B \cdot b}{C} + h, \dfrac{\pm A \cdot a}{C} + k\right)$ and $\left(\dfrac{\pm A \cdot b}{C} + h, \dfrac{\pm B \cdot a}{C} + k\right)$

are on the ellipse.

We proceed to the two hyperbolas. The left-right hyperbola $\dfrac{(x-h)^2}{a^2} - \dfrac{(y-k)^2}{b^2} = 1$,

when given the Pythagorean triple (A, B, C), has the following eight rational points

$\left(\dfrac{\pm C \cdot a}{A} + h, \dfrac{\pm B \cdot b}{A} + k\right)$ and $\left(\dfrac{\pm C \cdot a}{B} + h, \dfrac{\pm A \cdot b}{B} + k\right)$. Lastly, the up-down hyperbola

$\dfrac{(y-k)^2}{a^2} - \dfrac{(x-h)^2}{b^2} = 1$, when given the Pythagorean triple (A, B, C), has the eight

rational points $\left(\dfrac{\pm A \cdot b}{B} + h, \dfrac{\pm C \cdot a}{B} + k\right)$ and $\left(\dfrac{\pm B \cdot b}{A} + h, \dfrac{\pm C \cdot a}{A} + k\right)$.

We turn our attention now to denesting radicals. Since denesting radicals is usually not taught anymore in precalculus, and we will have a need for this capability, we will take some pains to show how to do it—at least in some easier types of cases since more advanced cases requires knowledge of higher mathematics. It will not be assumed that the student already knows how to do denesting of these easier cases. Thus, the first chapter in this book needs to discuss denesting radicals along with the already discussed fraction concepts and rational points concepts. We begin the denesting discussion with the concept of a surd.

A surd is a number $\sqrt[n]{k}$, which is the unique real root of the equation $x^n = k$, where n is a positive integer and k is a rational number. Thus, $\sqrt{2}$ is a surd since $n = 2$, which is a positive integer, and $k = 2$ is a rational number. Also, $\sqrt[12]{\dfrac{91}{119}}$ is a surd since $n = 12$, which is a positive integer, and $k = 91/119$ is a rational number. Under this definition, all integers and rational numbers are surds because $n = 1$ is a positive integer. Thus, we are not wrong when we say that $\sqrt{\dfrac{81}{196}}$ is a surd because it simplifies to 9/14, which is a surd.

Our first task will be to attempt to simplify the square root of a three-term sum of surds. We need the three-term sum to be of the format $a^2 \pm 2ab + b^2$ because of the fact that $\sqrt{a^2 \pm 2ab + b^2} = \sqrt{(a \pm b)^2} = (a \pm b)^{\frac{2}{2}} = |a \pm b|^1 = |a \pm b| = a + b, a - b$ if $a > b$, or $b - a$ if $b > a$. Observe that when the fractional exponent is 2/2, we do not need the absolute value signs; but when we reduce the fractional exponent to 1, we change the index of the radical from two to one. The expression under an even indexed radical cannot be negative if we wish to stay in the real numbers. Thus, the fractional exponent reduction from 2/2 to 1 must block the potential of the radical going into the imaginary numbers. This is why we need to place the absolute value signs here. If our three-term sum is not of this format, we will not attempt to reduce it because we do not have a way to generalize our procedure to encompass a larger set of three-term sum problems.

Suppose we are asked to denest $\sqrt{9 + 12\sqrt[3]{3} + 12\sqrt[3]{9}}$. We should not just assume that the procession of terms in the three-term sum are in the order of $a^2, \pm2ab, b^2$. A nice way to go about this is to assign x, y, and z to the terms in the order as they appear in the radical. Thus, we let $x = 9$, $y = 12\sqrt[3]{3}$, and $z = 12\sqrt[3]{9}$. Then, looking for zero, compute the following quantities just below this paragraph. Upon finding a zero, compute the expression below the one yielding the zero for the denested answer. The reason this works is because if $x = a^2$ and $y = b^2$, then $4xy = 4a^2b^2 = (\pm2ab)^2 = z^2$. So if a zero is not found with these expressions, then the three-term sum is not in the $(a \pm b)^2$ format. Note that the sgn(x) function returns just three values: +1, if x is > 0; 0, if $x = 0$; and −1, if $x < 0$. It returns the sign of its input value.

$$4xy - z^2$$
$$\left| \sqrt{x} + \text{sgn}(z) \cdot \sqrt{y} \right|$$
$$\qquad\qquad 4xz - y^2$$
$$\left| \sqrt{x} + \text{sgn}(y) \cdot \sqrt{z} \right|$$
$$\qquad\qquad 4yz - x^2$$
$$\left| \sqrt{y} + \text{sgn}(x) \cdot \sqrt{z} \right|$$

Using this template, we compute $4xy - z^2 = 4 \cdot 9 \cdot 12\sqrt[3]{3} - (12\sqrt[3]{9})^2 = 432\sqrt[3]{3} - 144\sqrt[3]{81} = 432\sqrt[3]{3} - 144 \cdot 3 \cdot \sqrt[3]{3} = 0$. Thus, the first one we tried worked. We now write the denested answer using the first expression $|\sqrt{9} + (+1) \cdot \sqrt{12\sqrt[3]{3}}| = |3 + 2 \cdot \sqrt[6]{81}| = |3 + 2 \cdot 3^{\frac{4}{6}}| = \left| 3 + 2 \cdot |3|^{\frac{2}{3}} \right| = 3 + 2\sqrt[3]{9}$. Of course, the student may skip some of these steps. If z were negative, we would see a (−1) in place of sgn(z).

Example: Denest $\sqrt{20 - 12\sqrt[6]{3125} + 9\sqrt[3]{25}}$. Here, the minus sign is a giveaway. We go right to the answer:

$$\left|\sqrt{20}+(-1)\cdot\sqrt{9\sqrt[3]{25}}\right| = \left|2\sqrt{5}-3\cdot\sqrt{\sqrt[3]{25}}\right| = \left|2\sqrt{5}-\sqrt[6]{5^2}\right| = \left|2\sqrt{5}-\sqrt[3]{5}\right| = 2\sqrt{5}-\sqrt[3]{5}$$

However, we would be remiss if we did not approximate both of the expressions with a grapher to check that the approximations were equal. The original problem approximates to 0.6578, and the derived answer approximates to 2.762, which does not agree with the original problem's approximation. Hence, we conclude that we do not have the correct middle term in our three-term sum. Had we computed the three expressions above by assigning to x, y, and z, we would not have found a zero. The moral to this example is to show that the reader should not make blind assumptions because that leads to trouble. It is a good idea to check one's work whenever possible.

Of course, expressions under a square root can have a different number of terms than three. So we next present a theorem that will help us handle these other cases.

Theorem (Denesting Radicals): Given $x, y \in \Re$ with $x \geq y \geq 0$, then

$$\sqrt{x \pm y} = \sqrt{\frac{1}{2}x + \frac{1}{2}\sqrt{x^2 - y^2}} \pm \sqrt{\frac{1}{2}x - \frac{1}{2}\sqrt{x^2 - y^2}}$$

We will use this theorem to denest a few examples. We will also use the theorem to tell us when we are not getting anywhere, and this tells us that our problem cannot be denested.

Example: Denest $\sqrt{3 + 2\sqrt{2}}$. We identify x and y in the theorem as 3 and $2\sqrt{2}$, respectively, since x needs to be greater than or equal to y; and $3 \geq 2\sqrt{2}$. It remains to plug the numbers in the correct places and simplify.

$$\sqrt{3 + 2\sqrt{2}} = \sqrt{\frac{1}{2}\cdot 3 + \frac{1}{2}\cdot\sqrt{3^2 - (2\sqrt{2})^2}} + \sqrt{\frac{1}{2}\cdot 3 - \frac{1}{2}\cdot\sqrt{3^2 - (2\sqrt{2})^2}}$$

$$= \sqrt{\frac{3}{2} + \frac{1}{2}\cdot\sqrt{9 - 8}} + \sqrt{\frac{3}{2} - \frac{1}{2}\cdot\sqrt{9 - 8}}$$

$$= \sqrt{2} + 1 \quad \text{Thus, this example denested.}$$

Example: Denest $\sqrt{2\sqrt{10} + 7}$. Here, x is 7 and y is $2\sqrt{10}$ since $7 > 2\sqrt{10}$. We proceed to use the theorem.

$$\sqrt{2\sqrt{10}+7} = \sqrt{\frac{1}{2}\cdot 7 + \frac{1}{2}\cdot\sqrt{(7)^2-(2\sqrt{10})^2}} + \sqrt{\frac{1}{2}\cdot 7 - \frac{1}{2}\cdot\sqrt{(7)^2-(2\sqrt{10})^2}}$$

$$= \sqrt{\frac{7}{2}+\frac{1}{2}\cdot\sqrt{49-40}} + \sqrt{\frac{7}{2}-\frac{1}{2}\cdot\sqrt{49-40}}$$

$$= \sqrt{\frac{10}{2}} + \sqrt{\frac{4}{2}} = \sqrt{5}+\sqrt{2}$$

Example: Denest $\sqrt{5\sqrt{5}+111}$. Here, clearly x is 111, and y is $5\sqrt{5}$. So we plug in the values and compute.

$$\sqrt{5\sqrt{5}+111} = \sqrt{\frac{1}{2}\cdot 111 + \frac{1}{2}\cdot\sqrt{111^2-(5\sqrt{5})^2}} + \sqrt{\frac{1}{2}\cdot 111 - \frac{1}{2}\cdot\sqrt{111^2-(5\sqrt{5})^2}}$$

$$= \sqrt{\frac{111}{2}+\frac{\sqrt{12196}}{2}} + \sqrt{\frac{111}{2}-\frac{\sqrt{12196}}{2}} = \sqrt{\frac{111}{2}+\sqrt{3049}} + \sqrt{\frac{111}{2}-\sqrt{3049}}$$

At this point, we observe that we now have two nested radicals with a rational number and a square root term, and we started off with just one nested radical with a rational number and a square root term. Thus, we are worse off now, so the denesting procedure will not work for this one. We conclude that the original expression cannot be denested.

Example: Denest $\sqrt{37+6\sqrt{14}-2\sqrt{210}-4\sqrt{15}}$. Note that here we have four surds under the radical. It turns out that we have to divide the expression up into two groups of two, and it does not matter how we do this as long as the pair chosen for x is greater than the pair chosen for y to satisfy the requirements of the theorem. Nevertheless, it makes sense to use the two positive terms as x and the two negative terms as y. Thus, $x = 37 + 6\sqrt{14}$ and $y = 2\sqrt{210}+4\sqrt{15}$. We assume this is OK because otherwise the negatives would outweigh the positives, and the radicand would have an overall negative quantity, which implies that the radical is imaginary. The reader is encouraged to try a different set of pairings for x and y that satisfy the theorem. Now, we plug in the values into the formula and compute.

$$\sqrt{37+6\sqrt{14}-2\sqrt{210}-4\sqrt{15}} = \sqrt{\frac{1}{2}\cdot(37+6\sqrt{14})+\frac{1}{2}\cdot\sqrt{(37+6\sqrt{14})^2-(2\sqrt{210}+4\sqrt{15})^2}} -$$

$$\sqrt{\frac{1}{2}\cdot(37+6\sqrt{14})-\frac{1}{2}\cdot\sqrt{(37+6\sqrt{14})^2-(2\sqrt{210}+4\sqrt{15})^2}}$$

$$= \sqrt{\frac{37+6\sqrt{14}}{2} + \frac{1}{2}\cdot\sqrt{444\sqrt{14}+1873-(240\sqrt{14}+1080)}} -$$

$$\sqrt{\frac{37+6\sqrt{14}}{2} - \frac{1}{2}\cdot\sqrt{444\sqrt{14}+1873-(240\sqrt{14}+1080)}}$$

$$= \sqrt{\frac{37+6\sqrt{14}}{2} + \frac{1}{2}\cdot\sqrt{204\sqrt{14}+793}} - \sqrt{\frac{37+6\sqrt{14}}{2} - \frac{1}{2}\cdot\sqrt{204\sqrt{14}+793}}$$

We have here a couple of nested radical situations. We need to check to see if the inner nested radical can be denested. Thus, we have a subproblem to the main problem. Since 793 is greater than $204\sqrt{14}$, we choose x to be 793 with y being the square root term.

$$\sqrt{204\sqrt{14}+793} = \sqrt{\frac{1}{2}\cdot793 + \frac{1}{2}\cdot\sqrt{793^2-(204\sqrt{14})^2}} + \sqrt{\frac{1}{2}\cdot793 - \frac{1}{2}\cdot\sqrt{793^2-(204\sqrt{14})^2}}$$

$$= \sqrt{\frac{793}{2} + \frac{1}{2}\cdot\sqrt{46225}} + \sqrt{\frac{793}{2} - \frac{1}{2}\cdot\sqrt{46225}} = \sqrt{\frac{793}{2} + \frac{215}{2}} + \sqrt{\frac{793}{2} - \frac{215}{2}}$$

$$= \sqrt{\frac{1008}{2}} + \sqrt{\frac{578}{2}} = \sqrt{504} + \sqrt{289} = 6\sqrt{14}+17$$

Now that the subproblem is finished, we turn back to the main problem.

$$\sqrt{\frac{37+6\sqrt{14}}{2} + \frac{1}{2}\cdot\sqrt{204\sqrt{14}+793}} - \sqrt{\frac{37+6\sqrt{14}}{2} - \frac{1}{2}\cdot\sqrt{204\sqrt{14}+793}} =$$

$$\sqrt{\frac{37+6\sqrt{14}}{2} + \frac{1}{2}\cdot(6\sqrt{14}+17)} - \sqrt{\frac{37+6\sqrt{14}}{2} - \frac{1}{2}\cdot(6\sqrt{14}+17)} =$$

$$\sqrt{\frac{37+6\sqrt{14}}{2} + \frac{6\sqrt{14}+17}{2}} - \sqrt{\frac{37+6\sqrt{14}}{2} - \frac{6\sqrt{14}+17}{2}} = \sqrt{27+6\sqrt{14}} - \sqrt{10}$$

We're almost there! Once we split the last nested radical, we are finished.

$$\sqrt{27+6\sqrt{14}} = \sqrt{\frac{1}{2}\cdot27 + \frac{1}{2}\cdot\sqrt{27^2-(6\sqrt{14})^2}} + \sqrt{\frac{1}{2}\cdot27 - \frac{1}{2}\cdot\sqrt{27^2-(6\sqrt{14})^2}} =$$

$$= \sqrt{\frac{27}{2} + \frac{\sqrt{225}}{2}} + \sqrt{\frac{27}{2} - \frac{\sqrt{225}}{2}} = \sqrt{\frac{27}{2} + \frac{15}{2}} + \sqrt{\frac{27}{2} - \frac{15}{2}} = \sqrt{21} + \sqrt{6}$$

Thus, the expression $\sqrt{27 + 6\sqrt{14}} - \sqrt{10} = \sqrt{21} + \sqrt{6} - \sqrt{10}$.

We can square this result to obtain the original four surds under the radical to check our work.

This last example required applying the theorem three times, but it did denest the original radical. Thus, we were able to conquer a difficult problem. Incidentally, my grapher (a TI-89) failed to denest the radical given in this example. So the reader can now do something that a nice grapher cannot do.

Exercises

1. Writing in mathematics: Using the concepts of *lcm* and *gcf*, explain through a patterned argument why the product of n positive integers is equal to the *lcm* of the n positive integers multiplied by the *gcf* of the n integers raised to the $(n-1)$th power. [Hint: Start with $n = 2$, and try a few examples with a *gcf* $\neq$ 1. Then, move to $n = 3$. Attempt in this fashion to establish the pattern.]

2. Reduce the fraction 98861/130587 by finding the *gcf* by the method explained in the chapter.

3. Reduce the fraction 38556/238595 by finding the *gcf* by the method explained in the chapter.

4. Simplify the expression $\dfrac{121}{49} \cdot (11 - 5\frac{3}{4})^2 + \dfrac{21}{64} - 1.23$ by hand. Show all work.

5. Compute the expression $-\dfrac{1^2}{2} + \dfrac{2^2}{3} - \dfrac{3^2}{5} + \dfrac{4^2}{7} - \ldots + \dfrac{n^2}{p}$ where $n = 1, 2, 3, \ldots$ (the positive integers), and $p = 2, 3, 5, 7, 11, 13, \ldots$ (the prime numbers) for $p = 11$ and 13.

6. Use Midy's theorem on the following fractions to generate a string of nines.

a) $\dfrac{4}{13}$ b) $\dfrac{11}{17}$ c) $\dfrac{11}{23}$ d) $\dfrac{13}{31}$

7. Why don't we get a string of nines when we use Midy's theorem on the fraction $\dfrac{11}{21}$ since the numerator 11 is prime?

8. Find another counterexample (besides the one given in the text) to demonstrate that Midy's theorem doesn't work in reverse.

9. Construct the complete extended Midy's theorem table as demonstrated in the text for the following fractions.

   a) $\dfrac{3}{7}$ b) $\dfrac{5}{13}$ c) $\dfrac{12}{17}$ d) $\dfrac{18}{29}$ e) $\dfrac{13}{31}$ f) $\dfrac{19}{23}$

10. Add, subtract, and multiply the following dyadic fractions. Verify in each case that the result is another dyadic fraction. Show that the results have terminating decimal expansions.

   a) $\dfrac{3}{8}, \dfrac{5}{16}$ b) $\dfrac{1}{2}, \dfrac{1}{8}$ c) $\dfrac{5}{32}, \dfrac{7}{16}$ d) $\dfrac{21}{128}, \dfrac{21}{256}$

11. Add, subtract, and multiply the following fractions with denominators of powers of five. Verify in each case that the result is another fraction having a denominator with a power of five. Show that the results have terminating decimal expansions.

   a) $\dfrac{1}{5}, \dfrac{3}{5}$ b) $\dfrac{3}{25}, \dfrac{4}{5}$ c) $\dfrac{2}{5}, \dfrac{19}{125}$ d) $\dfrac{6}{625}, \dfrac{124}{125}$

12. Use the corresponding fractions in parts a – d from the previous two problems. Add, subtract, and multiply them. Show that the results have terminating decimal expansions.
[i.e. a) $\dfrac{3}{5}, \dfrac{1}{5}$ and $\dfrac{5}{16}, \dfrac{3}{5}$ etc. Part a will have 6 such pairings, and 3*6 = 18 computations.]

13. Construct two members of the set of fractions $\left\{ \dfrac{a}{2^{b_1} \cdot 5^{b_2}} \right\}$ with neither b_1 nor b_2 equal to zero, and b_1 not equal to b_2. Add, subtract, and multiply them. Verify that the result is another member of the set $\left\{ \dfrac{a}{2^{b_1} \cdot 5^{b_2}} \right\}$. Show that the decimal expansions terminate.

57

14. Generate a three to six digit decimal number between zero and one. Show that it is a member of the set of fractions depicted in the previous question by identifying a, b_1, and b_2. Point out that no other factors are present in the denominator.

15. Convert the following decimals into fractions. Some of them are repeating, and some of them are terminating. Show your work.

a) $0.\overline{3}$ b) $1.\overline{23}$ c) 0.99 d) 1.23 e) $11.\overline{18}$ f) $0.5\overline{4}$

g) 0.125 h) $9.\overline{9}$ i) $0.1\overline{54}$ j) $0.6\overline{647}$ k) $0.1\overline{234}$ l) $0.12\overline{385}$

16. Compute the mediant of the following pairs of fractions. Show by comparison of decimal expansions that the mediant is strictly between the two originating fractions. Identify which of the given pairs is a Farey pair (and why), the lowest order of the Farey sequence they belong, and the lowest order of the Farey sequence in which their mediant will first appear. If the given pair of fractions is not a Farey pair, state the order of the Farey sequence that their mediant *first* appeared or will appear. Conjecture a formula for the order that it *first* appeared or will appear, but do not attempt to prove or disprove this conjectured formula.

a) $\dfrac{3}{5}, \dfrac{6}{7}$ b) $\dfrac{1}{4}, \dfrac{5}{6}$ c) $\dfrac{2}{5}, \dfrac{3}{7}$ d) $\dfrac{5}{8}, \dfrac{2}{3}$ e) $\dfrac{4}{9}, \dfrac{6}{13}$

17. Approximate how many terms are in F_1 through F_8. Generate the Farey sequences F_1 through F_8. Show all work (i.e. explain why some mediants get tossed out).

18. Between what two orders should we expect a Farey sequence to exceed 1000 terms?

    For exercises 19 – 32, find the rational points on the graph. If there are no rational points, then state that there are none. If there are infinitely many, then state the conditions under which those infinitely many points can be found. For example, "Given a rational x, y will also be rational." Note that this statement suggests the possibility that some given rational y's will produce x's that are not rational. Thus, be careful in your wording.

19. $x = 4$ 20. $y = 3.2$ 21. $x = \pi$ 22. $y = \sqrt{7}$

23. $y = 2x + 5$ 24. $x = 3y$ 25. $xy = 2$ 26. $xy = \sqrt[3]{9}$

27. $y = 4(x+3)^2 - 2$ 28. $x = (y+5)^2 + 5$ 29. $y = \sqrt{8} \cdot x^2 + 5x - \sqrt{2}$ 30. $3x = 2y$

31. $x = \sqrt{27} \cdot y^2 + \sqrt{12} \cdot y - 5$ 32. $\dfrac{e-1}{7} x - e(1-x)^2 = y - 3e$

For exercises 33 – 38, use the Pythagorean triple given for parts a, b, and c. For each part, draw a graph and label the eight rational points. The student can think of a circle as an ellipse with $a = b = r$ to aid in the computations.

$\quad$ a) $(A, B, C) = (3, 4, 5)$
$\quad$ b) $(A, B, C) = (5, 12, 13)$
$\quad$ c) $(A, B, C) = (8, 15, 17)$

33. $x^2 + y^2 = 1$ 34. $\dfrac{(x+2)^2}{16} - \dfrac{(y-5)^2}{25} = 1$

35. $\dfrac{(y+6)^2}{9} - \dfrac{(x+6)^2}{4} = 1$ 36. $\dfrac{x^2}{25} + \dfrac{(y-3)^2}{36} = 1$

37. $\dfrac{(x+3/4)^2}{100} + \dfrac{y^2}{0.81} = 1$ 38. $(x+1)^2 + (y-2)^2 = 9$

39. Writing in mathematics: Describe how to get eight integer points using a single Pythagorean triple for an ellipse or hyperbola. Confine your discussion to the Pythagorean triple (3, 4, 5). As part of the discussion, provide an example of a tall ellipse, a wide ellipse, an up-down hyperbola, and a left-right hyperbola that each has eight integer points. In each case, provide a detailed diagram with the relevant features labeled.

40. Denest $\sqrt{3\sqrt{15}+9}$.   41. Denest $\sqrt{12+2\sqrt{35}}$.   42. Denest $\sqrt{5+\sqrt{28\sqrt{35}+189}}$.

43. Denest $\sqrt{\sqrt[3]{36}+\sqrt[3]{4}+\sqrt[3]{96}}$.   44. Denest $\sqrt{\sqrt{4\sqrt{2}+9}+4\sqrt{2}+8}$.

45. Denest $\sqrt{2\sqrt{10}+16-6\sqrt{2\sqrt{10}+7}}$.   46. Denest $\sqrt{47-4\sqrt{105-30\sqrt{10}}-10\sqrt{10}}$.

47. Denest $\sqrt{10\sqrt[3]{3}+25+\sqrt[3]{9}}$.   48. Denest $\sqrt{64-24\sqrt[3]{6}+\sqrt[3]{36}}$.

49. Denest $\sqrt{43+16\sqrt{3}+8\sqrt{15}+12\sqrt{5}}$.

50. Let d_1 and d_2 be the two segments of a focal chord. For the three conics capable of having focal chords, it is a fact that $\dfrac{1}{d_1} + \dfrac{1}{d_2}$ is a constant. For the parabola $y = ax^2$, this constant is $4|a|$. For this problem, use the parabola $y = 3x^2$.

    a) Find the equation of the focal chord with a slope of two.
    b) Construct a system of equations involving the equation of the focal chord and the parabola to yield the intersection points of the focal chord with the parabola. Note that these two points have radicals.
    c) Find d_1 and d_2 by direct computation using the distance formula. This step produces nested radicals. Use the denesting techniques in this chapter to denest the nested radicals.
    d) Plug these values into the equation $\dfrac{1}{d_1} + \dfrac{1}{d_2} = 4\,|\,a\,|$ to yield an identity.

51. Using the ideas in problem 50, for a tall ellipse centered at the origin, determine the constant in terms of a and b that is equal to $\dfrac{1}{d_1} + \dfrac{1}{d_2}$. [Hint: Use $m = 0$ since we are after a constant.]

52. Using the ideas in problem 50, for a left-right hyperbola centered at the origin, determine the constant in terms of a and b that is equal to $\dfrac{1}{d_1} + \dfrac{1}{d_2}$.

53. Redo problem #50 with the results from problem 51 for the ellipse $\dfrac{x^2}{9} + \dfrac{y^2}{16} = 1$ with a focal chord through the top focus with a slope of four.

54. Redo problem #50 with the results from problem #52 for the left-right hyperbola $\dfrac{x^2}{25} - \dfrac{y^2}{16} = 1$ with a focal chord through the left focus with a slope of three.

The Deluxe Toolkit

When solving problems in mathematics, we use tools that we've picked up along the way. Some of these tools are the slope formula, the perpendicular slope formula, the distance formula, and the midpoint formula. Other devices are things like knowing how to set up a system of equations that yield a desired outcome. We do this when we want to find the equation of a circle through three points, the equation of a parabola through three points, or the equation of a rectangular hyperbola through three points. We have a lot of standard equations for objects as well. Some examples are the following: a few equations for a line; equations for a circle, ellipse, parabola, and hyperbola; and equations for the area and perimeter of an assortment of polygons. The further we go in mathematics, the more tools we pick up along the way, which we put into our toolkit. Having more tools enables us to solve more and more kinds of problems.

We will encounter more kinds of problems in this textbook. So we need some more tools to handle them. To that end, this chapter is a collection of tools to put into your toolkit. You may find a need for them from time to time in this course. Also, do not forget about the tools already present in your toolkit. Just because you get some new tools in this chapter does not mean that the old tools have no value. On the contrary, many times you will find that you need several tools from the toolkit to handle a problem. Some of them might be new, but some of them that were necessary were older tools.

Our first concept in this chapter is an extension of the slope formula. We call it the slope function. This function takes a number of forms. For the parabola $y = ax^2 + bx + c$, the slope function is $m(x) = 2ax + b$. This is to say that if we want to find the slope of the parabola at any given value of x, we just plug that value of x into the slope function. Thus, for example, suppose we want to find the slope of the parabola $y = 2x^2 - 5x + 1$ at the point (2, -1), we compute $m(2) = 2(2)(2) - 5 = 3$. So the equation of the tangent line through (2, -1) is $y = 3x - 7$. If we want to find the equation of a normal line through a point on the parabola $y = ax^2 + bx + c$, we compute the required slope with the function $m_\perp(x) = \dfrac{-1}{2ax + b}$ and then proceed with finding the line with this slope through the indicated point. The slope function can be extended further to other conic sections, but we should really have calculus under our belts before we attempt that task. For now, these two functions for a parabola will serve us well enough.

Our next section of tools in this toolkit chapter deals with angles. You may already know about these tools. If so, then you will have an easier time of it than other students who may not have known about these tools.

If we have two intersecting lines that are not perpendicular, then two of the angles will be acute angles while the other two angles will be obtuse. An acute angle is smaller than 90 degrees, and an obtuse angle is more than 90 degrees. The two acute angles will be opposite one another, and the two obtuse angles will be opposite one another. It is a fact that the two acute angles will be equal. It is also a fact that the two obtuse angles will be equal. We call the angles in this opposite position *vertical*. We say that vertical angles are equal. This covers both the acute and obtuse angles with one statement.

If two angles add up to 90 degrees, then we say that they are complementary angles. An angle of 90 degrees is called a right angle. Thus, complementary angles look like the corner angle of a regular sheet of paper when they are butted together sharing the vertex. Suppose we have two intersecting lines that are not perpendicular. We can use the perpendicular slope formula for either line's slope to find the slope of a line that creates a complementary angle to the acute angle. We send a line of this slope through the intersection point of the two given lines. Then, the new angle created is complementary to the original acute angle. The obtuse angle cannot have a complementary angle because it is already more than 90 degrees.

Example: When 20.25 degrees is added to the square of one complementary angle, the result is one-half of the difference of the two complementary angles. Find the two angles. Solution: We let x = one of the complementary angles, and y = the other. Next, we see some ambiguity in the language statement of the problem. We are just told about the difference of the two complementary angles. We have no way of differentiating between the two angles. Thus, we will just have to set up two systems of equations. One will have the difference as $x - y$, and the other will have the difference as $y - x$. We construct both systems with this understanding.

$$\begin{cases} x^2 + 20.25 = \dfrac{1}{2}(x - y) \\ x + y = 90 \end{cases} \qquad \begin{cases} x^2 + 20.25 = \dfrac{1}{2}(y - x) \\ x + y = 90 \end{cases}$$

We proceed to solve the left system, but we encounter imaginary solutions there. Hence, the left system yields no solution. Our only chance, then, lies in the other system. We solve that system to yield $(x, y) = (4.5, 85.5)$ and another solution with negative degrees. We rule out the negative degrees solution as extraneous. Thus, the two angles are 4.5 degrees and 85.5 degrees.

We proceed to our next topic. If two angles add up to 180 degrees, then we say that they are supplementary angles. An angle of 180 degrees is a straight line. If we have two intersecting lines, then by the above statements, any two adjacent angles will be

62

supplementary. The acute angle added to the obtuse angle yields 180 degrees. If the initial pair of intersecting lines is perpendicular, then all four of the angles are right angles. Any two adjacent right angles will still yield a straight line.

Example: Two non-perpendicular lines intersect. Twice the sum of the acute angles is one-third of one of the obtuse angles. What are all of the angle measures?

Solution: We know that the vertical angles are equal. Hence, the two acute angles are equal, and the two obtuse angles are equal. Let x = the measure of an acute angle. Let y = the measure of an obtuse angle. The sum of the acute angles is $x + x$, and twice this sum is $2(x + x) = 2(2x) = 4x$. We know from the problem that this quantity, $4x$, is one-third of one of the obtuse angles. Hence, we get the following system of equations.

$$\begin{cases} 4x = \dfrac{1}{3} y \\ x + y = 180 \end{cases}$$

We solve this system to yield $(x, y) = (180/13, 2160/13)$. Thus, we would say that the acute angles measure 180/13 degrees or about 13.85 degrees, and the obtuse angles measure 2160/13 degrees or about 166.15 degrees.

Next, we have the parallel lines cut by a transversal situation. This creates two pairs of vertical acute angles and two pairs of vertical obtuse angles if the transversal line is not perpendicular to the parallel lines. All four of the acute angles are equal. All four of the obtuse angles are equal. Names exist for differing pairs of these angles, such as alternate interior angles, etc. However, we only have two angle sizes present—the acute angle size, and the obtuse angle size. If the transversal is perpendicular to the parallel lines, then all eight of the angles are right angles. A counterpart to parallel lines exists, and it is presented in the *Concyclic Points* chapter for the student who wants to peek at a coming attraction.

We move the discussion to polygons. The sum of the angles in an n sided polygon is $180(n - 2)$ degrees. This is a useful tool because we often deal with polygons in our work. If n is 3, then the polygon is a triangle. The sum of the angles in a triangle is $180(3 - 2) = 180$ degrees. If we want the triangle to be regular, meaning that all sides and all angles are equal, then we divide the sum of the degrees by the number of sides. Hence, a regular triangle has $180/3 = 60$ degrees for each angle. It is not possible for a triangle having three 60-degree angles not to be regular. However, it is possible for polygons with $n > 3$ sides. A regular triangle is also called an equilateral triangle. A regular quadrilateral is known as a square. A rectangle is an example of a polygon having all angles equal, but it is not regular. A rhombus is an example of a polygon having all sides equal, but it is not regular. We have names for five, six, seven, eight, nine, and ten sided polygons, which are pentagon, hexagon, heptagon, octagon, nonagon, and decagon, respectively. These can be regular or not.

Example: How many degrees does each corner of a regular decagon have?

Solution: We use the polygon angle sum formula to determine that the total of the angles is $180(n-2) = 180(10-2) = 180(8) = 1440$ degrees. Then, we divide this result by the number of angles present, which is 10. Hence, $1440/10 = 144$. Therefore, we conclude that each angle in a regular decagon is 144 degrees.

The number of diagonals that a polygon has is determined as follows: From a vertex, a diagonal may be extended to another vertex as long as it is not an adjacent vertex. This restriction means that we may not count the originating vertex, nor either adjacent vertex. Thus, the number of diagonals extended from a vertex is three less than the number of vertices n, which is $n-3$. Since there are n vertices, we then get $n(n-3)$ diagonals. However, a diagonal running from vertex n_1 to say n_5 gets counted twice: once from n_1's perspective and again from n_5's perspective. Therefore, we must cut the amount in half. Let N be the number of distinct diagonals of a polygon having n vertices. Then, the equation relating N and n is the following: $N = \dfrac{n(n-3)}{2}$.

Our next section of tools in this chapter deals with circles. Again, some or all of these tools may be familiar to you. However, some students may have not heard of these before. We just want to be sure that the students have the tools they need to solve the later problems in this book.

Thales of Miletus (c. 625 BC – c. 547 BC) was an ancient Greek mathematician. He proved that an angle in a semicircle is a right angle. This result is known as Thales' theorem. If we choose a point on a circle as the vertex for a right angle, then the other two points of intersection of the right angle with the circle are on a diameter of the circle. This is the converse of the preceding, which is also Thales' theorem. A diameter of a circle passes through the center of the circle.

Suppose we are given two points on a diameter to make a circle. We solve this by finding the midpoint of the two given points, which would yield the center of the circle. The radius of the circle is the distance between the center of the circle and either of the two given points. We get the radius by using the distance formula. We then plug these values into the equation for a circle: $(x-h)^2 + (y-k)^2 = r^2$.

Suppose we want to reflect a point P across the center of a circle. We usually denote the reflected point as P', and we say P prime. To do this, we find the equation of the line between P and the center of the circle. Then, we solve the system of equations between this line and the circle. One solution is P, and the other solution is P'. We call the point P' the diametric opposite of P. We also may say that P' is diametrically opposed to point P.

Example: Find the diametrically opposed point to $P(6, 2)$ on the circle with equation $(x-3)^2 + (y+2)^2 = 25$.

Solution: We will label the point diametrically opposite P as P'. Then, P' lies on the line through P and the center $(h, k) = (3, -2)$ of the circle. We find the equation of the line through $P(6, 2)$ and $(h, k) = (3, -2)$, which is $y = \dfrac{4}{3}x - 6$. Next, we set up the system of equations involving this line and the circle. This yields the following:

$$\left\{ \begin{array}{l} (x-3)^2 + (y+2)^2 = 25 \\ y = \dfrac{4}{3}x - 6 \end{array} \right\}$$

The solutions to this system are $P(6, 2)$ and $P'(0, -6)$. Another way exists to find the diametrically opposed point to P on a circle. This other method will be discussed in *The Circle* chapter.

A property of circles is that the tangent line to a circle is perpendicular to the radius at that point. A tangent line just touches or kisses another curve and stays on the same side of the curve after the intersection. A tangent line does not go through to the other side of the curve. This property enables us to find the equation of the tangent line to a circle quite easily. We have other methods to find the equation of the tangent line to a circle at a given point. We will explore some of these other methods in this book. However, the easiest and most direct route to obtaining this tangent line is by employing this property.

The midpoint formula is useful in many contexts. However, we may have occasion to use a midpoint formula extension. This is a weighted average between two given points. We assign the weights in the following manner:

$$0 \le w_1, w_2 \le 1 \text{ and } w_1 + w_2 = 1$$

The weight w_1 gets assigned to point 1, and the weight w_2 gets assigned to point 2. We compute the new x and y coordinates from these weights. We will work an example.

Example: Find the coordinates of the point that is five-sixths of the way from $P_1(2, -3)$ to $P_2(1, 4)$.

Solution: To solve this, we must first understand which point we will be closer to when we are done. If we are five-sixths of the way, we are most of the way to P_2. Hence, we will be closer to P_2 when we are done. This means that P_2 will get the weight of five-sixths—the heavier weight because this weight is greater than one half. Then, P_1 will get the weight of 1/6—the lighter weight because this weight is less than one half. We now compute $(w_1 P_{1x} + w_2 P_{2x}, w_1 P_{1y} + w_2 P_{2y}) = \left(\frac{1}{6} \cdot 2 + \frac{5}{6} \cdot 1, \frac{1}{6}(-3) + \frac{5}{6} \cdot 4 \right) = (7/6, 17/6)$. Note that we do not assign the weight w_1 to any P_2 coordinate, and we do not assign the weight w_2 to any P_1 coordinate. We should observe that the goal of being five-sixths of the way is 5/6 of the distance from P_1 to P_2. The distance from P_1 to $(7/6, 17/6)$ is five sixths of the distance from P_1 to P_2. In symbols, we write $d(P_1, (7/6, 17/6)) = \frac{5}{6} d(P_1, P_2)$. We would check this distance relationship to see if our work was correct using the distance formula twice. As a remark, we do not require that w_1 be lighter than w_2. However, we do require that the heavier weight is assigned to the coordinates of the point that we will be closer to when we are done.

The next thing we would like to have in our toolkit is how to find the distance from a point to a line. We have two versions of the formula. The first version is the following: Given the line $y = mx + b$ and the point $P(x_0, y_0)$, the distance from the point to the line is $d = \frac{|y_0 - mx_0 - b|}{\sqrt{m^2 + 1}}$. The second version is the following: Given the line $ax + by + c = 0$ and the point $P(x_0, y_0)$, the distance from the point to the line is $d = \frac{|ax_0 + by_0 + c|}{\sqrt{a^2 + b^2}}$. We can also find the distance between two parallel lines simply by selecting a point P on one of these lines to transform the original situation into one of these two situations.

We should remark that an infinite quantity of distances exist between a given point P and a line ℓ. We may choose a point on ℓ, and then we can compute the distance between P and this point. We can repeat this for another point on ℓ. It should be clear that we would never run out of selections for choosing a point on ℓ. Hence, we will never run out of differing distances from P to ℓ. However, we may notice that some symmetry is at work here. We get an equal distance for two different points on ℓ that are symmetric with respect to the axis of symmetry, which is the perpendicular line to ℓ through P. Because of this axis of symmetry, one point on ℓ provides a unique distance to P that no other point on ℓ provides. This special point is the intersection point of ℓ

and the axis of symmetry that we just discussed. This unique distance is also the shortest possible distance from P to any point on ℓ. We call this unique distance *the distance*. Thus, when we ask for the distance from a point P to a line, we are requesting this special, unique distance. We also sometimes call this distance *the perpendicular distance*.

Two special triangles exist that are typically taught in a high school geometry course. The first is the 45-45-90 triangle (a square cut in half along a diagonal), and the second is the 30-60-90 triangle (an equilateral triangle cut in half along an altitude). Since both have a 90 degree angle, they are both examples of right triangles, so the Pythagorean theorem applies. In the 45-45-90 triangle, both legs are equal, so $a^2 + a^2 = c^2$ implies that $2a^2 = c^2$. When we solve for c, we get $c = \sqrt{2} \cdot a$. Thus, the three sides of a 45-45-90 triangle are a, a, $\sqrt{2} \cdot a$. With a 30-60-90 triangle, the short leg is half the hypotenuse because the altitude of the equilateral triangle splits the base in half creating the short leg of the 30-60-90 triangle. We say that the hypotenuse is a and that the short leg is $\frac{1}{2}a$.

The height of the altitude is b, and we use the Pythagorean formula to compute b:

$$\left(\frac{1}{2}a\right)^2 + b^2 = a^2 \Rightarrow b^2 = a^2 - \left(\frac{1}{2}a\right)^2 \Rightarrow b = \sqrt{a^2 - \left(\frac{1}{2}a\right)^2} = \sqrt{\frac{4a^2 - a^2}{4}} = \frac{\sqrt{3}}{2}a$$

Thus, the three sides of a 30-60-90 triangle are $\frac{1}{2}a$, $\frac{\sqrt{3}}{2}a$, a.

Next, we will discuss four reflection situations involving points and lines. The first situation among the four covered in this chapter is that we can reflect a point P across another point Q. To do this we let the point Q that is doing the reflecting be the center of a circle with radius equal to the distance between the two given points P and Q. That distance can be found with the distance formula. Then, we find the diametrically opposed point P' to the point P that is being reflected as in a previous example in this chapter.

Another reflection situation is when we have a point P reflected across a line ℓ. To solve this, we send a line perpendicular to the given line through the point P. We then find the intersection point Q between these two lines. Then, we reflect the given point P across this intersection point Q as in the previous paragraph by using the circle idea.

What do we do when we want to reflect a line ℓ across a point P? The result is a parallel line to the given line. We can do this a couple of ways. The way we will discuss is to find the intersection point between the given line and a perpendicular line $\ell_\perp$ sent through the given point, P. Then, reflect this intersection point across the given point to create the reflected point P'. Finally, send a parallel line $\ell_{//}$ to the given line through the reflected point P'.

The last reflection we will discuss in this chapter is to reflect a line across another line. The line doing the reflecting we will call the mirror. We choose a point P on the line to be reflected that is not the intersection point X. Then, run a line perpendicular to the mirror through P. Find the coordinates of the intersection point Q of the mirror with this perpendicular line. Reflect the point P across the point Q to get the point P'. The line $P'X$ is the reflected line that we seek.

We have two reflection situations that we will present. The first is to reflect each of the three vertices across its opposite side. This results in three new vertices that make a new triangle. This new triangle is called the *reflection triangle*. It is possible for the reflection triangle to be degenerate, which means the three new vertices all line on the same line.

Our second reflection situation is called *reflecting a point across a triangle*. We present it with a theorem statement.

Triangle Reflection Point Theorem: Given $\triangle ABC$ and any point P not on the triangle's lines or their extensions. Reflect the point P across the lines of $\triangle ABC$ to create three points. The labeling of these three reflections matters, so follow this procedure. When reflecting P across the line BC, label the reflection point as X. When reflecting P across the line AC, label the reflection point as Y. When reflecting P across the line AB, label the reflection point as Z. Next, find the equations of the four circles AYZ, XBZ, XYC, and ABC. If these conditions are satisfied, then these four circles will all pass through a common point P', the triangle reflected point of P.

We will discuss the isotomic transversal line next. It involves reflecting points across the midpoints of the sides of a triangle. Given a triangle ABC and a line ℓ not parallel to any of the triangle's sides, we extend the sides where necessary to obtain three intersection points. The line ℓ intersects side BC at point A'. The line ℓ intersects side AC at point B'. The line ℓ intersects side AB at point C'. Reflect the point A' across the midpoint of side BC to obtain the point A''. Reflect the point B' across the midpoint of side AC to obtain the point B''. Finally, reflect the point C' across the midpoint of side AB to obtain the point C''. Then, the points A'', B'', and C'' are collinear, and this line is called the isotomic transversal of line ℓ with respect to triangle ABC. We denote the isotomic transversal with the symbol ℓ'. Note that if we were given ℓ', we would have found ℓ through the very same isotomic transversal process.

The theorem highlight chapter discussing the symmedian point is an important situation in mathematics when we reflect a line across another line. This special point in triangle geometry is something akin to one of the seven wonders of the ancient world. We can also think of it in modern terms along the lines of viewing the Grand Canyon or the Great Wall of China. The symmedian point is a geometric marvel. The student can peek at it now, of course, but we have much to discuss before the student can truly grasp it. A number of other reflection situations come up in other places throughout this book as well. Thus, this is an important concept to understand.

Our next discussion in this chapter is about Heron's formula. This formula is named after Heron of Alexandria ($c.\ 10 - c.\ 75$), an ancient Greek mathematician. Given the three side lengths a, b, and c of a triangle, we first compute the semi-perimeter $s = \dfrac{a+b+c}{2}$. Then, the area of the triangle is $A = \sqrt{s(s-a)(s-b)(s-c)}$. We use two other forms of Heron's formula (without the semi-perimeter), which are the following:

$$A = \frac{1}{4}\sqrt{(a+b+c)(-a+b+c)(a-b+c)(a+b-c)}$$

$$A = \frac{1}{4}\sqrt{2(a^2b^2 + a^2c^2 + b^2c^2) - (a^4 + b^4 + c^4)}$$

The form to use depends on the way the lengths look. One form may be more receptive than another to different levels of sophistication in the lengths of the triangle's sides.

The presence of the square root is troublesome if we desire an integral or a rational area. This problem has led to the concept of a Heronian triangle. This is a triangle with integer side lengths such that the triangle also has integer area. Any triangle having a Pythagorean triple (a, b, c) where $a^2 + b^2 = c^2$ for its side lengths is also a Heronian triangle. However, an infinite quantity of other Heronian triangle lengths exist, such as (5, 5, 6) and (9, 10, 17). Two Pythagorean triple triangles can be butted together to form a Heronian but non-Pythagorean triangle. We can be quite clever in resizing one or both of the Pythagorean Triple triangles to obtain a new Heronian triangle. Web sites exist that catalogue tables of these triples. A small catalog is presented for the reader.

In order for a Heronian triangle to yield a rational area that is not integral, we must multiply the triple by a positive rational scalar k. This method also works for Pythagorean triples to get us a triangle of rational area. Thus, for example, a modified Heronian triangle is $k(17, 17, 30)$ where $k = 0.1$ yields the triangle (1.7, 1.7, 3), which has rational area. Another example is a modified Pythagorean triangle $k(20, 21, 29)$ where $k = 1/12$ is

$\left(\dfrac{5}{3}, \dfrac{7}{4}, \dfrac{29}{12}\right)$, which also has rational area. Triangles such as these are friendly to us when we work with the concepts presented in the *Triangle Geometry* chapter. Triangles not based on a Heronian or Pythagorean triangle will cause the algebra work to exhibit radicals, which makes it harder.

A Few Heronian Triangles

(3, 4, 5)	(3, 25, 26)	(4, 13, 15)	(5, 5, 6)	(5, 5, 8)	(5, 12, 13)
(6, 8, 10)	(6, 25, 29)	(7, 15, 20)	(7, 24, 25)	(8, 15, 17)	(9, 10, 17)
(9, 12, 15)	(10, 10, 12)	(10, 10, 16)	(10, 13, 13)	(10, 17, 21)	(10, 24, 26)
(11, 13, 20)	(11, 25, 30)	(12, 16, 20)	(12, 17, 25)	(13, 13, 24)	(13, 14, 15)
(13, 20, 21)	(14, 25, 25)	(15, 15, 18)	(15, 15, 24)	(15, 20, 25)	(16, 17, 17)
(16, 25, 39)	(17, 17, 30)	(17, 25, 26)	(17, 25, 28)	(18, 20, 34)	(18, 24, 30)
(19, 20, 37)	(20, 20, 24)	(20, 20, 32)	(20, 21, 29)	(25, 25, 30)	(25, 25, 40)
(25, 25, 48)					

Our final discussion in this chapter is a unified definition of the conic sections. We do not consider the circle a conic section in this unification. If we are told the coordinates of the focus F, the directrix d, some point $P(x, y)$ in the plane, and the eccentricity e of the conic section, then we can find the equation of the conic section. We locate the point D on the directrix d such that the distance between P and d is minimized (i.e. the distance that was discussed earlier in this chapter). For a non-rotated conic section, this will either be the x-coordinate of point P or the y-coordinate of point P with the other coordinate the value of the constant in the directrix equation. We may then declare that the perpendicular distance from the point P to the directrix d is PD. We will usually find PD by taking the absolute value of a subtraction, which is how we find a distance on a vertical or horizontal line, but the distance PF will require the full distance formula because we will not in general have an equal coordinate for P and F. The unified definition is that the ratio of the distances $\dfrac{PF}{PD} = e$. We note that if $0 < e < 1$, then the conic section is an ellipse. If $e = 1$, then the conic section is a parabola. If $e > 1$, then the conic section is a hyperbola.

Exercises

1. Graph the lines $y = 2x$ and $y = 3x$ on the same set of axes. Identify the acute vertical angles and the obtuse vertical angles.

2. Two angles are complementary. Thrice the first is two more than half the second. Find the angles.

3. Two angles are supplementary, but the sum of the reciprocals in degrees is complementary. Find the angles.

4. Two angles are supplementary. Twice the first is twenty less the reciprocal of the other. Find the angles. Answer exactly, and approximate to the nearest hundredth.

5. Identify an adjacent angle pair in problem 1. Are these angles complementary, supplementary, or neither?

6. What angle size is present in a regular hexagon?

7. Solve $S = 180(n - 2)$ for n. Suppose S is 1800. Compute n.

8. How many diagonals does a nonagon have?

9. Solve $N = \dfrac{n(n - 3)}{2}$ for n. Analyze the resulting discriminant.

10. The points (1, 5) and (3, -4) are the endpoints of the diameter of a circle. Find the equation of the circle.

11. Reflect the point $P(1, 8)$ across the center of the circle with center (2, 4). What is the equation of the circle?

12. Use the property of a circle that a tangent line is perpendicular to the radius at the point of contact to find the equations of the tangents to the circle at P and P' in problem 11.

13. Find the coordinates of the point that is one-third of the distance from $P_1(4, 4)$ to $P_2(9, 11)$. Note: We can trisect a line segment with the extended midpoint theorem, but we need better stuff than this if we want to trisect an angle—one the ancient Greek problems.

14. Find the point that is 28% of the way from $P_1(1, 3)$ to $P_2(7, 8)$.

15. Solve $d = \dfrac{|y_0 - mx_0 - b|}{\sqrt{m^2 + 1}}$ for m.

16. Solve $d = \dfrac{|ax_0 + by_0 + c|}{\sqrt{a^2 + b^2}}$ for a.

17. A line has a y-intercept of $(0, 4)$. The point $(x_0, y_0) = (4, 9)$ is not on this line. We request a distance between the point (x_0, y_0) and the line to be two units. What slope or slopes of the line make this possible. Use the result from problem 15.

18. Reflect the point $(1, 1)$ across the point $(2, 2)$.

19. Reflect the point $(4, 1)$ across the line $y = -2x - 4$.

20. Reflect the line $y = -2x - 4$ across the point $(4, 1)$.

21. Reflect the line $y = 2x$ across the line $y = 3x$.

22. Triangle ABC has vertices at $A(0, 0)$, $B(3, 0)$, $C(0, 4)$. The point $P(1, 1)$ is to be reflected across this triangle. Find the reflection point P'. Use the *Triangle Reflection Point* theorem presented in the text. Be sure to follow its guidelines in labeling.

23. Find the three vertices of the reflection triangle for triangle ABC in problem 22.

24. For the triangle with vertices at $A(4, 5)$, $B(8, 8)$, and $C(4, 1/2)$, find the isotomic transversal for the line $\ell : y = 2x - 5$.

25. Solve $A = \sqrt{s(s - a)(s - b)(s - c)}$ for a.

26. Show that $(9, 10, 17)$ is a Heronian triangle by showing it has integer area.

27. In the Heronian triangle $(9, 10, 17)$, find the two primitive Pythagorean triples contained in it. Note: This task will be asked again in the *Pythagorean Triples* chapter when the student knows more about them.

28. Write a computer program to generate all Heronian triangles up to the smallest side length of 25. Note: Answers vary, but the author's C program in the appendix may be used as a guide.

29. Use the unified definition of a conic section to find the equation of the conic section with focus $F(5, 0)$, directrix d: $y = 4$, the point $P(x, y)$, and $e = 1/3$. Repeat this for $e = 1$ and $e = 3$.

Desargues' Theorem

Gérard Desargues (1591 – 1661), a French mathematician, discovered this theorem, which bears his name.

Desargues' Theorem: If two nondegenerate triangles (i.e. having positive area) are perspective at a point, then the intersection points of corresponding pairs of sides lie on a straight line as long as none of the pairs of sides are parallel. Furthermore, if the intersection points of corresponding pairs of sides of two nondegenerate triangles (implying that none of the pairs of sides are parallel) lie on a straight line, then the two triangles are in perspective at a point.

What does it mean for two triangles to be in perspective at a point? It means that the lines joining the corresponding vertices of the two triangles are concurrent (i.e. intersect at a common point). The intersection point of the lines through the vertices is called the center of perspectivity or the perspector. We usually denote the perspector with a P. The line that goes through the intersection points of the extensions of pairs of corresponding sides is called the axis of perspectivity or the perspectrix. Desargues' theorem says that if you have one, you automatically have the other. If you find that one pair of corresponding sides are parallel, then the perspectrix obtained from the other two intersection points will be parallel to the pair of sides that are parallel. We will not deal with the cases having more than one pair of sides parallel. We also will not deal with the cases involving degenerate triangles (i.e. the three vertices are collinear).

Thus, if you have two triangles $\triangle ABC$ and $\triangle abc$, and you can show that they are in perspective at a point, then you know that there is a perspectrix. Likewise, if you can show that two triangles have a perspectrix, then you know that the triangles have a perspector. Demonstration of having one of these things means that, through Desargues' theorem, you know that you automatically have the other.

Let us show a couple of triangles in perspective. We know that the perspectrix must exist by way of Desargues' theorem.

Let us talk about the picture in Figure 1. Observe the two triangles: $\triangle ABC$ and $\triangle abc$. They look like they are differently shaped, so we will not assume that they are similar triangles. The six vertices are $A(1, 6)$, $B(4, 0)$, $C(-1, 1)$, $a(8, 5)$, $b(12.4, 34/15)$, and $c(6, 2)$. We are just given the six vertices, so we must determine whether or not they are in perspective from a point or from a line. If we can show either one of these types of perspectivity, then we will have met the prerequisite conditions for Desargues' Theorem. Thus, we would know that the other type of perspectivity applied as well.

Let us determine if we have a perspector. If so, we will then look for the perspectrix. We need to know the equations of the three lines through the pairs of vertices *Aa*, *Bb*, and *Cc*. Then, we solve this over-determined system of equations. If we get no solution, we would know that we do not have a perspector, and thus, would not have a perspectrix either. If, however, we do get a single intersection point, then we would have found the perspector; and we could proceed with confidence in Desargues' theorem that we would successfully find the perspectrix.

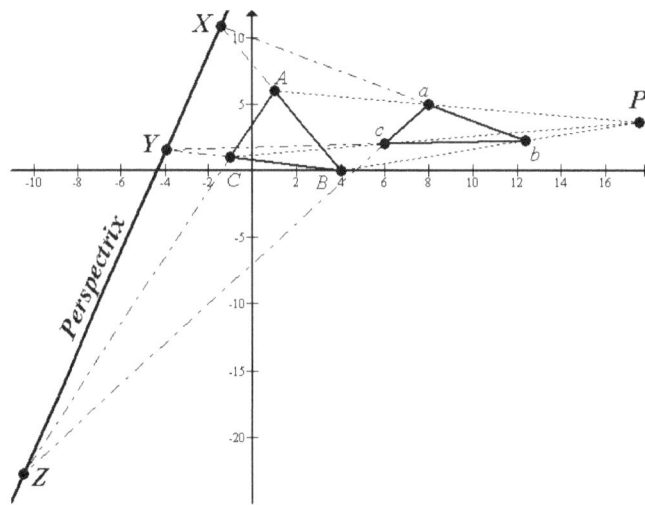

Figure 1: Perspective Triangles with Perspector *P* and Perspectrix

After performing the algebra, we arrive at the following three lines:

$$\textit{Aa}: \ y = \frac{-1}{7}x + \frac{43}{7}, \ \textit{Bb}: \ y = \frac{17}{63}x - \frac{68}{63}, \text{ and } \textit{Cc}: \ y = \frac{1}{7}x + \frac{8}{7}$$

We solve the over-determined system by choosing two of these equations to find the intersection point of the two chosen lines. Observe that none of the slopes are equal, so we know by inspection that these three lines will intersect at least in pairs. What we cannot tell by mere inspection is whether all three have a common intersection point. After we get the intersection point from the pair of equations we chose as our two equation system, we place that ordered pair into the third equation to see if it checks (i.e. if we get an identity). If it checks, then we have a common intersection point for the three lines, and this shows that the two triangles are in perspective from a point. If it fails to check, then we just have a couple of triangles plotted where Desargues' theorem does not apply. Well, solving this system shows us that the perspector is at the point

P(35/2, 51/14). So now we know that the perspectrix exists because of Desargues' theorem, and we can set out to find it.

The next thing to do is to find the equations of the pairs of corresponding sides of the two triangles. We will take them one pair at a time. The first pair of lines is *AB* and *ab*. We need to find the intersection point of these two lines.

$$AB: \ y = -2x + 8$$
$$ab: \ y = \frac{-41}{66}x + \frac{329}{33}$$

We solve this system of equations to find that the intersection point is *X*(-10/7, 76/7). We then proceed to the next pair of corresponding sides: *AC* and *ac*.

$$AC: \ y = \frac{5}{2}x + \frac{7}{2}$$
$$ac: \ y = \frac{3}{2}x - 7$$

We solve this system to find that the intersection point is *Z*(-21/2, -91/4). We find the third point on the perspectrix by setting up and solving the remaining pair of corresponding sides of the two triangles.

$$BC: \ y = \frac{-1}{5}x + \frac{4}{5}$$
$$bc: \ y = \frac{1}{24}x + \frac{7}{4}$$

This last system yields the remaining point on the perspectrix *Y*(-114/29, 46/29). It remains for us to use the three points *X*, *Y*, and *Z* that we have found to find the equation of the perspectrix. To that end, we can use two of the points to determine first the slope of the perspectrix with the slope formula. Then, we can use one of those points along with the slope to determine the equation of the perspectrix. However, we should all understand the importance of checking our work when it is possible to do so. Here, it is possible to do so. To check to see if we have the correct equation of the perspectrix, we place the third point into the perspectrix equation to see if we have an identity. If we do, then all is right with the world. If we do not, then at least one mistake has been made, and we hope that it is later in our work rather than earlier. And we definitely hope that the mistake was not as early on in our work as copying the problem down wrong, which I have seen happen. I must confess that I have done that once or twice along the way as well.

Well, we can pick the two points X and Y to find that the slope is $m = \dfrac{941}{254}$. We can then use the point X to find that the equation of the perspectrix is $y = \dfrac{941}{254}x + \dfrac{2051}{127}$. Finally, we insert the point Z into the equation to find that $\dfrac{-91}{4} = \dfrac{-91}{4}$. So we have found what Desargues' theorem assured us was there—the perspectrix.

We could have found the perspectrix first, as previously mentioned. Then, we would have known that the perspector existed by Desargues' theorem. Thus, we could have then proceeded to find the perspector.

A couple of remarks are in order. The first remark is that these Desargues' theorem problems really put the student's skills to the test. Fractions abound. The student has to find the equations of a lot of lines. With these equations, the student has to select appropriate systems of equations to solve. A lot of checking for correctness is also a must along the way because these problems are voluminous in terms of the amount of writing required to solve them. Making a mistake can produce of lot of wasted effort. It may be a prudent suggestion that when doing the homework, the student should use a graphing utility or a computer algebra system (CAS) to aid in the computations. A missed sign or hastily writing that $2 \cdot 3$ is five is a sad error to discover. Choosing an incorrect pair of equations is not something that the algebra technology can help with, however. That is an error in judgment. An extension of this remark is that the student should label the algebra. Conventions such as Aa: $y = \dfrac{-1}{7}x + \dfrac{43}{7}$ become very important when the work has to be checked for errors. Seeing a field of raw linear equations with a sentence at the bottom, "The perspectrix is _____." is frustrating to read later should the student have a need to borrow from that already completed work. It is also frustrating for the instructor to grade. If the work is neat and orderly, with the algebra and diagrams labeled, then it is much easier to read. It also shows discipline—the hallmark of a professional.

The other remark is that Figure 1 depicts just one type of way that two triangles can be in perspective. It is possible that the two triangles can overlap one another. One can be nested inside the other. They can be similar while not being parallel, so Desargues' theorem can still apply. The chapter on triangle geometry has a number of triangles in perspective, so these pairs of triangles have a perspector and a perspectrix. The perspector and perspectrix of these special cases of Desargues' theorem get named in honor of the mathematicians who discovered them because they were nontrivial examples of Desargues' theorem to find.

Exercises

1. a) Fill in the missing details of the example problem discussed in this theorem highlight chapter. The rest of the parts of this problem ask the student to move point b along BP further and further in one direction until the two triangles overlap.

   b) Find the coordinates of a new point b along the line BP such that the new line ab is parallel to the line AB. Explain in words what you expect to happen to the perspectrix. Find the equation of the new perspectrix to confirm your suspicions. Draw a graph of the complete situation and label all of the relevant parts.

   c) Move point b further along the line BP such that b is the same point as B. Thus, the two triangles now share a vertex. Find the new perspectrix.

   d) Move point b again to a place on the line BP such that the two triangles overlap. Find the new perspectrix for the situation you have chosen.

2. The first triangle has vertices $A(0, 0)$, $B(10, 0)$, and $C(3, 9)$. Let point a be the midpoint of side BC. Let point b be the midpoint of side AC. Let c be the midpoint of side AB.

   a) Find the coordinates of triangle abc by computing the midpoints of the three sides. This triangle is called the medial triangle of triangle ABC.

   b) Find the equations of the three lines Aa, Bb and Cc. These lines are called the median lines.

   c) Show that the lines Aa, Bb, and Cc are concurrent. This perspector is known as the centroid of triangle ABC. It is a fact that a triangle and its medial triangle share the same centroid because they share the same three median lines.

   d) Show that triangle ABC is similar to triangle abc by computing the distance relationships as follows: $\dfrac{d(A,B)}{d(a,b)} = \dfrac{d(B,C)}{d(b,c)} = \dfrac{d(A,C)}{d(a,c)}$.

   e) Observe that more than one pair of corresponding sides is parallel. Thus, we will not attempt to find the perspectrix as promised in the chapter.

3. The first triangle has vertices $A(0, 0)$, $B(7, 0)$, and $C(3, 6)$. We will find the second triangle, the perspector, and the perspectrix as follows:

a) Find the equation of the line perpendicular to *BC* traveling through *A*. We call this line the altitude through *A*. Point *a* is the intersection point of the line *BC* with the altitude through *A*. Observe that the line *Aa* is the altitude through *A*. Find the lines *Bb* and *Cc* in a similar manner.

b) Show that the lines obtained in part *a* are concurrent (i.e intersect in the same point).

c) The point of concurrency found in part *b* is called the orthocenter of the triangle. The new triangle *abc* is called the orthic triangle of triangle *ABC*. The orthocenter is the perspector for these two triangles. Since we have shown that the triangles are in perspective from a point, we know that they are in perspective from a line. Find the perspectrix for this situation. We call this perspectrix the orthic axis.

4. The first triangle is the same as in exercise 3. The second triangle will be obtained by following the steps outlined in the rest of this problem.

a) Find the midpoints of the three sides of the triangle. Find the equation of the circle going through these midpoints. Note that this circle is also the circumcircle of the orthic triangle—thus, the feet of the altitudes are on this circle, called the nine-point circle.

b) The center of the circle in part a is called the nine-point center. We will have more to say about this in the *Triangle Geometry* chapter. The orthic triangle of the previous problem will be used to find our second triangle for this problem. Reflect vertex *a* of the orthic triangle across the nine-point center to obtain point *a'*. Repeat for the other two vertices of the orthic triangle to obtain points *b'* and *c'*. The perspector of triangle *ABC* and *a'b'c'* is called the Prasolov point. Find the perspectrix and the Prasolov point.

Pythagorean Triples

Pythagorean triples are ordered triples of positive integers of the form (a, b, c), where the Pythagorean formula $a^2 + b^2 = c^2$ holds. This formula is for a right triangle, and it says that the sum of the squares of the legs of the right triangle is equal to the square of the hypotenuse of the triangle. Two Pythagorean triples are (3, 4, 5) and (5, 12, 13). We customarily write Pythagorean triples in ascending order. Thus, a and b are the short leg and long leg, respectively; and c is the hypotenuse. There are infinitely many Pythagorean triples, so it is impossible to list them all. If the student enters in the words "Pythagorean triples" into a search engine on the Internet, a large quantity of hits will come back. Thus, a wealth of information about Pythagorean triples exists. Our interest in Pythagorean triples in this course is to be able to construct nice problems with them. All too often, the problems found in textbooks involve the two Pythagorean triples cited above. We would like more variety in our computational tasks. One of the nice things about Pythagorean triples with conic sections is that we can have integers for the vertices and foci, and the equations that we compute with will have integer coefficients. With some work, we can show that circles, ellipses, and hyperbolas have an infinite quantity of rational points that we can use to make nice problems. The ellipses and hyperbolas seem to need perfect squares for their constants, and when that happens, we can find the rational points with the help of Pythagorean triples. We will also see problems that do not have nice features like this (i.e. without perfect squares for the constants or not bothering to use the rational points that are available), but we will see a Pythagorean triple featured at times. Hence, it is important for the student to know something about these.

Euclid's m and n Expressions

Euclid, in one of the books of *The Elements* (c. 300 B.C.), enunciates the triple of expressions $(m^2 - n^2, 2mn, m^2 + n^2)$ where $m, n \in Z^+$, $m > n$ will allow us to generate all possible Pythagorean triples (accounting for interchanging the roles of a and b—the two legs) because any Pythagorean triple (a, b, c) has an associated infinite family of Pythagorean triples (ka, kb, kc) where $k \in Z^+$. The symbol Z refers to the set of integers, and the superscript + means the positive integers. If no such common factor of k exists between the three members of a Pythagorean triple, then the Pythagorean triple is said to be primitive. The members a, b, and c in a primitive Pythagorean triple are relatively prime. The triple of expressions cited above will generate all primitive Pythagorean triples along with some nonprimitive ones. Often, $m^2 - n^2 > 2mn$; however, this is not always the case. To use these expressions, we just pick two positive integers m and n such that $m > n$ and compute what the Pythagorean triple is. For example, let us choose m as 4 and n as 1. Then, $m^2 - n^2 = 4^2 - 1^2 = 15$; $2mn = 2 \cdot 4 \cdot 1 = 8$; and $m^2 + n^2 = 4^2 + 1^2 = 17$. Thus, (8, 15, 17) should be a Pythagorean triple. We observe that $8^2 + 15^2 = 17^2$. We can factor each member of the Pythagorean triple (8, 15, 17) as $(2 \cdot 2 \cdot 2, 3 \cdot 5, 17)$. No common factor exists, so we conclude that the Pythagorean triple is a primitive one. Since we know that this is a primitive triple, we can construct an infinite sequence of Pythagorean triples $k(8, 15, 17)$, $k = 1, 2, 3, \ldots$ This generates the sequence of Pythagorean triples (8, 15, 17), (16, 30, 34), (24, 45, 51), (32, 60, 68), (40, 75, 85), ... Any one of these triples could be used in a conic section problem. We may choose another value of m and n to get another Pythagorean triple. We can factor its members to see if they have a common k value to create the associated primitive Pythagorean triple. Then, we can construct the infinite sequence of Pythagorean triples associated with that primitive one. I have written a computer program that will help the student in this cause because the program allows the user to enter in a range of values for m and n. It will find all of the Pythagorean triples falling into those ranges for m and n, and it weeds out all cases where $m \leq n$. The user can specify that the program only display primitive Pythagorean triples. Thus, with a ready-made list of primitive Pythagorean triples, the associated infinite sequences for each is a mere formality.

I used the program Pythtrip.exe with m ranging from 1 to 15 and n ranging from 1 to 15. I asked the program to give me only the primitive Pythagorean Triples. I also asked the program to flag the Pythagorean triples with a difference of 1 between the terms with an asterisk. I got the following printout.

```
*1)  [3, 4, 5]       *2) [5, 12, 13]    3) [8, 15, 17]
*4)  [7, 24, 25]     *5) [20, 21, 29]  *6) [9, 40, 41]
7)  [12, 35, 37]      *8) [11, 60, 61]  9) [28, 45, 53]
10) [33, 56, 65]    *11) [13, 84, 85]     12) [16, 63, 65]
13) [48, 55, 73]       14) [39, 80, 89]
*15) [15, 112, 113]      16) [36, 77, 85]
17) [65, 72, 97]      *18) [17, 144, 145]
19) [20, 99, 101]      20) [60, 91, 109]
21) [51, 140, 149]     *22) [19, 180, 181]
23) [44, 117, 125]      24) [88, 105, 137]
25) [85, 132, 157]      26) [57, 176, 185]
*27) [21, 220, 221]      28) [24, 143, 145]
*29) [119, 120, 169]      30) [95, 168, 193]
*31) [23, 264, 265]      32) [52, 165, 173]
33) [104, 153, 185]      34) [133, 156, 205]
35) [105, 208, 233]      36) [69, 260, 269]
*37) [25, 312, 313]      38) [28, 195, 197]
39) [84, 187, 205]      40) [140, 171, 221]
41) [115, 252, 277]      42) [75, 308, 317]
*43) [27, 364, 365]
```

With a printout such as this of all of the primitive Pythagorean triples within the specified ranges for m and n, which is almost in order by hypotenuse, we may make some interesting observations. The hypotenuse 65 has two primitive Pythagorean triple representations. However, 65 is a composite number, which factors into $5 \cdot 13$. We have hypotenuses of 5 and 13 on the printout. Thus, we have four ways to represent the square of 65 as a sum of squares of two smaller natural numbers. Thus, $65^2 = 16^2 + 63^2 = 33^2 + 56^2 = 25^2 + 60^2 = 39^2 + 52^2$. Based on this printout, can we say for sure that there are only four ways to write the square of 221 as the sum of squares of smaller natural numbers? No, we can't because the 40th triple on the printout has a hypotenuse of 221, and the printout stops at the 43rd triple. The triples are not ranked in strict order by hypotenuse— they are only more or less in order. The printout would have to be extended by a fair amount to enable us to see if there were more triples with a hypotenuse of 221. It turns out that there are only four ways to represent the square of 221 as the sum of squares of two smaller natural numbers because I extended the printout with identical ranges of m and n from 1 to 25 and saw no further occurrences of 221 as the hypotenuse. Only the extended printout gave me the confidence to say this with certainty.

The above argument may leave some students a bit unsettled. The argument against further sums of squares of smaller natural numbers yielding the square of 221 rests on the premise of an extended printout that is more or less in order by hypotenuse. To be able to say something with certainty in mathematics requires a proof. Another class beyond this one is called *Introduction to Advanced Mathematical Thinking*, and it deals with proofs— how to read them and construct them. We can agree that there are finitely many natural number solutions to the equation $x^2 + y^2 = 221^2$. We rearrange this equation to $y^2 = 221^2 - x^2$, and let x be 1, 2, 3, ..., 219, 220. We can solve each of these equations for y to see if it yields a natural number. We compile a list of all cases when this is so. This would constitute an exhaustive proof by cases. Nevertheless, the printout argument is sound due to logic because the expressions in m and n have been proven to yield all primitive Pythagorean triples. As m and n become ever larger, the hypotenuses generated become ever larger. That is evident in the printout. One will not see a hypotenuse of 221 on a printout of primitive Pythagorean triples with larger ranges for m and n than from 1 to 25 in, say, the 5000th position. Because the printout is exhaustive and hypotenuses of 221 have clearly been passed up, the argument above will not be overturned. Having said all this in this paragraph, proofs will not be required in this class. The aspiring math student should be aware, however, that this topic of proof exists and comes up in future courses.

Another course, even beyond the one mentioned above, is called *Introduction to Number Theory*. This course will show the student how to prove that there are only four cases where 221^2 can be written as the sum of two other squares. The student can look forward to that when he or she gets into that course.

Pythagorean Triples from a Fibonacci Sequence

A Fibonacci sequence is a recursive sequence such that the new term is the sum of the previous two terms. We normally use a capital F to denote a Fibonacci sequence. Thus, the recursive equation $F_{n+1} = F_n + F_{n-1}$ denotes a nonspecific Fibonacci sequence. We must specify the necessary previous terms to enable the recursive formula to generate a specific Fibonacci sequence. The usual case is to let $F_1 = 0$, and $F_2 = 1$. This generates the sequence 0, 1, 1, 2, 3, 5, 8, 13, 21, 34, ... This sequence is *the* Fibonacci sequence. Numbers in *the* Fibonacci sequence are called Fibonacci numbers. Choosing other values of F_1 and F_2 will generate *a* Fibonacci sequence. Numbers in *a* Fibonacci sequence are not distinguished with any special name. For example, a Fibonacci sequence is 3, 4, 7, 11, 18, 29, ..., where $F_1 = 3$ and $F_2 = 4$. Another Fibonacci sequence is generated when $F_1 = 92$ and $F_2 = 93$. We get an infinite quantity of unique Fibonacci sequences by choosing $F_1 = t$ and $F_2 = t + 1$ for any $t \in Z^+$. We may let F_1 and F_2 have larger differences than 1 as well, of course. We may even let $F_2 < F_1$; however, such a Fibonacci sequence immediately straightens out to its hallmark behavior of each term being greater than its predecessor.

A Fibonacci sequence has the property that it can produce Pythagorean triples. Take any four consecutive terms in a Fibonacci sequence. Let us label these terms w_1, w_2, w_3, and w_4. The product of the outer two and twice the product of the inner two yields two of the numbers (both legs) in the Pythagorean triple. The sum of the squares of the inner two numbers always yields the hypotenuse, which is the c term in the Pythagorean triple. Thus, $(w_1 \cdot w_4, \ 2 \cdot w_2 \cdot w_3, \ w_2^2 + w_3^2)$ will usually give us the ascending order for the triple, but we may have to switch the first two terms sometimes. Let us use a Fibonacci sequence with $F_1 = 8$ and $F_2 = 17$. We get the sequence 8, 17, 25, 42, 67, 109, ... We may choose any four consecutive terms from this sequence. Let us choose 25, 42, 67, and 109. Thus, $w_1 = 25$, $w_2 = 42$, $w_3 = 67$, and $w_4 = 109$. We compute the Pythagorean triple $(w_1 \cdot w_4, \ 2 \cdot w_2 \cdot w_3, \ w_2^2 + w_3^2)$, which yields (2725, 5628, 6253). The student can verify that this is a Pythagorean triple. We may factor each term in this Pythagorean triple to see if a common factor of k exists. Thus, we get $(5 \cdot 5 \cdot 109, 2 \cdot 2 \cdot 3 \cdot 7 \cdot 67, 13 \cdot 13 \cdot 37)$, and we see no common factor of k. We may now create an infinite sequence of triples $k \cdot$(2725, 5628, 6253), where $k = 1, 2, 3, ...$

The Pythagorean triple program, Pythtrip.exe, will generate Pythagorean triples from a Fibonacci sequence. I used the program with the first number as 3 and the second number as 9, and I told the program to give me 15 Pythagorean triples. The output from the program is the following:

```
The Fibonacci Sequence is the following:
3, 9, 12, 21, 33, 54, 87, 141, 228, 369, 597, 966, 1563,
2529, 4092, 6621, 10713, 17334

1) [ 63, 216, 225 ]       2) [ 297, 504, 585 ]
3) [ 648, 1386, 1530 ]      4) [ 1827, 3564, 4005 ]
5) [ 4653, 9396, 10485 ]      6) [ 12312, 24534, 27450 ]
7) [ 32103, 64296, 71865 ]      8) [ 84177, 168264, 188145 ]
9) [ 220248, 440586, 492570 ]
10) [ 576747, 1153404, 1289565 ]
11) [ 1509813, 3019716, 3376125 ]
12) [ 3952872, 7905654, 8838810 ]
13) [ 10348623, 20697336, 23140305 ]
14) [ 27093177, 54186264, 60582105 ]
15) [ 70930728, 141861546, 158606010 ]
```

Notice that these all have common factors of 9, and some have terms that are all even. Thus, none of these triples are primitive. Let us find the corresponding primitive Pythagorean triple for the ninth Pythagorean triple on this printout. Thus, the triple (220248, 440586, 492570) factors as

$(2 \cdot 2 \cdot 2 \cdot 3 \cdot 3 \cdot 7 \cdot 19 \cdot 23, 2 \cdot 3 \cdot 3 \cdot 3 \cdot 41 \cdot 199, 2 \cdot 3 \cdot 3 \cdot 5 \cdot 13 \cdot 421)$. We see that the common factor is $k = 2 \cdot 3 \cdot 3 = 18$. Thus, this ninth triple is the triple $18 \cdot (12236, 24477, 27365)$, where the triple $(12236, 24477, 27365)$ is the associated primitive Pythagorean triple. We may now generate the infinite sequence of Pythagorean triples $k \cdot (12236, 24477, 27365)$, where $k = 1, 2, 3, \ldots$

The Dye-Nickalls Method of Generating Pythagorean Triples

A third way to generate Pythagorean triples is a recently discovered way by R. H. Dye and R. W. D. Nickalls in the paper *A New Algorithm for Generating Pythagorean Triples*. This paper was published in The Mathematical Gazette (1998); volume 82 (March), pages 86-91. This is the method to generate Pythagorean triples with a constant gap between the short leg and long leg or between the long leg and the hypotenuse. The authors state in the paper that as far as they know, these algorithms have not been described before. This chapter will not delve into the mathematics as deeply as their paper. We will just take their equations as they are and use them.

The notation is a bit strange on the first pass, which makes it difficult to understand. To help the student overcome this, we will develop it here. Then, we will use their equations. The variable a measures the gap between the short leg and long leg. Thus, in the triple (3, 4, 5), the gap between the short leg and long leg is 4 – 3 = 1. Thus, a = 1 for this triple. The variable b measures the gap between the long leg and the hypotenuse. Thus, in the triple (3, 4, 5), the gap between the long leg and hypotenuse is 5 – 4 = 1. So, b in this triple is 1. In the triple (5, 12, 13), a = 7, and b = 1. In the triple (60, 91, 109), a = 31, and b = 18. I ran the program Pythtrip.exe in its Dye-Nickall's mode with a = 1 held constant starting with the triple (3, 4, 5) and asked for 7 triples. The program produced the following printout:

```
1) [ 3, 4, 5 ]        2) [ 20, 21, 29 ]
3) [ 119, 120, 169 ]     4) [ 696, 697, 985 ]
5) [ 4059, 4060, 5741 ]     6) [ 23660, 23661, 33461 ]
7) [ 137903, 137904, 195025 ]
```

The program is generating a sequence of Pythagorean triples, but this sequence is not like what we have seen before. In the first place, these are all primitive because the short leg and long leg are separated by just one number, which makes the legs relatively prime. Thus, we do not need to know what the hypotenuse is to determine if the three members are relatively prime together because it is established already with the first two members of the triple. Furthermore, no discernible pattern emerges other than the legs having the difference of one between them for each triple. Thus, the variable a has been preserved at 1 as it was supposed to be. However, b in the first triple is 1, and b in the second triple is 8. The variable b then becomes 49 and 288 in turn for the next two triples. It increases rapidly beyond that. So, while the variable a is pinned down to 1 for all triples in the sequence, the variable b increases with each new triple in the sequence. A similar phenomenon occurs in a Dye-Nickalls sequence of Pythagorean triples when the variable b is held constant although the variable a does not increase as rapidly as b does when the variable a is held constant.

We can use the terminology (X, Y, Z) to represent a term in a Dye-Nickalls sequence of Pythagorean triples. We can introduce subscripts to specify a particular member in a sequence. Thus, in the printout above, (X_1, Y_1, Z_1) is (3, 4, 5); and (X_7, Y_7, Z_7) is (137903, 137904, 195025). We are now ready to present the Dye-Nickalls recursive equations.

For a held constant (i.e. $X_n + a = Y_n$): $\left\{\begin{array}{l} X_{n+1} = 3X_n + 2Z_n + a \\ b_{n+1} = X_n + Z_n \end{array}\right\}$

For b held constant (i.e. $Z_n = Y_n + b$): $\left\{\begin{array}{l} Z_{n+1} = 2X_n + Z_n + 2b \\ a_{n+1} = X_n + Y_n \end{array}\right\}$

We will fully describe the a = constant one. We must give the system one previous triple to start the process rolling since the new term has a subscript of $n + 1$, and the previous terms only have subscripts of n. We do need all three elements of the previous triple because the program Pythtrip.exe checks to see that it is starting with a Pythagorean triple. In addition, the variables a and b are computed from the given initial Pythagorean triple by subtracting the X-value from the Y-value and by subtracting the Y-value from the Z-value, respectively. We first need to compute the new X-value from the given Pythagorean triple. This is accomplished with the top equation $X_{n+1} = 3X_n + 2Z_n + a$. To use a recursive equation, we need to render it in a form with specific values of n. Since we are constructing a sequence, and we are given the first triple in the sequence, it makes sense to let the subscripts referring to this first triple be 1. Thus, we need to start n at 1. Then, X_n and Z_n when n is 1 become X_1 and Z_1, respectively. The variable X_{n+1} when n is 1 becomes X_{1+1}, which is X_2. Therefore, the top equation when n is 1 becomes $X_2 = 3X_1 + 2Z_1 + a$. Well, X_1 and Z_1 are in the first triple and are 3 and 5, respectively. The variable a is the positive difference between the short and long leg, which is $4 - 3 = 1$. We have enough information to compute: $X_2 = 3X_1 + 2Z_1 + a = 3\cdot3 + 2\cdot5 + 1 = 20$. Look on the printout for the X-value for the second triple. It is 20. The second equation in the curly brace is $b_{n+1} = X_n + Z_n$. To make this equation a specific equation, we let $n = 1$, and compute the subscripts. Thus, when $n = 1$, this second equation becomes $b_2 = X_1 + Z_1 = 3 + 5 = 8$. This is the 8 that we observed earlier. The equation in parentheses before the curly brace is $X_n + a = Y_n$. This is not a recursive equation because we do not compute a next value of something. Rather, this equation tells us that for any triple in the sequence, the Y-value is the variable a more than the X-value in that triple. The variable a is already established as 1 from the first triple. We let $n = 2$ to access the second triple we are now working on. Thus, $X_n + a = Y_n$ becomes $X_2 + a = Y_2$, which means that $Y_2 = 20 + 1 = 21$. Look at the Y-value for the second triple on the printout. It is 21. To compute Z_2, we add the gap size between it and Y_2, which means that $Z_2 = Y_2 + b_2 = 21 + 8 = 29$. The third triple in the sequence will repeat these steps after incrementing n to 2. Then, increment n to 3 and repeat. The process continues with incrementing n and repeating until the desired triple in the sequence is reached or the process runs out of steam due to some external physical limit. The Dye-Nickalls equations for computing a sequence of Pythagorean triples by holding b constant operate similarly to what was just described.

For the special cases $a = 1$ (a = constant) and $b = 1$ (b = constant), these recurrence equations will generate all the corresponding Pythagorean triples with gaps of size 1. However, for all other cases of specified a and b, the above equations necessarily yield only some of the triples. One reason is that if the initial triple is primitive, then every member of the generated sequence is also primitive. Furthermore, if the initial triple has a common factor k, then all subsequent triples will have that common factor of k as well. Thus, if a triple exists such that its hypotenuse is between the hypotenuses of adjacent triples in a Dye-Nickalls Pythagorean triple sequence, and the triple has the specified gap *a* or *b* but does not have the common *k* value as the gcd of its three members (which is 1 for an initial primitive triple), then that triple will be missed by these Dye-Nickalls equations. These claims are proven in the Dye-Nickalls paper with a divisibility proof, so we may have confidence in these claims due to this proof. The student is not expected to construct such a divisibility proof, but the student is expected to know that these equations will miss some triples having the specified gap because the triples that are missed will not have the common factor of *k* (as the gcd) that the original Pythagorean triple fed to the system had. The student is also expected to be able to find any triples missed by the Dye-Nickalls equations. It is also evident that these equations do not even catch all of Pythagorean triples having that common factor of *k* present.

We will work an example to illustrate. Let us feed in the triple (5, 12, 13) holding *a* constant. Let us generate the next triple. We compute $a = 12 - 5 = 7$. Since 5 and 13 are both prime, the common factor of *k* between the three members of the triple is 1. (i.e. $k = \text{gcf}(5, 12, 13) = 1$). Next, we compute $X_2 = 3X_1 + 2Z_1 + a = 3(5) + 2(13) + 7 = 48$. Then, $Y_2 = X_2 + a = 48 + 7 = 55$. Now, we compute $b_2 = X_1 + Z_1 = 5 + 13 = 18$. Finally, $Z_2 = Y_2 + b_2 = 55 + 18 = 73$. Thus, our next triple in the sequence is (48, 55, 73). We would like to know which triples were missed by the Dye-Nickalls equations. Since $a = 7$, any triple in this sequence has the following form:

$$(x, x+a, \sqrt{x^2 + (x+a)^2}) = (x, x+7, \sqrt{2x^2 + 14x + 49})$$

We need to know the values of the *z* position over the interval $5 < x < 48$ to see if we get any integer solutions. Of course, one could compute all those *z* terms by hand, but let's be real. With a grapher, we have several ways to accomplish this feat. One way is to enter the square root function into the *y*-editor, and use the table feature with the table starting at 6 and the increment set to 1. Then, scroll down the list to see if there are any integer results stopping, of course, at $x = 47$. Another way is to use the statistical editor (or list editor). In one list, say L1, put the sequence of integers from 6 to 47 using the grapher's sequence command. This command will generate the integers from 6 to 47 in order, and is useful for large lists from something like 11 to say 5000. For the next list, say L2, we would type the command √(2*(L1)^2+14*(L1)+49) referencing the first list to generate the *z*-terms in that list. Then, scroll down the second list to see if there are any integers. A third way to accomplish the feat is to write a program, but only the most ambitious students will try this approach. However, if the range of *x*-values for which the student has to scroll through becomes large, the appeal of the program simply reporting the skipped triples grows. When we accomplish the feat, we find that the triples that were missed were (8, 15, 17) with $k = 1$ and (21, 28, 35) with $k = 7$. We see that one of these triples had the same value of *k* as the given triple, but the other missed triple had a different value of *k*.

Identity Equations

In the exercises, the student is asked to show when an equation is an identity. This may be something new for the student, so we will take some pains to show what this means here. The symbol for an identity is an equals sign with three bars instead of the usual two bars. Hence, an equals sign is =, and an identity sign is ≡. When solving an algebra equation, we may alter the value of both sides as long as what we do to one side is done to the other side of the equation. Thus, to solve the equation $2x + 5 = 9$, we first subtract 5 from both sides. However, when we do this, we clearly have changed the value of the right hand side of this equation from a 9 to a 4. This is OK because we are not concerned about preserving the value of each side—we are only concerned about keeping the sides equal to each other. Well, in doing identity problems, we are concerned about preserving the value of each side. Thus, we are not allowed to add to the sides or subtract from the sides of an identity equation. Having said this, we are allowed to add zero to either side (or both) because zero is the additive identity, and we will sometimes add very clever forms of zero to an expression (or side of an identity problem) to get it to reduce to something more manageable. We are also not allowed to multiply the sides or divide the sides by anything other than one since one is the multiplicative identity. The student who has had trigonometry is familiar with the process of identity equations. We usually transform the more complicated side into the less complicated side. The verification path is a straight line down the one side of transformations. However, in extreme situations, we may have to transform both sides until they look like each other. The verification path of analysis will then go down one side and bend up to the other side. In either case, the path of steps to perform to transform one side into the other is clear.

Example 1: Show that the equation $D = RT$ is an identity when $T = D/R$.

Solution: $D = RT$ (original equation)

$D \equiv RT$ (replace equals sign with identity symbol to create identity equation)

$D \equiv R \cdot \dfrac{D}{R}$ (substitute the equivalent value of T given)

$D \equiv D$ (after canceling, we see that the two expressions have the same <u>value</u>)

Example 2: Find an expression for z to make the equation $3xyz = a^2 + 2b^2$ an identity equation if it is known that $x = 2a - 5b$ and $y = \dfrac{1}{b}$.

Solution: The simplest way to do this is to solve the given equation for z. Then, plug in the given expressions for x and y, and reduce the result as much as possible.

$$z = \frac{a^2 + 2b^2}{3xy} = \frac{a^2 + 2b^2}{3(2a - 5b) \cdot \frac{1}{b}} = \frac{b \cdot (a^2 + 2b^2)}{3(2a - 5b)}$$

Let us now show that the equation $3xyz = a^2 + 2b^2$ is an identity equation with the expressions for x, y, and z. However, we need to be mindful of the denominators, so $b \neq 0$ and $2a \neq 5b$.

$3 \cdot x \cdot y \cdot z = a^2 + 2b^2$ (original equation)

$3 \cdot x \cdot y \cdot z \equiv a^2 + 2b^2$ (identity symbol inserted)

$3 \cdot (2a - 5b) \cdot \dfrac{1}{b} \cdot \dfrac{b(a^2 + 2b^2)}{3(2a - 5b)} \equiv a^2 + 2b^2$ (make the substitutions)

$a^2 + 2b^2 \equiv a^2 + 2b^2$ (simplify, and the identity is revealed)

Exercises

1. Write the infinite sequence of Pythagorean triples for the following Pythagorean triples. Be sure that the first triple for when $k = 1$ is the corresponding primitive Pythagorean triple (i.e. write out the prime factorizations). Construct these sequences in the same fashion as they are done in the text. Show all work. In each case, write out the first five terms of the sequence, and indicate that the sequence continues with ellipsis (i.e. the three dots: "..."").

    a) (3, 4, 5)   b) (5, 12, 13)   c) (48, 55, 73)   d) (234, 480, 534)
    e) (165, 532, 557) f) (279, 440, 521) g) (156, 495, 519) h) (252, 539, 595)

2. Compute the Pythagorean triples (a, b, c) using Euclid's m and n equations when m and n are the following:

    a) $m = 24$, $n = 11$ b) $m = 5$, $n = 4$ c) $m = 150$, $n = 99$ d) $m = 120$, $n = 175$

3. Use Euclid's m and n equations to construct all possible infinite sequences of Pythagorean triples for m ranging from 9 to 11 and n ranging from 8 to 10 by hand (i.e. without the benefit of the program Pythtrip.exe). Be sure that the first triple for when $k = 1$ is the corresponding primitive Pythagorean triple (i.e. write out the prime factorizations). Construct these sequences in the same fashion as they are done in the text. Show all work.

4. Use the program Pythtrip.exe to generate a printout with m ranging from 20 to 30 and n ranging from 18 to 28. Tell the program not to do just the primitive Pythagorean triples (i.e., it will find all of the triples in these ranges for m and n), and flag the Pythagorean triples with a difference of 1 between the sides. The instructor will assign enough triples to each student in the class so that all of the nonprimitive triples from the printout will get used for finding their corresponding primitive Pythagorean triples. The students will construct the infinite sequences of Pythagorean triples from the ones that they are assigned on the board. Compare corresponding primitive triples with the rest of the students in the class on the board. How many of the triples on this printout have the same corresponding primitive Pythagorean triple?

5. Generate by hand the first ten terms of the Fibonacci sequence when $F_1 = 4$ and $F_2 = 9$. Use the seventh, eighth, ninth, and tenth terms of this sequence to create a Pythagorean triple. Find the associated primitive Pythagorean triple, and write the infinite sequence of triples for it. Write the first five Pythagorean triples for this sequence.

6. Generate by hand the first six terms of the Fibonacci sequence when $F_1 = 6$ and $F_2 = 15$. Use the third, fourth, fifth, and sixth terms of this sequence to create a Pythagorean triple. Find the associated primitive Pythagorean triple, and write the infinite sequence of triples for it. Write the first five Pythagorean triples for this sequence.

7. a) Show that the m and n expressions yield an identity when plugged into the Pythagorean formula.

b) Show that any four consecutive terms in a Fibonacci sequence yield a Pythagorean triple under the conditions that the product of the outer two terms along with twice the product of the inner two terms yield the legs and the sum of the squares of the outer two terms yield the hypotenuse. [Hint: Let the four consecutive terms be $p, q, p + q$, and $p + 2q$. Then, subject these terms to the conditions to obtain the legs and hypotenuse in terms of p and q. Finally, show that the legs and hypotenuse yield an identity when plugged into the Pythagorean formula.]

8. In the text, the Pythtrip.exe program was used to generate 15 Pythagorean triples from a Fibonacci sequence. The first five triples on the printout were the following:
1) [63, 216, 225],   2) [297, 504, 585], 3) [648, 1386, 1530], 4) [1827, 3564, 4005], and 5) [4653, 9396, 10485]. The simple m and n formulas will generate all primitive Pythagorean triples, but they will not generate all possible Pythagorean triples. None of the five triples in this list are primitive.

a) Find the corresponding primitive Pythagorean triples for these five triples.

b) What are the values of m and n that would yield the five primitive triples in part a?

c) Will the m and n formulas find any of the first five triples on the printout? If so, find the values of m and n for these triples.

9. Use the Dye-Nickalls equations for holding the variable a constant to compute the next three triples by hand with the following initial triples:

a) (24, 45, 51)   b) (28, 45, 53)   c) (14, 48, 50)   d) (40, 42, 58)

10. Identify any triples that the Dye-Nickalls equations missed in question 9 (i.e. triples that have a gap size of the variable a, but were passed over by the Dye-Nickalls equations).

11. Use the Dye-Nickalls equations for holding the variable b constant to compute the next three triples by hand with the following initial triples:

a) (36, 77, 85)   b) (13, 84, 85)   c) (39, 80, 89)   d) (18, 80, 82)

12. Identify any triples that the Dye-Nickalls equations missed in question 11 (i.e. triples that have a gap size of the variable b, but were passed over by the Dye-Nickalls equations).

13. Regarding the printout having seven triples with $a = 1$, discuss whether or not any triples were missed up to that point. If triples were missed, identify them. If triples were not missed, provide an explanation why none were missed and state the circumstances required for the Dye-Nickalls equations to miss triples.

14. Use a grapher to verify that the twelfth and thirteenth members of the Dye-Nickalls sequence of Pythagorean triples with $a = 1$ and the first triple is (3, 4, 5) are indeed Pythagorean triples. What went wrong? Write an essay describing the problem and how we can repair the information in the 13th and 14th triples. Include in the essay the correct 13th and 14th triples. [Hint: The C language stores the integers generated by the Dye-Nickalls equations in memory using four 8-bit memory locations.]

1) [3, 4, 5]   2) [20, 21, 29]   3) [119, 120, 169]   4) [696, 697, 985]
5) [4059, 4060, 5741]   6) [23660, 23661, 33461]   7) [137903, 137904, 195025]
8) [803760, 803761, 1136689] 9) [4684659, 4684660, 6625109]
10) [27304196, 27304197, 38613965] 11) [159140519, 159140520, 225058681]
12) [927538920, 927538921, 1311738121]

13) [1111125707, 1111125708, 3350402749]
14) [1444248028, 1444248029, 1610809189]

15. a) Find the *m* and *n* values necessary to generate the first five triples listed in problem 14.

   b) Construct a Fibonacci sequence that yields 4 consecutive terms such that when these four terms are subjected to the conditions in problem 7b, the sixth Pythagorean triple in problem 14 is yielded.

16. Find the two primitive Pythagorean Triples contained in the Heronian triangle (9, 10, 17).

17. Find the two primitive Pythagorean Triples contained in the Heronian triangle (13, 14, 15).

The Circle

We have several things to go through in this chapter. We need to be able to locate rational points on a circle, but this will be done later in this chapter. With the ability to locate rational points on a circle, we have an application using that ability: plotting a triangle with given rational lengths on the Cartesian plane.

The student should already be able to find the equation of a circle given three points. The center-radius form of the equation of a circle is $(x-h)^2 + (y-k)^2 = r^2$. The center of the circle is (h, k), and the radius of the circle is $|r|$. Given three points that are not collinear, we can use this equation to find the equation of the circle going through the three points.

Example: Find the equation of the circle going through (1,-2), (5,-6), (-7, 8).

We replace the x and y coordinates for x and y, respectively, in the equation. This gives us three equations, one for each point. We have three unknowns: h, k, and r.

$$\begin{cases} (1-h)^2 + (-2-k)^2 = r^2 \\ (5-h)^2 + (-6-k)^2 = r^2 \\ (-7-h)^2 + (8-k)^2 = r^2 \end{cases}$$

We solve this system by expanding out the binomials. Observe that each equation is equal to r^2. This means that they are equal to each other. We can take these equations in pairs to create a system of two equations in two unknowns. Thus, set the left hand side of the first equation equal to the left hand side of the second equation. This is the first new equation. Set the left hand side of the first equation equal to the left hand side of the third equation. This is the second new equation. The third possible new equation is redundant, so we leave it out. The h^2 and k^2 terms drop out of our reduced system, and we have two linear equations left. From there, we find h and k. Then, plug h and k back into any of the three original equations to find r^2. With h, k, and r^2 known, plug them into the equation $(x-h)^2 + (y-k)^2 = r^2$ to obtain the equation of the circle through the three points.

When we do this, we find $h = 62$ and $k = 55$. We plug these into the first equation to find that $r^2 = 6970$. We construct the equation $(x-62)^2 + (y-55)^2 = 6970$. We check with the grapher to see if the three given points each yield an identity to be sure our work is correct, and we find that it is. Note that this example shows us that 6970 is expressible as a sum of two squares in three ways. Perhaps 6970 can be expressible as a sum of two squares in even more ways than this. It is possible to know because a deep connection exists between a number being expressible as a sum of squares in more than one way and its lack of primality. In fact, a formula exists that tells us how many ways a composite number can be written as a sum of two squares. However, this takes us too far into number theory, but the interested student can look forward to this in a future math class. We discussed this in the last chapter.

Another equation of a circle is from the reduced form of the general second-degree equation $Ax^2 + Bxy + Cy^2 + Dx + Ey + F = 0$. When $B = 0$, $A = C = 1$, and $\left(\dfrac{D}{2}\right)^2 + \left(\dfrac{E}{2}\right)^2 > F$, then we have a circle of nonzero radius. Its center is $\left(\dfrac{-D}{2}, \dfrac{-E}{2}\right)$, and its radius is $\sqrt{\left(\dfrac{D}{2}\right)^2 + \left(\dfrac{E}{2}\right)^2 - F}$. The reduced form is $x^2 + y^2 + Dx + Ey + F = 0$. An equation in reduced form can be converted to the center-radius form by completing the square on both x and y. We convert the other way by expanding out and collecting like terms.

An advantage of the reduced form is that we can use a matrix to find the equation of the circle. Using the same three points as in the first example, which were (1,-2), (5,-6), (-7, 8), we generate an augmented 3x4 matrix that can be put in reduced row echelon form (rref) with a graphing utility. The –5 is obtained by $1^2 + (-2)^2 = 5$ on the left hand side of the first equation, which becomes –5 on the right hand side of the equation.

$$\begin{bmatrix} 1 & -2 & 1 & -5 \\ 5 & -6 & 1 & -61 \\ -7 & 8 & 1 & -113 \end{bmatrix} \xrightarrow{\ rref\ } \begin{bmatrix} 1 & 0 & 0 & -124 \\ 0 & 1 & 0 & -110 \\ 0 & 0 & 1 & -101 \end{bmatrix}$$

The equation $x^2 + y^2 + Dx + Ey + F = 0$ after substitution becomes the following: $x^2 + y^2 - 124x - 110y - 101 = 0$. We can either complete the square on both x and y, or we can use the relationships established earlier. Let's do the latter. The center (h, k) is $\left(\dfrac{-D}{2}, \dfrac{-E}{2}\right) = (62, 55)$, and r^2 is $\left(\dfrac{D}{2}\right)^2 + \left(\dfrac{E}{2}\right)^2 - F = 62^2 + 55^2 - (-101) = 6970$. Thus, we can write the equation of the circle in center-radius form as the following: $(x - 62)^2 + (y - 55)^2 = 6970$, which is the same as earlier. Our situation is captured in the diagram on the next page. Observe that when the three points are close to being collinear, one gets a big circle!

Next, we will discuss the rational points on the unit circle part along with translation to circles of different radii and centers. The unit circle is centered at the origin with a radius of one. That's why it is called the unit circle. Trigonometry makes extensive use of the unit circle. We have an interest in it as well. The equation of the unit circle is the following: $x^2 + y^2 = 1$. Given a particular Pythagorean triple (a, b, c), we know from the Pythagorean theorem that $a^2 + b^2 = c^2$. If we divide both sides of the Pythagorean theorem by c^2, we get the following: $\dfrac{a^2}{c^2} + \dfrac{b^2}{c^2} = 1$. We observe that $x = |a/c|$ and $y = |b/c|$ yields four rational points on the unit circle by including the minus signs. We can switch the positions of a and b to get $x = |b/c|$ and $y = |a/c|$ to yield four more rational points on the unit circle by including the minus signs. Therefore, one Pythagorean triple gives us eight rational points on the unit circle.

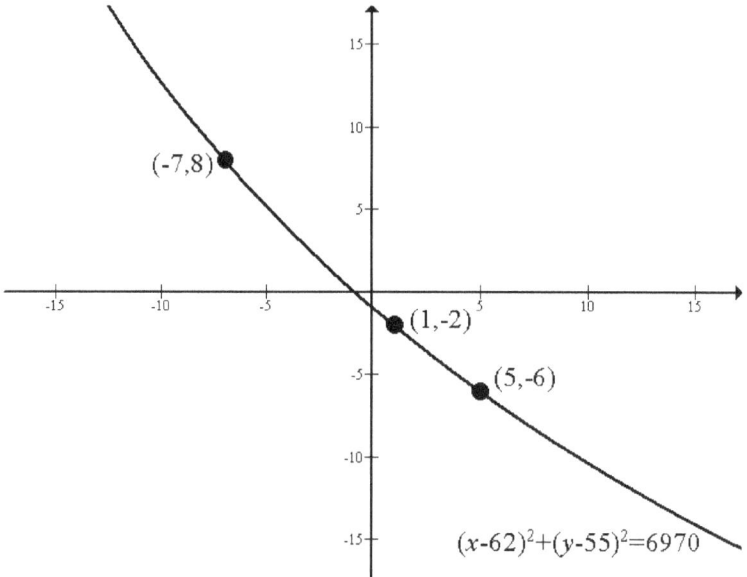

Figure 1: The Big Circle Through the Three Given Points

We will use the notation x_u and y_u to designate the x-coordinate and y-coordinate of a rational point on the unit circle. Thus, the point (x_u, y_u) is an ordered pair rational point on the unit circle.

All circles have the same shape. We know the circumference and area equations of a circle. It does not matter what size the circle is because these equations apply to any size circle. This notion is important because a rational point on the unit circle has a corresponding point on the circle $(x - h)^2 + (y - k)^2 = r^2$ if h, k, and r are all rational. If

94

one or more of h, k, and r are not rational, then the new circle may not have any rational points at all. We will not concern ourselves with finding rational points on a circle unless all three of h, k, and r are rational. There are higher-level math courses that will handle this concern when one or more of h, k, and r are not rational. We also mention in passing that a corresponding point between two circles exists whether or not those points are rational.

Let us extend our notation. We will call a new circle by C, and the variables x, y, and r will be subscripted with a C to show that these belong to the circle C. Thus, we can now say C: $(x_C - h)^2 + (y_C - k)^2 = r_C^2$ is a circle with radius $|r_C|$ and center (h, k).

We can find a rational point on C when $h = k = 0$ by taking a rational point on the unit circle and multiplying it by r_C. Thus, $x_C = r_C(x_u)$ and $y_C = r_C(y_u)$. We may construct the ordered pair $(x_C, y_C) = (r_C(x_u), r_C(y_u))$. This is true of circles of rational radius r centered at the origin.

Is it possible that either or both of h and k are nonzero? Of course. We then use h and k to move the resized rational point to its new location. Thus, $x_C = r_C(x_u) + h$ and $y_C = r_C(y_u) + k$. We may now construct the ordered pair $(x_C, y_C) = (r_C(x_u) + h, r_C(y_u) + k)$. Thus, these maneuvers tell us to resize the circle first before moving it. If we move the rational point first before we resize it, the multiplying by the resizing factor r_C would get distributed to h and k causing them to yield incorrect values. In other words, the center of the new circle would be $(r_C h, r_C k)$, which is not (h, k) like it was supposed to be, unless r_C is one (of course!). The point is stressed because some students make these types of errors, and we wish to bring these details to light for those students to help them. Note that the equations $x_C = r_C(x_u) + h$ and $y_C = r_C(y_u) + k$ are called linear transformations.

Next, we will apply our linear transformations to plot some triangles of given rational lengths. A procedure is given in the *Triangle Geometry* chapter to plot a triangle of lengths a, b, and c. The advice there says to plot $(0, 0)$ and $(c, 0)$ as two of the vertices, and it then proceeds to find a suitable third point. The problem with that procedure is that one of the sides is always horizontal. We can take the coordinates and reverse them to get $(0, 0)$ and $(0, c)$, but now one of our sides is vertical. In the same vein, we can vertically or horizontally translate the obtained coordinates, but we will still have a vertical or horizontal side. The procedure we will describe here gets a triangle onto the axes in numerous orientations; however, we may wind up with a vertical or horizontal side with this procedure. It depends on the triangle and the choices made during the procedure. Note that we may not get rational points to go along with our rational lengths unless we have the nice triangle lengths provided by Pythagorean and Heronian triangles, which includes those that are resized by a rational scalar factor k described in the *Deluxe Toolkit* chapter. Our first task, then, will be to plot the simplest Pythagorean triple triangle: a 3-4-5 triangle. Thus, we have $a = 3$, $b = 4$, and $c = 5$.

We begin by selecting a rational point on the unit circle. For this example, we select the ordered pair (-3/5, 4/5). We choose the rational point based on the steepness of the angle the point provides because that steepness will carry over to one of the sides of our finished plotted triangle. Next, we need to select (h, k)—the center of our new circle C.

A remark is in order here. Note that we will have the vertices A, B, and C on our triangle, and this new circle C that we are utilizing is called by the same letter C. This happens in mathematics sometimes that we use the same letter for different things in the same problem. I remember being confused on one problem in one of my homework problems because I did not realize that a letter was being used twice for different things in the same problem. I reached the incorrect conclusion that there was a conflict in the problem statement's variables, so I reported that as my answer when I submitted my homework. I got a zero on the problem, and the instructor wrote the unsympathetic, "There is no conflict." I am reporting this to students to show that I sympathetic when things like this come up—using the same letter in the same problem for different things. We now return to the task of selecting the center of the circle C.

For this example, we select the point $(h, k) = (2, 6)$. This will be the vertex B in our plotted triangle for this example, but note that it doesn't always have to be this vertex. When we get to select something in math, this means that we are free to choose whatever we would like. We could just as well have chosen $(9^{(9^9)}, 10^{1000} - 1)$, a couple of serious heavyweight coordinates. However, since we have to do math with these coordinates, we select a nice, small integer ordered pair like (2, 6). If we were so inclined, we could get a little more ambitious than (2, 6) with fractional coordinates such as (1/2, 4/9). Next, we select one of the three given triangle lengths as the radius of our new circle C, and for this example, we choose $r = 3$. Hence, our variables are $r = 3$, $h = 2$, $k = 6$, $x_u = -3/5$, and $y_u = 4/5$.

With these values, we translate the ordered pair (x_u, y_u) to the new circle C, so we compute the point $C(x_C, y_C) = (rx_u + h, ry_u + k) = (3(-3/5) + 2, 3(4/5) + 6) = (0.2, 8.4)$. So now we have vertex C. Because $r = 3$, the distance between the two points $B(2, 6)$ and $C(0.2, 8.4)$ should be $a = 3$, and the student is encouraged to verify this with the distance formula before proceeding with their own independent problems because an error here leads to wasted effort upon proceeding. A simple little check here can let the student proceed on sure footing.

The slope through B and C is $m = \dfrac{y_2 - y_1}{x_2 - x_1} = \dfrac{6 - 8.4}{2 - 0.2} = \dfrac{-4}{3}$. Since a Pythagorean triangle has a right angle, we can use the negative reciprocal to get a perpendicular line, which yields a right angle. At this point, we need to decide on which vertex to place it. So we will decide to send our right angle through vertex C because convention dictates that the right angle is usually labeled C. Hence, our new problem now is to find the equation of a line with a slope of $m_\perp = \dfrac{3}{4}$ through $C(0.2, 8.4)$. This yields the equation

$y = \dfrac{3}{4}x + \dfrac{33}{4}$. The vertex A is somewhere on this line. We can get this vertex by setting up a system of equations involving this line and either a circle of radius 4 centered at vertex C or a circle of radius 5 centered at vertex B. We will depict the latter.

$$\begin{cases} y = \dfrac{3}{4}x + \dfrac{33}{4} \\ (x-2)^2 + (y-6)^2 = 25 \end{cases}$$

We get the choice of $x = 17/5 = 3.4$ and $x = -3$. We select the former. Hence, we compute the y-coordinate using either the line equation or the circle equation in the above system. Of course, it is easier to use the line equation. This yields point $A(3.4, 10.8)$. It would be good to verify that the distances are correct, especially if this triangle is to be used for further work. We show our plotted triangle in a square-viewing window in Figure 2.

Our next triangle that we will plot has lengths 5-5-8. This is called a Heronian triangle because a Heronian triangle has integer side lengths and integer area. If we view the base as 8, we would soon see that the height has to be 3 with the Pythagorean formula. We would also observe that this Heronian triangle is two 3-4-5 Pythagorean triangles butted together on their $a = 3$ sides. Note that every Pythagorean triangle is also a Heronian triangle, but we have many other Heronian triangles that are not Pythagorean.

As before, we will translate a rational point on the unit circle to a new circle C. This point on C and the center of C will yield two vertices of our triangle. We will select a different rational point on the unit circle this time: $(x_u, y_u) = (4/5, -3/5)$. We will let $r = 8$, the longest side of our 5-5-8 Heronian triangle. We will choose (h, k) this time to be $(2, 11)$. We do not have the Pythagorean convention of $(a, b, c) = $ (short leg, long leg, hypotenuse) with a Heronian triangle. Hence, we can let our selection of $(h, k) = (2, 11)$ be vertex A without being mindful of a convention. We perform the same type of computations as before with our new circle C to get $B(x_C, y_C) = (8.4, 6.2)$. Again, it is prudent to check the distance between the vertices $A(2, 11)$ and $B(8.4, 6.2)$ to ensure that it is 8.

We need the midpoint M of the side AB of this triangle. This is so that we can swing a line through it representing altitude of the Heronian triangle. Our third vertex will lie on this line. We compute $M = \left(\dfrac{1}{2}(2 + 8.4), \dfrac{1}{2}(11 + 6.2) \right) = M(5.2, 8.6)$. We find that the slope through A and B is $-3/4$, so the perpendicular line for the altitude through M needs to be $m_\perp = 4/3$. Thus, the desired line is $y = \dfrac{4}{3}x + \dfrac{5}{3}$.

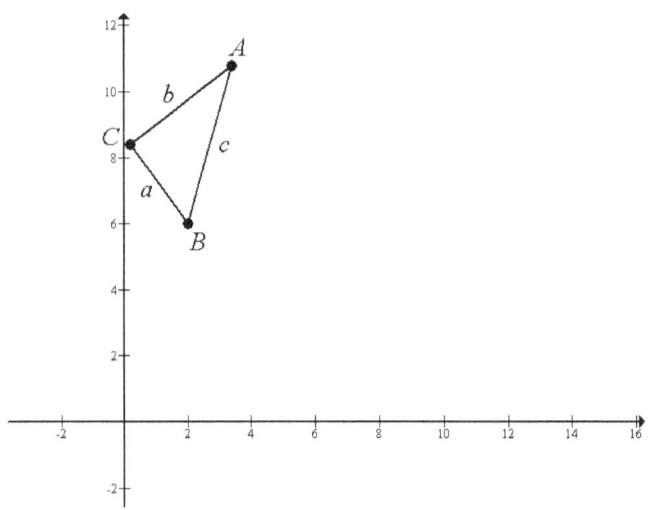

Figure 2: Our Plotted 3-4-5 Triangle

We now need to pair up this line to create a system of equations. The other equation in this system is a circle, but we have three choices. We can use the circle of radius 5 centered at either A or B, or we can use the circle of radius 3 centered at M. We will depict the first of these.

$$\begin{cases} y = \dfrac{4}{3}x + \dfrac{5}{3} \\ (x - 8.4)^2 + (y - 6.2)^2 = 25 \end{cases}$$

Solving this system yields the pair of points $(3.4, 6.2)$ and $(7, 11)$ either of which can be vertex C. Note that either of these choices yields a side of the Heronian triangle that is horizontal, which is not bad per se. However, if we want a triangle that has tilted slopes

98

for all three of its sides, then we would have to change something in our procedure here. For example, we could use a different Pythagorean triple with a 5 or an 8 in it, say (5, 12, 13) or (8, 15, 17). Pick a rational point on the unit circle from one of these triples, and we would then not see a horizontal or vertical side for the 5-5-8 Heronian triangle. We show both of our Heronian triangles in Figure 3. Observe that both of these triangles together form a rhombus.

Suppose we want to plot a triangle that is not so nice as a Pythagorean or Heronian triangle? For example, suppose we want to plot a nasty 6-7-8 triangle. With our side of length 8, we can reuse some of our earlier work. Thus, we can keep the $A(2, 11)$ and $B(8.4, 6.2)$ vertices from the previous problem.

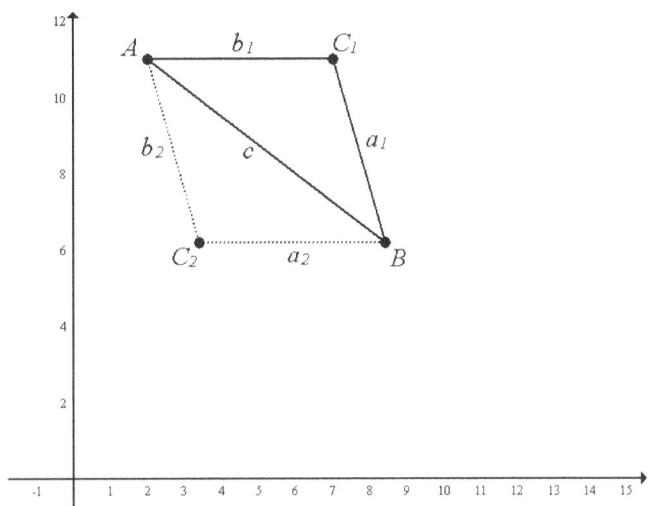

Figure 3: Both 5-5-8 Heronian Triangles Plotted Form a Rhombus

What if we attempt to set up a system of equations involving a pair of circles? Let's say one with radius 6 centered at $A(2, 11)$ and one with radius 7 centered at $B(8.4, 6.2)$. This yields the below system.

$$\begin{cases} (x-2)^2 + (y-11)^2 = 36 \\ (x-8.4)^2 + (y-6.2)^2 = 49 \end{cases}$$

Some computer algebra systems will not return a result for this system. Others will. Doing this by hand is certainly problematic because we will wind up with an unfriendly fourth degree polynomial. This route of two circles is, therefore, not the best choice. Let us try another route, then.

Because of the asymmetry, we cannot use the midpoint M of AB now like we did in the Heronian triangle example. What will we do? Recall the extended midpoint tool in the *Deluxe Toolkit* chapter. We have a need for it now. However, what will we call the weights w_1 and w_2? We will have to solve for them. Look at Figure 4 to see how we will accomplish this feat.

We observe that the side of length 8 is split into two parts: x_1 and x_2. These are not coordinates, but rather, these are lengths. Thus, $x_1 + x_2 = 8$. We have a couple of right triangles in the figure as well. By the Pythagorean theorem, we have $x_1^2 + h^2 = 36$ and $x_2^2 + h^2 = 49$. These three relationships form a system of three equations in three unknowns. The following system is consistent.

$$\begin{cases} x_1 + x_2 = 8 \\ x_1^2 + h^2 = 36 \\ x_2^2 + h^2 = 49 \end{cases}$$

The solution to this system is $x_1 = 51/16$ and $x_2 = 77/16$. Remember that these two values are lengths, and w_1 and w_2 are proportions where $0 \le w_1, w_2 \le 1$ and $w_1 + w_2 = 1$. So how do we convert these lengths to proportions? We construct the ratio of the part over the whole. Thus, $w_1 = \dfrac{x_1}{x_1 + x_2} = \dfrac{51/16}{8} = \dfrac{51}{128}$ and $w_2 = \dfrac{x_2}{x_1 + x_2} = \dfrac{77/16}{8} = \dfrac{77}{128}$.

Observe that both of these weights are between zero and one and that they add to one.

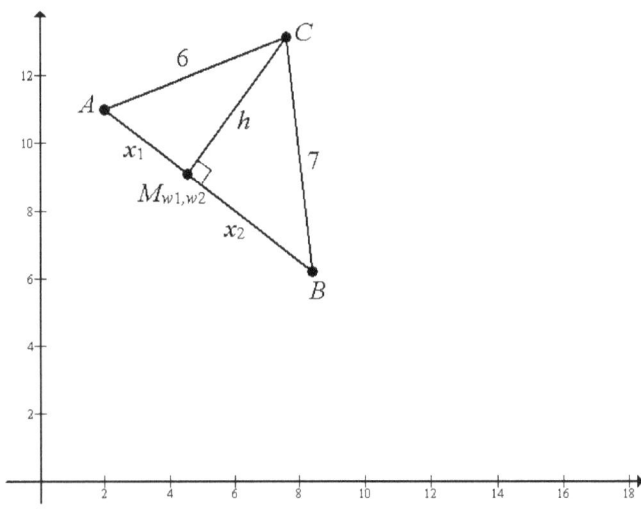

Figure 4: Our Sketch to Find w_1 and w_2 Algebraically

We did this work so as to be able to find the extended midpoint $M_{w1,w2}$ between A and B. Viewing the figure, we see that $M_{w1,w2}$ is closer to A than to B. This tells us that the coordinates for point A get the bigger weight. We may now compute the following:

$$M_{\frac{51}{128},\frac{77}{128}}(B,A) = \left(\frac{51}{128} \cdot (8.4) + \frac{77}{128} \cdot 2, \frac{51}{128} \cdot (6.2) + \frac{77}{128} \cdot (11) \right) = \left(\frac{91}{20}, \frac{727}{80} \right)$$

We would like to send a perpendicular line to AB through M. Thus, we find the slope through AB as $m = -3/4$. Then, $m_\perp = 4/3$. Of course, $M = (91/20, 727/80)$. Solving this produces the line $y = \frac{4}{3}x + \frac{145}{48}$. Since we have three points, we have three choices of the system to create. We can pair this line with any of three circles: a circle of radius 6 with center at A, a circle of radius h centered at M, and a circle of radius 7 centered at B. With x_1 and x_2 at our disposal, it wouldn't be much trouble to find h. However, why do that extra step when it isn't necessary? We can just use the first circle from our selection of three. Thus, we solve the following system.

$$\begin{cases} (x-2)^2 + (y-11)^2 = 36 \\ y = \dfrac{4}{3}x + \dfrac{145}{48} \end{cases}$$

The computer CAS that couldn't solve the two-circle system of equations above could solve this system involving the line and circle. In addition, a person can manually solve this system involving the line and circle. Yes, it would take some work, but it can be done. Hence, this route of solution involving the extended midpoint is a better route. We get two solutions to this system. Thus, we have a choice for where to put vertex C for our 6-7-8 triangle. We will choose the plus version for x, which yields the coordinates for our third vertex C as $\left(\dfrac{7(9\sqrt{15} + 52)}{80}, \dfrac{84\sqrt{15} + 727}{80} \right)$. As before, the distances AC and BC should be checked to ensure that they are 6 and 7, respectively. Observe that these coordinates have a radical in them. Hence, these are not rational coordinates. Finding some of the things described in the *Triangle Geometry* chapter with this triangle could become overwhelming mathematically. On a positive note for what we just accomplished, we should be happy that we plotted a difficult triangle with sides of 6, 7, and 8. It involved a lot of concepts.

Our final topic in this chapter is to develop a method to find the diametrically opposed point to a given point on a circle. We promised to do this in the *Deluxe Toolkit* chapter.

The linear transformations $x_C = r_C(x_u) + h$ and $y_C = r_C(y_u) + k$ are equations. As such, they can be solved for their other variables. By solving each of these for their unit circle coordinates, we get that $x_u = \dfrac{x_C - h}{r}$ and $y_u = \dfrac{y_C - k}{r}$. Thus, if we have a point P on a circle C, we can find the coordinates of its corresponding point on the unit circle.

Our next step is to observe that a point on the unit circle (x_u, y_u) has as its diametrically opposed point the point $(x'_u, y'_u) = (-x_u, -y_u)$. This is because the unit circle has a nice $r = 1$ and $(h, k) = (0, 0)$.

Combining these ideas, the procedure is evident. We take the point P, and find out its location on the unit circle with the equations $x_u = \dfrac{x_C - h}{r}$ and $y_u = \dfrac{y_C - k}{r}$. Then, we switch the signs on this point to get its diametrically opposed point (x'_u, y'_u) on the unit circle. Then, we find out the coordinates of (x'_u, y'_u) with the original linear transformation equations $x_C = r_C(x_u) + h$ and $y_C = r_C(y_u) + k$. With these two equations, we will insert the coordinates from (x'_u, y'_u) into the places calling for x_u and y_u. The new point P' is (x'_u, y'_u). We will remark that it does not matter if a point on C is irrational on the unit circle. This procedure will work regardless.

In the *Desargues' Theorem* chapter, we studied the Prasolov point in the exercises. We needed to reflect the orthic triangle across the nine-point center to yield a new triangle that was in perspective with the original triangle *ABC*. The job of reflecting those vertices could now be accomplished using the procedure outlined just now with the unit circle.

Exercises

1. Find the equation of the circle passing through the points (1, 3), (1, 11), and (-1, -8) by using the center-radius form of the equation of a circle.

2. Find the equation of the circle passing through the points (2, 2), (5, 7), and (-9, 2) by using the reduced general second-degree equation for a circle. Use the rref command on a graphing utility once the augmented matrix is obtained.

3. For the rational point ($7/13, -12/13$) on the unit circle, find its corresponding point on the circles in questions 1 and 2.

4. The point (1, 3) is a rational point on the circle in question 1. What is its corresponding point on the unit circle? Identify the primitive Pythagorean triple used to generate this point on the unit circle.

5. For the example circle in this chapter $(x - 62)^2 + (y - 55)^2 = 6970$, find the exact and approximate (to the nearest tenth) coordinates of the intercepts. Also, find the exact and approximate area and circumference of this circle. How long would a particle take to travel around this circle at 5 units per second (to the nearest tenth)? How long would it take to paint the interior of this circle if it is painted at 40 square units per second (to the nearest tenth)?

6. Write the equation $(x - 5)^2 + (y + 2)^2 = 36$ in its general second-degree form.

7. Write the equation of the circle $4x^2 + 4y^2 - 16y + 15 = 0$ in center-radius form.

8. Write the equation of the circle $x^2 + y^2 - 4x + 6y - 23 = 0$ in center-radius form.

For the remaining problems, make sure that the resulting triangle sides are tilted.

9. Use the procedures described in this chapter to plot a 6-8-10 Pythagorean triangle.

10. Use the procedures described in this chapter to plot a 8-15-17 Pythagorean triangle.

11. Use the procedures described in this chapter to plot a 5-5-6 Heronian triangle.

12. Use the procedures described in this chapter to plot a 15-15-16 Heronian triangle.

13. Use the procedures described in this chapter to plot a 3-4-6 triangle.

14. Use the procedures described in this chapter to plot a 7-10-11 triangle.

15. Use the procedures described in this chapter to plot a $\dfrac{41}{7} - \dfrac{91}{11} - \dfrac{37}{3}$ triangle.

16. Find the diametrically opposed point to $P(3, 6)$ on C: $(x+1)^2 + (y-3)^2 = 25$ using the advice in the *Deluxe Toolkit* chapter. Next, find it using the method outlined at the end of this chapter. Comment on each method.

17. Redo question 4 in the *Desargues' Theorem* chapter, but reflect the orthic triangle across the nine-point center using the advice outlined at the end of this chapter that utilizes the unit circle.

Theorem Highlight: The Wallace-Simson Line

Two mathematician's names are on this theorem. However, only one of them did the mathematical legwork for this line. Wallace is the mathematician deserving the credit for this line. Simson is improperly credited with this line. Thus, the line is known as the Simson line, but other writers call it the Wallace-Simson line to honor Wallace. I will also refer to it as the Wallace-Simson line. William Wallace (1768 – 1843) was a Scottish mathematician. Robert Simson (1687 – 1768) was also a Scottish mathematician.

The theorem highlighted in this discussion states that for any point P on the circumcircle of a triangle, the intersection points of the three perpendicular lines drawn from P to the sides of the triangle (extending the sides of the triangle when necessary) lie on a straight line.

We need to be given the three points of the triangle. The sides of the triangle are lines, and the equations of those lines can be found with standard techniques. We can determine the unique circle that goes through these three points, and it is the circumcircle. In *The Circle* chapter, we saw that the equation of the circle can be found by considering an equivalent form of the equation of a circle that is reduced for the general second-degree equation. From there, we can then select any point P on the circle that is not a vertex of the triangle because the ⊥ line from P to two of the sides of the triangle meet those sides at the vertex that is P. The only other ⊥ line is to the third side, which meets at some point other than P. Thus, we only have two points, which determine a line—in this case the Wallace-Simson line. However, we are not impressed with two points determining a line. Getting three points on a line is where the action is. So we choose a point P on the circumcircle that is not a vertex of the triangle. Next, we draw lines from P that perpendicularly intersect the sides of the triangle. These ⊥ lines that go through P can be found with standard techniques. Then, we compute the intersection points. Finally, we use two of those points to find the Wallace-Simson line, and we use the third point to verify our work.

Let us make up a situation to find the Wallace-Simson line. Given the three points of a triangle A(-4, 3), B(1, 1), C(2, -1) find the Wallace-Simson line for P being the upper point on the circumcircle with an *x*-coordinate of 0. Before we start, we need to be aware that we are going to do a lot of computation. Therefore, it behooves us to be neat and organized with our work. We should be fastidious about labeling the items that we find. That way, if we should make an error, we can trace back through the work with minimal headache.

The first thing we need to do is to find the equations of the three sides of the triangle. Part of being fastidious entails checking that these line equations actually go through the points.

Side with A(-4, 3) and B(1, 1):

$$m = \frac{y_2 - y_1}{x_2 - x_1} = \frac{1-3}{1-(-4)} = \frac{-2}{5}$$

$$y - y_1 = m(x - x_1) \Rightarrow y = m(x - x_1) + y_1 = \frac{-2}{5}(x-1) + 1 = \frac{-2}{5}x + \frac{7}{5}$$

Side with B(1, 1) and C(2, -1):

$$m = \frac{y_2 - y_1}{x_2 - x_1} = \frac{-1-1}{2-1} = \frac{-2}{1} = -2$$

$$y - y_1 = m(x - x_1) \Rightarrow y = m(x - x_1) + y_1 = -2(x-1) + 1 = -2x + 3$$

Side with A(-4, 3) and C(2, -1):

$$m = \frac{y_2 - y_1}{x_2 - x_1} = \frac{-1-3}{2-(-4)} = \frac{-4}{6} = \frac{-2}{3}$$

$$y - y_1 = m(x - x_1) \Rightarrow y = m(x - x_1) + y_1 = \frac{-2}{3}(x-2) + (-1) = \frac{-2}{3}x + \frac{1}{3}$$

Next, we need to find the equation of the circumcircle. Two forms of the equation of a circle exist. The usual equation $(x - h)^2 + (y - k)^2 = r^2$ is the better-known version. However, the other equation of a circle comes from a reduced form of the general second-degree equation $ax^2 + bxy + cy^2 + dx + ey + f = 0$. The equation of a circle does not have an xy term, and the coefficients of the x^2 and y^2 term are equal. Thus, the reduction of the general second-degree equation to represent a circle is $x^2 + y^2 + dx + ey + f = 0$. Of course, we discussed these matters in *The Circle* chapter. Using this equation of a circle and inputting the three points into the equation can accomplish our goal of finding the equation of the circle we seek. We construct one equation for each point that we have. Since we have three points, we will get a system of three equations in three unknowns.

Equation of Circle: $x^2 + y^2 + dx + ey + f = 0$

The points A(-4, 3), B(1, 1), and C(2, -1), generate the following system of equations:

106

$$(-4)^2 + 3^2 + d \cdot (-4) + e \cdot 3 + f = 0 \Rightarrow -4d + 3e + f = -25$$
$$1^2 + 1^2 + d \cdot 1 + e \cdot 1 + f = 0 \Rightarrow d + e + f = -2$$
$$2^2 + (-1)^2 + d \cdot 2 + e \cdot (-1) + f = 0 \Rightarrow 2d - e + f = -5$$

$$\begin{bmatrix} -4 & 3 & 1 & | & -25 \\ 1 & 1 & 1 & | & -2 \\ 2 & -1 & 1 & | & -5 \end{bmatrix} \xrightarrow{rref} \begin{bmatrix} 1 & 0 & 0 & | & 13/2 \\ 0 & 1 & 0 & | & 19/4 \\ 0 & 0 & 1 & | & -53/4 \end{bmatrix}$$

We input the problem as a matrix into the grapher, and compute the rref of that matrix. The command rref stands for reduced row echelon form, and this allows us to read the solution directly from the matrix. Of course, the matrix can be solved by hand as well, and the student will be asked to do this in the exercises. Thus, $d = 13/2$, $e = 19/4$, and $f = -53/4$. We place these coefficients into the circle equation we are using, and we can then convert that equation into the standard form for a circle.

$x^2 + y^2 + dx + ey + f = 0$ Substitute in the computed values.

$x^2 + y^2 + \dfrac{13}{2}x + \dfrac{19}{4}y - \dfrac{53}{4} = 0$ Commute terms to set up for completing the square.

$x^2 + \dfrac{13}{2}x + y^2 + \dfrac{19}{4}y = \dfrac{53}{4}$ Add in the required amounts on both sides.

$x^2 + \dfrac{13}{2}x + \dfrac{169}{16} + y^2 + \dfrac{19}{4}y + \dfrac{361}{64} = \dfrac{53}{4} + \dfrac{169}{16} + \dfrac{361}{64}$ Finish completing the square.

$\left(x + \dfrac{13}{4}\right)^2 + \left(y + \dfrac{19}{8}\right)^2 = \dfrac{1885}{64}$ Thus, $(h, k) = \left(\dfrac{-13}{4}, \dfrac{-19}{4}\right)$ and $r^2 = \dfrac{1885}{64}$.

Again, verify that this circle goes through the three given points before proceeding. Note that we used some formulas for h and k in *The Circle* chapter. Now that we have the equation for the circumcircle, we can compute the point P, which is given as the upper point on the circle with an x-coordinate of zero.

$\left(x + \dfrac{13}{4}\right)^2 + \left(y + \dfrac{19}{8}\right)^2 = \dfrac{1885}{64}$ Plug in the x-value of zero.

$\left(0 + \dfrac{13}{4}\right)^2 + \left(y + \dfrac{19}{8}\right)^2 = \dfrac{1885}{64}$ Solve for y.

$$\frac{169}{16} + \left(y + \frac{19}{8}\right)^2 = \frac{1885}{64}$$

We can use the square root method to solve.

$$\sqrt{\left(y + \frac{19}{8}\right)^2} = \pm\sqrt{\frac{1885}{64} - \frac{169}{16}}$$

The upper point means to take the positive root.

$$y + \frac{19}{8} = \sqrt{\frac{1209}{64}} = \frac{\sqrt{1209}}{8}$$

$$y = \frac{-19}{8} + \frac{\sqrt{1209}}{8} = \frac{-19 + \sqrt{1209}}{8}$$

The circumcircle point is P =

$$\left(0, \frac{-19 + \sqrt{1209}}{8}\right).$$

Let us think about the slopes of the three sides of the triangle. The negative reciprocal of any slope yields a perpendicular slope. Thus, we take the negative reciprocal of the slope in the equation for a side, and find the equation of the line with that slope through the given point P.

Equation of side of triangle (side $\overline{AB}$): $y = \frac{-2}{5}x + \frac{7}{5}$

$\perp$ line through P:

$$m = \frac{-2}{5} \Rightarrow m_\perp = \frac{5}{2} \Rightarrow y - \frac{-19 + \sqrt{1209}}{8} = \frac{5}{2}(x - 0) \Rightarrow y = \frac{5}{2}x + \frac{-19 + \sqrt{1209}}{8}$$

Equating: $\frac{-2}{5}x + \frac{7}{5} = \frac{5}{2}x + \frac{-19 + \sqrt{1209}}{8}$

Solving yields $x = \frac{151 - 5\sqrt{1209}}{116}$.

Plugging in yields: $y = \frac{-2}{5}x + \frac{7}{5} = \frac{-2}{5} \cdot \frac{-151 - 5\sqrt{1209}}{116} + \frac{7}{5} = \frac{51 + \sqrt{1209}}{58}$

Of course, the perpendicular line equation yields the same y-coordinate.

These two lines intersect at $(\frac{-151 - 5\sqrt{1209}}{116}, \frac{51 + \sqrt{1209}}{58})$, a point on the Wallace-Simson line.

108

Equation of side of triangle (side $\overline{BC}$): $y = -2x + 3$

$\perp$ line through P:

$$m = -2 \Rightarrow m_\perp = \frac{1}{2} \Rightarrow y - \frac{-19+\sqrt{1209}}{8} = \frac{1}{2}(x-0) \Rightarrow y = \frac{1}{2}x + \frac{-19+\sqrt{1209}}{8}$$

These two lines intersect at $\left(\dfrac{43-\sqrt{1209}}{20}, \dfrac{-13+\sqrt{1209}}{10} \right)$, another point on the

Wallace-Simson line.

Equation of side of triangle (side $\overline{AC}$): $y = \dfrac{-2}{3}x + \dfrac{1}{3}$

$\perp$ line through P:

$$m = \frac{-2}{3} \Rightarrow m_\perp = \frac{3}{2} \Rightarrow y - \frac{-19+\sqrt{1209}}{8} = \frac{3}{2}(x-0) \Rightarrow y = \frac{3}{2}x + \frac{-19+\sqrt{1209}}{8}$$

These two lines intersect at $\left(\dfrac{65-3\sqrt{1209}}{52}, \dfrac{-13+\sqrt{1209}}{26} \right)$, which makes the third

point on the Wallace-Simson line.

We will use the first two points we found on the Wallace-Simson line to establish the line. The third point can be used to verify our work. We just plug the third point into the Wallace-Simson line equation. If we get an identity, then all is well. If we get a contradiction, then at least one mistake was made along the way.

$$m = \frac{\dfrac{-13+\sqrt{1209}}{10} - \dfrac{51+\sqrt{1209}}{58}}{\dfrac{43-\sqrt{1209}}{20} - \dfrac{151-5\sqrt{1209}}{116}} = \dots = \frac{-21+\sqrt{1209}}{12}$$

$$y - \frac{51+\sqrt{1209}}{58} = \frac{-21+\sqrt{1209}}{12}\left(x - \frac{151-5\sqrt{1209}}{116} \right)$$

$$y = \frac{-21+\sqrt{1209}}{12}\left(x + \frac{-151+5\sqrt{1209}}{116} \right) + \frac{51+\sqrt{1209}}{58}$$

This is the Wallace-Simson line. This equation does simplify a fair amount mainly because the only square root term involved is $\sqrt{1209}$. After simplification, the line becomes

$$y = \frac{-21+\sqrt{1209}}{12}x + \frac{45-\sqrt{1209}}{6}.$$

Of course, we could have used either of the first two points. The form prior to simplification would look a little different, but it would be the same line upon simplification. Lastly, we use the third point to check our work. Getting an identity validates all of our work; however, a contradiction would mean we pore over our work to find the error or errors causing our troubles.

$$\frac{-13+\sqrt{1209}}{26} \equiv \frac{-21+\sqrt{1209}}{12}\cdot\frac{65-3\sqrt{1209}}{52} + \frac{45-\sqrt{1209}}{6} = \dots = \frac{-13+\sqrt{1209}}{26} \checkmark$$

The below diagram captures our situation, so the student can picture how the algebra and geometry interrelate. Observe side BC needs to be extended to intersect with the perpendicular line from P, but the other two sides of the triangle do not require an extension like this.

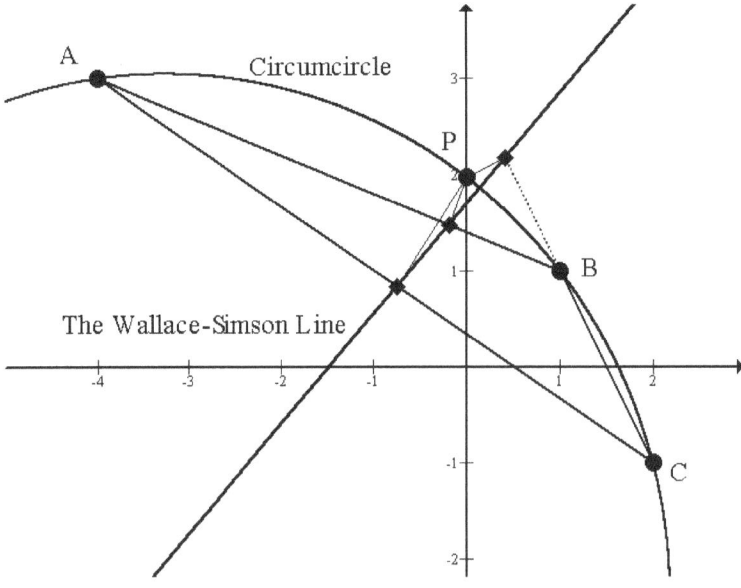

Figure 1: The Wallace-Simson Line

A theorem about Wallace-Simson lines states that the Wallace-Simson lines for two diametrically opposed points P and P' are perpendicular. The student will be asked to find the Wallace-Simson line for the point P' in the exercises and to verify that it is perpendicular to the Wallace-Simson line for point P. Another theorem states that the intersection point of the Wallace-Simson lines for two diametrically opposed points P and P' lies on the nine-point circle.

The student will be asked to find the equation for the nine-point circle for this triangle and to verify that the intersection of the two Wallace-Simson lines lies on it. The nine-point circle has the following nine points: the midpoints of the sides of the triangle, the feet of the three altitudes, and the midpoints from a vertex to the mutual intersection point of the three altitudes. The mutual intersection point of the three altitudes is also known as the orthocenter and is usually denoted by H. We will see the orthocenter again in the *Triangle Geometry* chapter among other places. Computationally, it is easiest to find the equation of the nine-point circle by finding the equation of the circle going through the midpoints of the three sides of the triangle. Be advised that the center of the nine-point circle is neither the orthocenter nor the circumcenter.

Exercises

1. Show that $\begin{bmatrix} -4 & 3 & 1 & | & -25 \\ 1 & 1 & 1 & | & -2 \\ 2 & -1 & 1 & | & -5 \end{bmatrix} \xrightarrow{rref} \begin{bmatrix} 1 & 0 & 0 & | & 13/2 \\ 0 & 1 & 0 & | & 19/4 \\ 0 & 0 & 1 & | & -53/4 \end{bmatrix}$ is true by performing elementary row operations on the matrix by hand to achieve the reduced row echelon form.

2. We were computing $m = \dfrac{\dfrac{-13+\sqrt{1209}}{10} - \dfrac{51+\sqrt{1209}}{58}}{\dfrac{43-\sqrt{1209}}{20} - \dfrac{151-5\sqrt{1209}}{116}} = \ldots = \dfrac{-21+\sqrt{1209}}{12}$ in the buildup to the Wallace-Simson line. Multiply the top and bottom by the LCD of all the denominators, and rationalize the denominator to obtain the value of m used in the Wallace-Simson line equation.

3. When we were checking to see if our Wallace-Simson line equation was correct, we had

$$\frac{-13+\sqrt{1209}}{26} \equiv \frac{-21+\sqrt{1209}}{12} \cdot \frac{65-3\sqrt{1209}}{52} + \frac{45-\sqrt{1209}}{6} = \ldots = \frac{-13+\sqrt{1209}}{26}$$. Fill in the missing steps by hand to complete the verification.

4. Find the point P' in the example, and use it to find the equation of the Wallace-Simson line for P'. Verify that the slopes of this pair of Wallace-Simson lines are perpendicular. Also, find the equation of the nine-point circle for the given triangle in the example. Solve the system of two Wallace-Simson equations for P and P' to get an intersection point. Plug this point into the nine-point circle equation to get an identity. This shows that the intersection point of the pair of Wallace-Simson lines for the points P and P' lie on the nine-point circle, according to the theorem.

5. Find the Wallace-Simson line for the situation in the diagram labeled for this problem below. Observe that P is an integer point, which will make all of the computations far easier than the example in the text.

6. Find the Wallace-Simson line for the point P' in exercise 5. Verify that the Wallace-Simson lines in exercises 5 and 6 are perpendicular. Find the equation of the nine-point circle for the triangle in exercise 5. Solve the system of equations of the two Wallace-Simson lines in exercise 5 and 6. Place the coordinates of the intersection point of these two Wallace-Simson lines into the equation for the nine-point circle to obtain an identity. This shows that these two Wallace-Simson lines intersect on the nine-point circle.

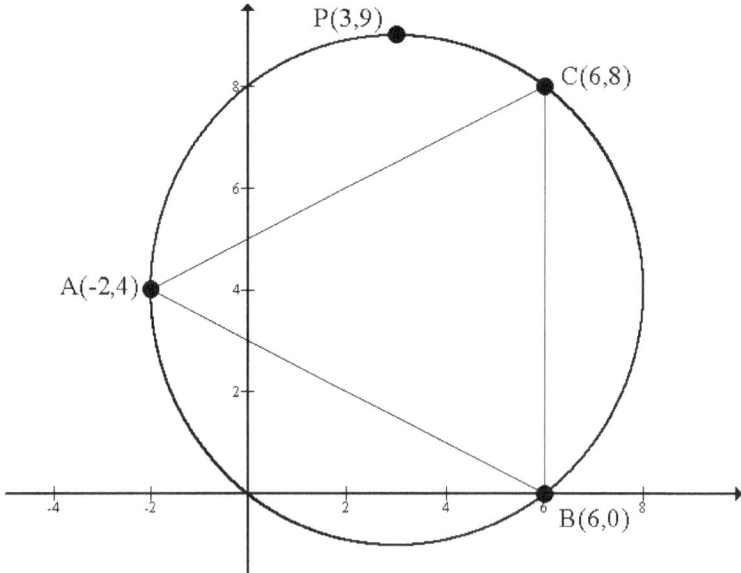

Diagram for #5

7. Find four integer points on the circle $(x+12)^2 + (y-5)^2 = 169$ with at most two with coordinates matching h or k from the center of the circle. Draw the circle with your four points labeled as A, B, C, and P. Repeat exercise 5 with this new situation. Note: Answers will vary.

8. Repeat exercise 6 with the situation in exercise 7. Note: Answers will vary.

Lines and Parabolas: Part 1

This is the first chapter with an asterisk in the *Table of Contents*, which means that this chapter is an original contribution to mathematics as far as the author knows. As such, this chapter begins the main body of material in this textbook—the lines and conic sections. For our first exposure to these new ideas, I ask you to attempt to answer the following two questions before reading any further. You may consult any text you like (with the exception of this one, of course) to refresh your memory of conic sections or algebra techniques. These two little exercises should demonstrate that this material is nontrivial. As a side note, this chapter comprises the first creative spark I had for this book. I drew a line threw the vertex of a parabola, and I was curious to find the distance between the vertex and the other intersection point. I thought that the slope of the line would obviously affect that assessment. It was just a short hop from there to asking myself, "Is there an equation that relates the slope, distance, and bend of the parabola constants?"

Incidentally, I have heard that someone once said that a person may think that any math problem is either trivial or impossible depending on whether or not the person can do it. If that person could do the math problem, then it was trivial. If the person could not do the math problem, then it was impossible. Of course, a problem that falls in the impossible category of problem could be *trivial* for someone else who knows how to do it, or it could be impossible for anyone regardless of how clever that person may be. In that case, that problem is really impossible because it asks something that cannot be done, which is able to be done in many ways (i.e. asking impossible questions can be done in many ways)—however, the many categories of impossible problems would require a book of its own, and it isn't this one. Let it suffice for us to say that everything that we attempt in this book is possible, but we just need to be clever enough to do it. With that in mind, please try the following two problems:

1)   Find the equation of the positively sloped line passing through the vertex of the parabola $y = 2x^2 - 5x + 4$ that intersects the parabola again nine units away.

2)   Find the equation of the downward facing parabola with a vertex of (-1, 3) that has a line of slope two hitting the vertex and intersecting the parabola again seven units away.

If you attempted these problems before you continued reading, congratulations. I trust that you found them to be nontrivial. If you solved them, then you are worthy of high praise indeed. However, most students will not be able to solve these problems—even those students who have strong math skills. However, if you are a Gauss or a Hilbert, I would hope that even you were not bored with these problems. I consider them real challenges and not mere child's play.

I absolutely must tell the story of how I got to do this project as my senior project at Lawrence Technological University. The Math Department Chair and one of my instructors were in one of the offices on campus. They asked me, "What do you plan to do for your senior project?" They informed me that the college was adjusting the requirements of the degree for math to include an independent senior project. I was the first to perform the deed, according to them. I said, "Well, I have been working on conic sections—specifically, the parabola." They had a board in the room, so I approached it and drew the parabola $y = x^2$. I then said, "Take a line of slope m through the vertex of the parabola." I drew said line. Then, I said, "If we want a certain distance d between these intersection points, then that affects the slope of the line m." I then turned back around to see them both have their arms folded across their chests with stern looking faces. They were not amused at all.

I recovered very quickly from this. I said, "Wait! There really is something here. The equations that are generated are non-trivial and not easy to work with at all!" I had a lot of excitement and passion in my voice because this project was on the line. I don't know what moved their hearts, but they consented for me to do the project. Thinking back on it, I guess they must have wondered what could a senior math major in college do with a parabola. They must have been sold solely on my passion because that was all I drew on that board—I did not show them the equation that related those three things. Thus, I conclude that they must have had enough respect for me that when I said that there really was something here that they believed me. Thank you both for that.

Continuing on with the line of thought began in the *Introduction*, perhaps, that is another clue as to why no one else pursued this idea of relating the three quantities. The mathematical people in the past that chanced to look at the lowly parabola with its a constant and the lowly line with its m constant and the lowly distance formula with its d—how could these three things form anything of value? Thus, the mathematical people turned their attention toward more worthy challenges. It might be that I had just enough mathematical ability to ask the question of relating the three quantities but not too much ability to dismiss it before I began to find the answers. The Math Department Chair and my instructor had already dismissed the idea of relating these three quantities as evidenced by their folded arms and stern faces. How many authors of algebra books have put these three equations together in their textbooks have not put these three ideas together in their minds? I am only saying that many many qualified people have had the opportunity to have this Eureka moment, but they simply have not. I put this paragraph in this book for the psychologists who study creativity. Perhaps making the link to put these equations together isn't so obvious after all. We now press on.

The first item on our agenda is the shape and location discussion. This discussion is important because our answers are found in the properties of the shape of things—not their location. In the above two questions, the information about the vertex of the parabola is irrelevant. Only the shape of the parabola has interest as far as making progress toward the solution. Thus, the general equation of the up-down parabola is $y = a(x-h)^2 + k$, and it has shape and location information in it. The variable a measures the bend of the parabola. This bending affects how far up the arc of the parabola we need to go to get to the desired distance along the line. Regardless of the where the vertex of the parabola is located, we still have to travel that same corresponding distance up (or down as the case may be) the arc to achieve the linear distance desired on the line. We may thus move the vertex of the parabola to the origin. This is accomplished by setting h and k in the parabola equation to zero. Now, our parabolic equation is $y = ax^2$.

For our first derivation, we will derive the equation for the slope of a line passing through the vertex of $y = ax^2$ that touches the parabola again d units away. We will thus have a general equation to find the slope for any line passing through a parabola having any shape (i.e. a is a real number) with its vertex at the origin that intersects the parabola again any distance d units away. This equation will enable us to answer the two specific questions posed at the beginning of this chapter.

The following graph (Figure 1 on the next page) is a line through the vertex of an up-down parabola having its vertex at the origin with a general distance d between intersection points on the parabola.

Let us discuss this picture. The parabola is $y = ax^2$, and the line is $y = mx$. The intersection points between the parabola and line are A and B. Of course, point A is the vertex of the parabola, which is the origin. The other intersection is point B, which has coordinates (x_1, ax_1^2), and x_1 is a generic (meaning arbitrary) x-value representing the x-coordinate of the intersection point between the line and the parabola. The value of x_1 is shown on the x-axis. The distance between point A and point B is labeled with a d to show it is a generic distance. The y-value of point B is shown on the y-axis as y_1 meaning it is a generic value.

We know that the distance d between the two points of intersection is the following:

$$d = \sqrt{(x_1 - 0)^2 + (ax_1^2 - 0)^2} = \sqrt{x_1^2 + a^2 x_1^4}$$

This equation can be solved for x_1 by squaring both sides and using the quadratic formula.

$$d^2 = x_1^2 + a^2 x_1^4$$
$$a^2 x_1^4 + x_1^2 - d^2 = 0$$

Now, from the quadratic formula, A is a^2, B is 1, and C is $-d^2$. Next, we can use the quadratic formula to solve for x_1^2.

$$x_1^2 = \frac{-1 \pm \sqrt{1 - 4a^2(-d^2)}}{2a^2} = \frac{-1 \pm \sqrt{1 + 4a^2 d^2}}{2a^2}$$

Thus, x_1 is $\pm$ the square root of this last expression.

$$x_1 = \pm \sqrt{\frac{-1 \pm \sqrt{1 + 4a^2 d^2}}{2a^2}} \cdot \sqrt{\frac{2}{2}} = \pm \sqrt{\frac{-2 \pm 2\sqrt{1 + 4a^2 d^2}}{4a^2}} = \frac{\pm \sqrt{-2 \pm 2\sqrt{1 + 4a^2 d^2}}}{2a}$$

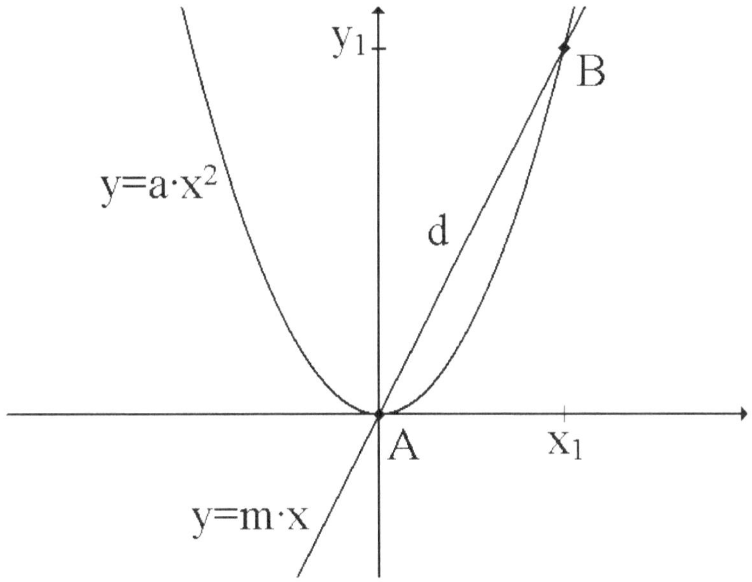

Figure 1: Our Situation with the Parabola

At this point, let us look at the expression $1 + 4a^2 d^2$ under the radical. If d is near zero, then $4a^2 d^2$ is near zero and positive, regardless of the sign of a. Thus, $1 + 4a^2 d^2$ is near one but larger than one. The square root of a number larger than one is a smaller number, but it is always larger than one. Therefore, $2\sqrt{1 + 4a^2 d^2}$ is a number greater than two because the 2 is multiplied by some number larger than one. These considerations are

117

important because $\sqrt{-2 \pm 2\sqrt{1 + 4a^2d^2}}$ needs to be interpreted. The expression $\sqrt{-2 - 2\sqrt{1 + 4a^2d^2}}$ is obviously complex or imaginary. Since $2\sqrt{1 + 4a^2d^2} > 2$, the expression $\sqrt{-2 + 2\sqrt{1 + 4a^2d^2}}$ must be a real number. Since we seek real values for x_1 on the intersection points, we may discard the complex or imaginary portion of this derivation. So, we have the following:

$$x_1 = \frac{\pm\sqrt{-2 + 2\sqrt{1 + 4a^2d^2}}}{2a}$$

Next, we can find the slope of the line with the slope formula. This is easy because the vertex of the parabola is at (0, 0). Note that x_1 in the slope formula is not the same as x_1 as we are using it in this derivation. Thus, the slope formula just wants two points—two x-coordinates and two y-coordinates. The x-coordinate called x_2 in the slope formula is our generic x-coordinate from point B, which is x_1.

$$m = \frac{y_2 - y_1}{x_2 - x_1} = \frac{ax_1^2 - 0}{x_1 - 0} = ax_1$$

It may be better to label points 1 and 2 in the slope formula as points A and B in our above picture. Then, we get the derived slope below, which should be less confusing.

$$m = \frac{y_B - y_A}{x_B - x_A} = \frac{ax_1^2 - 0}{x_1 - 0} = ax_1$$

Finally, we can use the slope intercept form of the equation of a line, $y = mx + b$, to find the equation of the line we seek.

$y = mx + b$ Here, b is zero because the y-intercept is zero. The slope m is ax_1.

$y = ax_1 \cdot x + 0$ We found x_1 earlier, so we can substitute that in now.

$$y = a \cdot \frac{\pm\sqrt{-2 + 2\sqrt{1 + 4a^2d^2}}}{2a} \cdot x$$

Of course, we can cancel the a in the numerator and denominator to yield our goal.

$$y = \frac{\pm\sqrt{-2+2\sqrt{1+4a^2d^2}}}{2} \cdot x$$

This is the derived equation of the line. Notice that there is no discriminant, and m is the entire mess in front of the x. Observe again that b in $y = mx + b$ is zero for this equation. We may also note that depending on the values of a and d in specific situations, the radical may be able to be denested.

We know that the general equation of a parabola with the vertex at the origin is $y = ax^2$. If we know p, the directed distance from the vertex to the focus of the parabola, we have the equivalent description of the parabola as $x^2 = 4py$. This becomes $y = x^2/(4p)$. Thus, a is $1/(4p)$. This means that a^2 is $1/(16p^2)$. We can use this to find an alternative equation of the line in terms of p. The algebra is a little involved, but it is doable. So we will just show the equivalence.

$$y = \frac{\pm\sqrt{-2+2\sqrt{1+4a^2d^2}}}{2} \cdot x = \frac{\pm\sqrt{-2p^2+\sqrt{4p^4+p^2d^2}}}{2p} \cdot x$$

The strategy required to solve the two problems shown at the outset of this chapter is to move the vertex of the parabola to the origin. This action will cause the h and k in the formula $y = a(x-h)^2 + k$ to go to zero. Then, the formula becomes $y = ax^2$. Whether you are given a or p initially will determine which form of the equation to use. However, converting p to a first may make the hand computations easier as a cursory inspection of the two versions will reveal.

We are now in a position to solve a problem of the first type asked at the outset. Find the equation of the positively sloped line passing through the vertex of the parabola $y = 7x^2 - 11x + 4$ that intersects the parabola again five units away.

We start by identifying a as 7 and d as 5. Then, we plug these values into the equation that we have derived.

$$y = \frac{\pm\sqrt{-2+2\sqrt{1+4a^2d^2}}}{2} \cdot x = \frac{\pm\sqrt{-2+2\sqrt{1+4\cdot7^2\cdot5^2}}}{2} \cdot x = \frac{\pm\sqrt{-2+2\sqrt{1+4900}}}{2} \cdot x =$$

$$\frac{\pm\sqrt{-2+2\sqrt{4901}}}{2} \cdot x = \frac{\pm\sqrt{-2+2\sqrt{13^2\cdot29}}}{2} \cdot x = \frac{\pm\sqrt{-2+26\sqrt{29}}}{2} \cdot x$$

This equation still does not quite achieve our goal because the problem asked us to find the equation of the positively sloped line. Thus, we have the following:

$$y = \frac{\sqrt{-2+26\sqrt{29}}}{2} x$$

But, the given parabola has a vertex that is not at the origin. So, we will need to account for that, but at least our line does have the correct slope now. The point-slope version of the equation of a line is ideal. We need to find the coordinates of the vertex first. The equation of the parabola was $y = 7x^2 - 11x + 4$. Thus, $a = 7$ and $b = -11$. We use the vertex formula of the parabola.

$$(x_1, y_1) = \left(\frac{-b}{2a}, y\left(\frac{-b}{2a} \right) \right) = \left(\frac{11}{14}, \frac{-9}{28} \right)$$

Of course, the point (x_1, y_1) is the coordinates of the vertex of the parabola.

$$y - y_1 = m(x - x_1)$$

$$y - \left(\frac{-9}{28} \right) = \frac{\sqrt{-2+26\sqrt{29}}}{2} \left(x - \frac{11}{14} \right)$$

$$y = \frac{\sqrt{-2+26\sqrt{29}}}{2} \left(x - \frac{11}{14} \right) - \frac{9}{28}$$

This is probably the best form to put the answer because if we try to distribute the numerical value of the slope and combine the numerical terms, we would not have a better form because of the presence of square roots and rational numbers together. In the next example, we will verify our work; however, the student is asked to verify the answer that we obtained analytically and graphically in the exercises.

It is interesting to show that the equation of a parabola and a line that passes through the vertex of the parabola share a similar format. The point-slope equation of a line has $x_1 = h$ and $y_1 = k$.

Parabola Equation: $y = a(x-h)^2 + k$

Line Equation: $y = m(x-h) + k$

The leading coefficient a, which determines the amount of bend that the parabola will have, has a similar position to m, the slope of the line in the respective equations. The $(x - h)$ term is squared in the parabola equation, and the $(x - h)$ term is to the first power in the line equation. Other than that, these two equations are identical. We know that $y = mx + b$ is the slope-intercept form of the equation of a line. So, b in the line equation above computes to $-mh + k$. If $m, h,$ and k have unusual formats of radicals and fractions, it is best to leave the equation in the $y = m(x - h) + k$ form because it will be more compact. This was done in the *Pappus Line* highlight to avoid even larger integers from appearing in the fractions.

Well, we asked one other question at the outset (question #2). Let us ask the same type of question but with more difficult numbers. Suppose we need to find the equation of an upward facing parabola with its vertex at $(-7\sqrt{6}, 8.3)$ with a line of slope $-3/11$ going through the vertex and intersecting the parabola again $\sqrt{15}$ units away. Attempting to answer a question like this before having this material would be extremely difficult. However, now it is reachable with the ideas presented in this chapter.

The first thing we need to do is to solve the derived slope equation for a. It requires some algebra, but the task is not too difficult. Thus, we will just present the original equation and final results solved for a and d.

$$m = \frac{\pm\sqrt{-2 + 2\sqrt{1 + 4a^2 d^2}}}{2}$$

$$a = \frac{\pm m}{d}\sqrt{m^2 + 1} \qquad\qquad d = \left|\frac{m}{a}\right|\sqrt{m^2 + 1}$$

Notice that there is still no discriminant in the formulas for a and d. If $m = 0$, then $a = 0$. And, if $m \neq 0$, then $m^2 + 1 > 0$, which means that the expression under the radical is positive and a is real. Similar remarks hold for d.

Well, we are now ready to proceed. We see from the given information that $m = -3/11$ and that $d = \sqrt{15}$. So, we just plug these values into the formula for a above.

$$a = \frac{\pm(-3/11)\sqrt{(-3/11)^2 + 1}}{\sqrt{15}}$$

After further simplification, which the student should have no trouble with, we arrive at the following:

$$a = \frac{\sqrt{78}}{121}$$

Note that we need the positive value of a because we want the upward facing parabola. We now know a, h, and k. So, we can insert those values into the parabolic equation above.

$$y = a(x-h)^2 + k = \frac{\sqrt{78}}{121}(x-(-7\sqrt{6}))^2 + 8.3 = \frac{\sqrt{78}}{121}(x+7\sqrt{6})^2 + \frac{83}{10}$$

We may want to verify our work, and this is always a good thing to do if it is possible. So we would want to know the equation of the line. It is given by the h and k form of the line above.

$$y = m(x-h) + k = \frac{-3}{11}(x-(-7\sqrt{6})) + 8.3 = \frac{-3}{11}(x+7\sqrt{6}) + \frac{83}{10}$$

To verify our work, we want to know the distance between the intersection points in the following system of equations:

$$\begin{cases} y = \dfrac{\sqrt{78}}{121}(x+7\sqrt{6})^2 + \dfrac{83}{10} \\ y = \dfrac{-3}{11}(x+7\sqrt{6}) + \dfrac{83}{10} \end{cases}$$

We have two methods available to us to do this. They are analytical, which yields exact results, and graphical, which yields approximate results. Using a grapher, put the two functions in the y-editor of your grapher. Find an appropriate viewing window. Use the intersection command to get the coordinates of the non-vertex intersection point. Store the x-coordinate in a variable. Lastly, use the distance command with one x-coordinate as the variable just obtained and the other x-coordinate as $-7\sqrt{6}$ to get an approximate distance between the two points. Then, ask the grapher to approximate $\sqrt{15}$, and compare the two. The results should agree to a large number of decimal places. We will now perform an analytical verification.

$$\frac{\sqrt{78}}{121}(x+7\sqrt{6})^2 + 83/10 = \frac{-3}{11}(x+7\sqrt{6}) + 83/10$$

$$\frac{\sqrt{78}}{121}(x+7\sqrt{6})^2 + \frac{3}{11}(x+7\sqrt{6}) = 0$$

$$(x+7\sqrt{6})\left(\frac{\sqrt{78}}{121}(x+7\sqrt{6}) + \frac{3}{11}\right) = 0$$

$$x = -7\sqrt{6}, \quad \frac{\sqrt{78}}{121}(x+7\sqrt{6}) + \frac{3}{11} = 0 \Rightarrow x = \frac{-11\sqrt{78}}{26} - 7\sqrt{6}$$

$$x = -7\sqrt{6} \Rightarrow y = \frac{-3}{11}(x+7\sqrt{6}) + \frac{83}{10} = \frac{-3}{11}(-7\sqrt{6}+7\sqrt{6}) + \frac{83}{10} = \frac{83}{10}$$

$$x = \frac{-11\sqrt{78}}{26} - 7\sqrt{6} \Rightarrow y = \frac{\sqrt{78}}{121}(x+7\sqrt{6})^2 + \frac{83}{10} = \frac{\sqrt{78}}{121}\left(\frac{-11\sqrt{78}}{26} - 7\sqrt{6} + 7\sqrt{6}\right)^2 + \frac{83}{10}$$

$$= \frac{15\sqrt{78} + 1079}{130}$$

$$d = \sqrt{\left(\frac{-11\sqrt{78}}{26} - 7\sqrt{6} + 7\sqrt{6}\right)^2 + \left(\frac{15\sqrt{78}+1079}{130} - \frac{83}{10}\right)^2} = \sqrt{\left(\frac{-11\sqrt{78}}{26}\right)^2 + \left(\frac{3\sqrt{78}}{26}\right)^2}$$

$$= \sqrt{\frac{121 \cdot 78}{26^2} + \frac{9 \cdot 78}{26^2}} = \sqrt{\frac{78 \cdot (121+9)}{26 \cdot 26}} = \sqrt{\frac{3 \cdot 26 \cdot 5}{26}} = \sqrt{15}$$

The two equations are each equal to y, so they are equal to each other. This creates a one-variable equation, which is quadratic. The two x-values are found by factoring the equation and setting each factor equal to zero. Then, each x-value is substituted back into one of the original equations (either equation works, but both equations are used here for completeness). Finally, the two points obtained are set into the distance equation and simplified. The result of $\sqrt{15}$ shows we did our work correctly because that was the given distance in the problem.

The three formulas we have derived thus far have their counterparts with p, the directed distance from the vertex to the focus, instead of a, the bend of the parabola. We omit the algebra and just present these results.

123

$$p = \frac{\pm d}{4m\sqrt{m^2+1}} \qquad\qquad d = 4|mp|\sqrt{m^2+1}$$

$$m = \frac{\pm\sqrt{-2p + \sqrt{4p^2 + d^2}}}{2\sqrt{p}}$$

Another situation exists for a line traveling through the vertex of a parabola intersecting the parabola a set distance d away. Of course, this other situation is for the left-right parabola. To be factually accurate, a technique exists called translation and rotation of axes, which allows a parabola to be at any angle; however, we will not get into that topic in this book. The reason is because the translation and rotation of axes topic is usually reserved for the second term of calculus, which is too advanced for this book. Perhaps, a future edition of this book will have an optional chapter in it with this material. For the moment, the left-right parabola situation will be as far as we take it in this chapter. Our left-right parabola situation is shown in the next diagram, which is Figure 2.

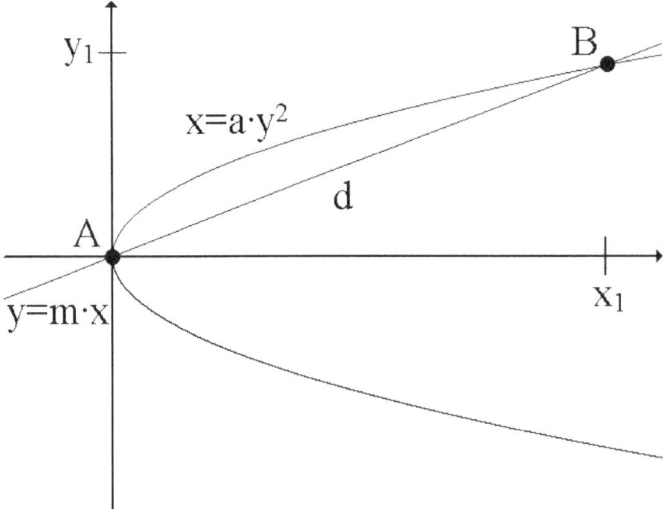

Figure 2: Our Situation with a Left-Right Parabola

Let us look at the diagram carefully. The parabola opens to the right, so $a > 0$ in the parabola equation $x = ay^2$. If $a < 0$, then the parabola would open leftward. Since h and k are zero, the vertex of the parabola is at the origin. The line $y = mx$ has a positive slope because it is an ascending line. However, m could just as well be negative with a descending line. Since $b = 0$, the line travels through the origin. Since both the line and the parabola have a point at the origin, point A must be the origin. Point B is the other intersection point between the line and the parabola, which is some distance d away from point A. Projecting the y-coordinate of point B onto the y-axis yields the point $(0, y_1)$. We can treat the left-right parabola as a function if we consider the domain as the y-axis. Hence, x is a function of y, or $x = f(y) = ay^2$. When we plug y_1 into the function f, we get ay_1^2. Thus, the coordinates of point B are (ay_1^2, y_1).

The first thing we need to do is to compute the slope of the line in terms of y_1 and a.

$$m = \frac{y_1 - 0}{ay_1^2 - 0} = \frac{1}{ay_1}$$

Next, we solve this equation for y_1 so that we can substitute into the distance equation for y_1.

$$y_1 = \frac{1}{am}$$

We can get a preliminary distance in terms of y_1.

$$d = \sqrt{(y_1 - 0)^2 + (ay_1^2 - 0)^2} = \sqrt{y_1^2 + a^2 y_1^4}$$

Now, we can substitute the expression for y_1 derived from the slope equation.

$$d = \sqrt{\left(\frac{1}{am}\right)^2 + a^2\left(\frac{1}{am}\right)^2} = \sqrt{\frac{1}{a^2 m^2} + \frac{a^2}{a^4 m^4}} = \sqrt{\frac{1}{a^2 m^2} \cdot \frac{m^2}{m^2} + \frac{1}{a^2 m^4}} = \sqrt{\frac{m^2 + 1}{a^2 m^4}} = \frac{\sqrt{m^2 + 1}}{|a| m^2}$$

We now have an equation in terms of a, d, and m. It remains to solve this equation for its other variables. The formula solved for a and m are the following:

$$a = \frac{\pm\sqrt{m^2 + 1}}{dm^2} \qquad\qquad m = \frac{\pm\sqrt{2 + 2\sqrt{1 + 4a^2 d^2}}}{2ad}$$

The directed distance from the vertex to the focus is p (but some textbooks use the letter c). The relationship between a and p is $a = 1/(4p)$. Using this relationship, we obtain versions of these formulas with p instead of a, which are the following:

$$d = \frac{4|p|}{m^2} \cdot \sqrt{m^2 + 1}$$

$$p = \frac{\pm\, dm^2 \sqrt{m^2 + 1}}{4(m^2 + 1)} \qquad\qquad m = \frac{\pm 2\sqrt{2p^2 + |p|\sqrt{4p^2 + d^2}}}{d}$$

Exercises

The following exercise set is for the student to try. Any good math exposition is always followed by exercises to test the student's grasp of the material. To that end, here are some exercises. These are difficult problems, but persevere to win the day.

1. Verify analytically and graphically that the distance between the intersection points of the parabola $y = 7x^2 - 11x + 4$ and the line $y = \frac{\sqrt{-2 + 26\sqrt{29}}}{2}\left(x - \frac{11}{14}\right) - \frac{9}{28}$ is five units. Sketch the image on your calculator, and label the parabola, line, and any important points.

2. Find the equation of the up-down parabola passing through the following three points: (1, 6), (4, 0), and (3, 4).

3. Find the equation of the up-down parabola that has a line of slope two intersecting the parabola at the vertex and hits the parabola again nine units away. The coordinates of the vertex are (-5, 8).

4. What is the distance between the intersection points of the parabola $y = -3x^2 - 2x + 11$ and the line parallel to $y = 5x + 2$ such that the line goes through the vertex of the parabola? [Hint: Use the slope equation solved for d.]

5. The upward facing parabola $y = ax^2 - 11x - 9$ has a line perpendicular to the line $5x - 3y = 2$ going through its vertex. The distance between the points of intersection between the parabola and this line going through its vertex is seven units. Find the coordinates of the vertex of the parabola. Use these coordinates to write the equation of the line. Discuss the differences in the solution to this problem if we make the parabola a downward facing one.

6. The coordinates of the vertex of an up-down parabola are (1, 2). The directed distance to the focus of this parabola is one unit. Find the equation of the negatively sloped line that goes through the vertex and hits the parabola again twelve units away.

7. A line goes through the end-point of the square root function $y = 2 - 7\sqrt{8 - 5x}$ and intersects the function again 21 units away. Find the equation of this line.

8. A square root function has its end-point at (-2, 4). A line with a slope of six goes through this end-point and intersects the square root function again eleven units away. Find the equation of the square root function.

9. Find the distance between the intersection points between the square root function $y = -2 + 3\sqrt{5x + 2}$ and the line of slope -4/7 going through the end-point of the square root function.

10. Solve the left-right parabola distance equation for the other two variables a and d.

11. Substitute $1/(4p)$ for a, and derive the last three equations listed in this chapter.

12. The parabola $x = -4y^2 + 3y - 7$ has a line perpendicular to the line $y = 5.7x - 1.9$ going through its vertex. Find the distance between the intersection points. Find the exact locations of the intersection points. Use a graphing utility to depict the situation, and use it to approximate the distance between the intersection points.

13. A left-right parabola has its vertex at (–8, 19). The directed distance from the vertex to the focus is –0.88 units. A negatively sloped line passes through the vertex of this parabola intersecting the parabola again a hundred units away. Find the slope of this line. Draw a diagram of this situation including all relevant details.

14. The parabola $x = 7y^2 - ty + 9$ has a line of slope two going through its vertex intersecting the parabola again forty units away. What value(s) of t make this happen? Draw a diagram for each value of t found.

15. The parabola $\dfrac{2x - 1}{1 - y} = 5y - 2$ has the line $y = 6x + b$ going through its vertex and intersecting it again $3b$ units away. What value of b makes this happen? Draw a diagram of this situation. Find and label the coordinates of all the important points.

16. Two perpendicular lines intersect at a point somewhere on the line $x = 5$, which happens to be the vertex of the parabola $x = 9y^2 - 6y + t$. The positively sloped line is $5 + t$ more than its perpendicular counterpart. Each line must intersect the parabola again some distance d away. Find these two distances. In addition, explain why even though there are two sets of perpendicular lines, each set has the same two distances.

17. (Modified Putnam Exam Question) Given the parabola $y = ax^2$, find the length of the shortest chord that is normal to the parabola at one end. Note that the length will be in terms of a. Write the equation of the line containing this chord. Now, let $a = 2$, and draw a graph of this parabola and chord. Find and label the intersection points.

Lines and Parabolas: Part 2

For our next derivation, we ask the basic question, "What is the slope of the line that travels through the focus of a parabola that has a distance d between the intersection points?" The situation is captured in Figure 1.

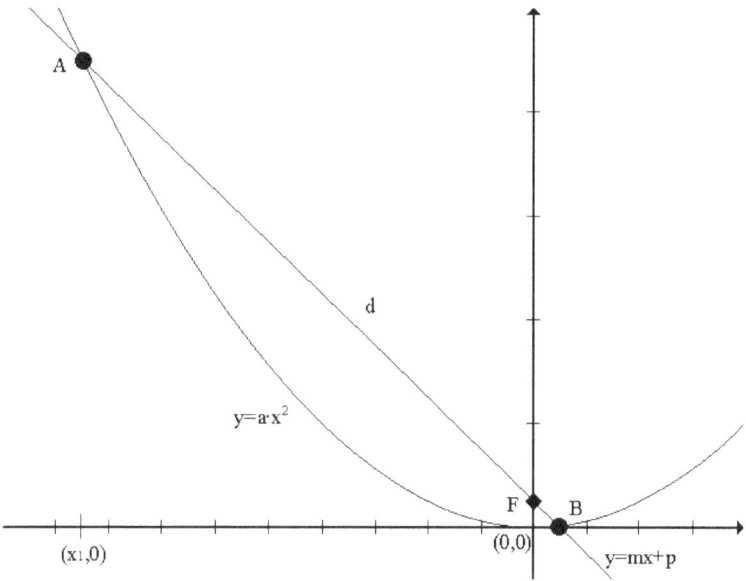

Figure 1: Through the Focus of an Up-Down Parabola

The equation of the parabola is $y = ax^2$, and the equation of the line is $y = mx + p$. The points A and B are the intersection points between the line and the parabola. The point F is the focus of the parabola, and the line goes through it. The d is the distance between the points A and B. In this graph, the vertex of the parabola is at the origin, which is labeled. However, in the actual problems, the vertex will most often not be at the origin. We place the vertex at the origin to help develop the equation for the slope of the line, according to the shape and location discussion in Part 1 of this chapter. The point A has been projected onto the x-axis, which means that its x-coordinate is preserved. Observe that in this picture, then, $x_1 < 0$. The coordinates of the focus F are $(0, p)$ or $(0, 1/(4a))$; however, the rendering with the variable a is better because of other computations that we will need to do.

The first thing that we need to do is to find the coordinates of the point B in terms of x_1 and a. To that end, we need to do some algebra. We do have knowledge of point A and F in terms of x_1 and a, so we can use it to find a preliminary slope of the line.

129

$$m = \frac{y_2 - y_1}{x_2 - x_1} = \frac{1/(4a) - ax_1^2}{0 - x_1} = \frac{ax_1^2 - 1/(4a)}{x_1} = \frac{4a^2 x_1^2 - 1}{4ax_1}$$

Next, we can use this slope in the slope-intercept equation of a line to find a preliminary equation of the line. Of course, the y-intercept is $(0, p)$.

$$y = mx + b = \frac{4a^2 x_1^2 - 1}{4ax_1} x + p = \frac{4a^2 x_1^2 - 1}{4ax_1} x + \frac{1}{4a}$$

The original parabola is $y = ax^2$. The line is $y = \frac{4a^2 x_1^2 - 1}{4ax_1} x + \frac{1}{4a}$. These two

equations give us a system of equations that we can use. The solutions of this system are the coordinates of the intersection points between the line and the parabola. We already knew one of the solutions to this system. It is the ordered pair (x_1, ax_1^2), which is point A. Of course, we seek the other ordered pair, which is point B.

Since both of the equations are equal to y, we can set them equal to each other. Then, we can solve for x. Note that we must treat a and x_1 as constants, but a plain x is the independent variable.

$$ax^2 = \frac{4a^2 x_1^2 - 1}{4ax_1} x + \frac{1}{4a}$$

$$ax^2 - \frac{4a^2 x_1^2 - 1}{4ax_1} x - \frac{1}{4a} = 0$$

This is quadratic in x. Since we know one solution already, we know one of the factors of this equation. Hence, we can construct the other factor. The student should check the middle term to make sure that this construction method yields the correct other factor or that no mistake has been made in the construction of this factor because that leads to erroneous results.

$$(x - x_1)\left(ax + \frac{1}{4ax_1}\right) = 0$$

$$x - x_1 = 0, \quad ax + \frac{1}{4ax_1} = 0$$

$$x = x_1, \quad x = \frac{-1}{4a^2 x_1}$$

130

Well, this gives us the x-coordinate for point B in terms of the constants a and x_1. Each factor has been set equal to zero because the zero-factor theorem states that if a product of factors is zero, then either one or both of them is zero. The student should note that in modular arithmetic, the zero-factor theorem does not hold. For example, if we use the regular clock digits 1-12, at noon or midnight if we add one hour, we get 1 o'clock, not 13 o'clock. In modular arithmetic, this reset point becomes zero. Thus, 12 in mod 12 is zero. Multiplying in mod 12, we have things like $2 \cdot 4 = 8$ and $0 \cdot 7 = 0$. However, we also have $3 \cdot 4 = 12 = 0$, and neither 3 nor 4 is 0, which violates the zero-factor theorem. All of the equations in this book are in a system of numbers (the real numbers) where the zero-factor theorem holds, so we do not have to worry about this sort of thing. The student is getting a small preview of coming attractions in future math courses with this mention of modular arithmetic.

Next, we can plug these x values into the original system to get the y values in the ordered pairs. However, we already knew one of the ordered pairs. Thus, we only need to find the corresponding y value for that other x-coordinate, which is point B.

$$y = ax^2 = a \cdot \left(\frac{-1}{4a^2 x_1} \right)^2 = \frac{1}{16a^3 x_1^2}$$

This is a big moment for us because we now know that $\left(\frac{-1}{4a^2 x_1}, \frac{1}{16a^3 x_1^2} \right)$ is the other solution. It does not matter in the abstract which of the ordered pairs is point A and which is point B. However, it does matter about the sign of x_1, which for point A was made explicitly less than zero from its position on the diagram. Thus, the picture is driving the mathematics as we proceed through the derivation. So, we now have three points on the line. They are $A = (x_1, ax_1^2)$, $F = \left(0, \frac{1}{4a} \right)$, and $B = \left(\frac{-1}{4a^2 x_1}, \frac{1}{16a^3 x_1^2} \right)$.

Next, we need to find the distance between the points A and B. However, if one tries to plug the points A and B into the distance formula, he or she would be going astray as I originally did, which will be explained shortly. Thus, in order to find the distance between the points A and B, one would add the distance between point A and the focus to the distance between point B and the focus. I will use the notation $d(A, F)$ to represent the distance between point A and the focus.

$$d(A, B) = d(A, F) + d(F, B) = d(A, F) + d(B, F)$$

$$d(A,F) = \sqrt{\left(\frac{1}{4a} - ax_1^2\right)^2 - (0 - x_1)^2} = \sqrt{\frac{16a^4x_1^4 - 8a^2x_1^2 + 1 + 16a^2x_1^2}{16a^2}}$$

$$= \frac{\sqrt{16a^4x_1^4 + 8a^2x_1^2 + 1}}{4\,|a|} = \frac{4a^2x_1^2 + 1}{4\,|a|}$$

$$d(B,F) = \sqrt{\left(\frac{1}{4a} - \frac{1}{16a^3x_1^2}\right)^2 + \left(0 - \frac{-1}{4a^2x_1}\right)^2} = \frac{4a^2x_1^2 + 1}{16\,|a|^3\,x_1^2}$$

Some of the algebra has been omitted, but the student is welcome to fill in the missing steps. Also, an easier derivation than using the distance formula here is to use the definition of a parabola. That is to say that the distance from the focus to the parabola is the same as the distance from that point on the parabola to the directrix. Then, we could just take the absolute value of the subtraction of the y-coordinates from point A and the corresponding point on the directrix. We could do the same for point B. This would yield the same two distances found above. This method will not work when going through the focus of an ellipse or hyperbola in the final two chapters of the book because the distance from the focus to the ellipse or hyperbola is not the same as the distance from that point on the ellipse or hyperbola to the associated directrix. However, the splitting up the distance as we have done here will work in all three of these cases, so it is the method used here. At this point, we can just add the two distances to get the total distance between the points A and B. An important point to observe about these two distances is that they are devoid of radicals, which makes it possible to add them nicely.

I promised to explain the problem about finding the distance between points A and B directly. The resulting distance when going directly from point A to point B with the distance formula is buried in an 8th-degree polynomial expression under the radical that is possible to factor, but we have to use the rational root theorem or a grapher to do it. That is why we are breaking up the distance into two pieces, which we will now add.

$$d(A,B) = d(A,F) + d(F,B) = d = \frac{4a^2x_1^2 + 1}{4\,|a|} + \frac{4a^2x_1^2 + 1}{16\,|a|^3\,x_1^2} = \frac{(4a^2x_1^2 + 1)\cdot(4a^2x_1^2 + 1)}{16\,|a|^3\,x_1^2}$$

$$d(A,B) = \frac{(4a^2x_1^2 + 1)^2}{16\,|a|^3\,x_1^2}$$

In the first chapter, we solved the distance equation for x_1 and plugged the expression into the preliminary slope equation. That approach will not work here because the resulting expression for x_1 is too cumbersome to work with. Hence, we will reverse our approach. That is to say, we will solve the preliminary slope equation for x_1 and substitute it into the distance equation we just obtained above.

$$m = \frac{4a^2 x_1^2 - 1}{4ax_1}$$

$$4amx_1 = 4a^2 x_1^2 - 1$$

$$4a^2 x_1^2 - 4amx_1 - 1 = 0$$

$$x_1 = \frac{4am \pm \sqrt{16a^2 m^2 - 4(4a^2)(-1)}}{2(4a^2)} = \frac{m \pm \sqrt{m^2 + 1}}{2a} \qquad \text{Note that we really need } x_1^2.$$

$$x_1^2 = \frac{2m^2 + 1 \pm 2m\sqrt{m^2 + 1}}{4a^2} \qquad \text{Recall the equation for } d.$$

$$d = \frac{(4a^2 x_1^2 + 1)^2}{16 |a|^3 x_1^2} \qquad \text{Now, we substitute.}$$

$$d = \frac{(4a^2 x_1^2 + 1)^2}{16 |a|^3 x_1^2} = \frac{\left(4a^2 \left(\dfrac{2m^2 + 1 \pm 2m\sqrt{m^2 + 1}}{4a^2} \right) + 1 \right)^2}{16 |a|^3 \left(\dfrac{2m^2 + 1 \pm 2m\sqrt{m^2 + 1}}{4a^2} \right)} = \frac{(2m^2 + 1 \pm 2m\sqrt{m^2 + 1} + 1)^2}{4 |a| (2m^2 + 1 \pm 2m\sqrt{m^2 + 1})}$$

$$d = \frac{(m^2 + 1 \pm m\sqrt{m^2 + 1})^2}{|a| (2m^2 + 1 \pm 2m\sqrt{m^2 + 1})}$$

At this point, we would seem to be at an impasse. However, the math technique that saves the day is called rationalizing the denominator. The student will be asked to do this in the exercises. Note that the pair of $\pm$ signs came from x_1 being in two places in the distance equation. Hence, there are only two versions of this distance equation—one with both $\pm$ signs in the plus position and one with both $\pm$ signs in the minus position. After rationalizing the denominator (through either version of the distance equation above), we arrive at the following:

$$d = \frac{m^2 + 1}{|a|}$$

Well, now this is a simple little equation. After all of that work, it's almost embarrassing that this is what we end up with; however, results in mathematics are what they are, not what we expect or wish them to be. Incidentally, looking at this d equation, when $m = 0$, then $d = \frac{1}{|a|}$, which is the length of the latus rectum. We still need to solve this equation for the other variables; however, the student will be asked to do this in the exercises. The results of solving for the other variables are the following:

$$m = \pm\sqrt{|a|d - 1} \qquad\qquad a = \frac{\pm(m^2 + 1)}{d}$$

We can now celebrate. The near incomprehensible formula for m (that was not depicted because of its ugliness) could not be simplified readily on its own. However, because it seemingly could not be solved in terms of its other variables, another pass had to be made through the derivation to get at d. One of my instructors at LTU observed that it could be factored to yield the collapse, but I did not see it. Happily, this reverse directional pass (that was depicted and was the route the author took through the derivation) yielded the incredible algebraic collapse of the distance equation after x_1 had been substituted. The collapse was possible through rationalizing the denominator. This collapse led to some very easy formulas for a, d, and m. So, the linked $\pm$ symbols in the distance equation have just dropped out of the picture. The equations of the two lines through the focus of the parabola with a given distance d between intersection points is the following:

$$y = \pm m(x - h) + k + \frac{1}{4a} = \pm\sqrt{|a|d - 1}(x - h) + k + \frac{1}{4a}$$

It was a long journey of analysis here; but the simple equations for a, d, and m were worth it. Let us use our new equations to answer a question. What is the equation of the downward parabola with a vertex at $(-5, 11)$ that has a line with a slope of seven going through its focus with a distance between intersection points of twenty units?

The equation for an up-down parabola is the following: $y = a(x - h)^2 + k$

The equation of the line going through the focus of the parabola is the following:

$$y = \pm m(x - h) + k + \frac{1}{4a}$$

We are given $m = 7$ and $d = 20$. Next, we use $a = \pm(m^2 + 1)/d = \pm(7^2 + 1)/20$ to find $a = \pm 5/2$. We are asked about a downward parabola, so $a = -5/2$. Now, all of the variable values are known. Thus, the equation of the parabola is $y = -(5/2)(x + 5)^2 + 11$. The equation of the line is $y = 7(x + 5) + 11 + 1/(4(-5/2)) = 7(x + 5) + 11 - 1/10 \Rightarrow y = 7(x + 5) + 109/10 = 7x + 459/10$

We need to verify our work, so we need to find the distance between the intersection points, which is supposed to be 20 units. Of course, this can be done analytically, which yields exact answers; or it can be done graphically, which yields approximate answers. The graphical approach should yield the exact answer in this case because it is a whole number. We will do the analytic approach first followed by the graphical approach.

$$-(5/2)(x + 5)^2 + 11 = 7x + 459/10$$

$$(-(5/2)(x^2 + 10x + 25) + 11) \cdot 10 = (7x + 459/10) \cdot 10$$

$$-25(x^2 + 10x + 25) + 110 = 70x + 459$$

$$-25x^2 - 250x - 625 + 110 = 70x + 459$$

$$25x^2 + 320x + 974 = 0$$

$$x = \frac{-320 \pm \sqrt{320^2 - 4(25)(974)}}{2(25)} = \frac{-32 \pm 5\sqrt{2}}{5}$$

$$x = \frac{-32 - 5\sqrt{2}}{5} \Rightarrow y = 7x + 459/10 = 7\left(\frac{-32 - 5\sqrt{2}}{5}\right) + 459/10 = \frac{11 - 70\sqrt{2}}{10}$$

$$x = \frac{-32 + 5\sqrt{2}}{5} \Rightarrow y = -(5/2)(x + 5)^2 + 11 = -(5/2)\left(\frac{-32 + 5\sqrt{2}}{5} + \frac{25}{5}\right)^2 + 11 = \frac{11 + 70\sqrt{2}}{10}$$

$$d = \sqrt{\left(\frac{-32 - 5\sqrt{2}}{5} - \frac{-32 + 5\sqrt{2}}{5}\right)^2 + \left(\frac{11 - 70\sqrt{2}}{10} - \frac{11 + 70\sqrt{2}}{10}\right)^2} = \sqrt{8 + 392} = \sqrt{400} = 20$$

Both equations are equal to y, so they are equal to each other, which creates a one-variable equation. Expand the binomial that is squared, and multiply both sides by the LCD = 10 to clear the fractions. Simplify the quadratic equation. We were able to factor the example in the last chapter, but that will not happen here. Therefore, we use the quadratic formula to solve the equation for x. Plug each value of x back into either equation to get its corresponding y-value. The same equation will work for both values of x, but both equations are used here for completeness. Finally, plug each (x, y) pair formed into the distance formula to compute the distance between the intersection points. We find that it is 20, and all is well.

Use a grapher for the graphical verification. Enter the two functions into the y-editor. Find an appropriate window setting for this problem. The x-coordinates of both intersection points need to be stored into other variables such as a for the lesser x-value and b for the greater x-value to use for the distance command. When using the distance command, input the lower bound as a and the upper bound as b. My grapher tells me that the distance is 20, so our graphical verification is complete.

This is a summary of all of the results for the up-down parabola for a line through the focus with a given distance d between intersection points.

$$m = \pm\sqrt{|a|\,d - 1}$$ Slope of line through focus of parabola.

$$d = \frac{m^2 + 1}{|a|}$$ Distance between intersection points on parabola.

$$a = \frac{\pm(m^2 + 1)}{d}$$ Leading coefficient of parabola.

$$y = m(x - h) + k + \frac{1}{4a}$$ Equation of line through focus of parabola.

We have one more stone to turn before we move to the next topic in this chapter. It has been known for a long time (at least to mathematicians anyway) that the sum of the reciprocals of the distances of the two focal chord pieces is constant. That constant can be expressed in terms of the length of the latus rectum of the parabola or in terms of its bend constant, $|a|$. When we do this, we get an equation. We will let d_1 and d_2 be the distances of the two focal chord pieces. Then, we get the following equation:

$$\frac{1}{d_1} + \frac{1}{d_2} = 4\,|a|$$

We may solve this equation for $|a|$ without much trouble to get the following:

$$|a| = \frac{d_1 + d_2}{4 d_1 d_2}$$

When we input this for $|a|$ and $d_1 + d_2$ for d in $m = \pm\sqrt{|a|\,d - 1}$ and simplify, we get the following:

$$m = \frac{\pm|d_2 - d_1|}{2\sqrt{d_2 d_1}}$$

This last equation reveals a major secret to us. Look at the radical in the denominator. It has a discriminant. If we want m to be rational when the two focal chord pieces are rational, then the product of the two focal chord pieces must be a perfect square. This is important enough to call it a theorem.

The Rational Slope and Focal Chord Pieces Theorem: Given d_1 and d_2 are the two rational focal chord pieces of a parabola, the slope m will be rational if and only if the product of d_1 and d_2 is a perfect square. Proof: Analysis of the discriminant of the slope equation just derived. It takes a perfect square to clear a square root.

Note: It is entirely possible that m is rational along with a and d, but that d_1 and d_2 are irrational. The above theorem just tells us the conditions to get a rational m when d_1 and d_2 are selected as rational numbers.

The three variables m, a, and d are linked together in a single equation as in the equation $m = \pm\sqrt{|a|\,d - 1}$. However, when we introduce the focal chord pieces d_1 and d_2, we get a system of three two-variable equations. We have learned that $y = f(x)$ before. The next step up is to say $z = f(x, y)$. We won't say much more than z depends on two inputs: x and y. Thus, x and y are independent variables, and z is the dependent variable in a standard two-variable function. Its graph requires three dimensions to graph. It's a place we will not delve into in this course, but multi-variable calculus deals with this idea extensively. We cast the three equations into this form so that we may choose d_1 and d_2 with impunity. This makes for nice algebra problems because not only will m, a, and d be rational, but so will d_1 and d_2 because we picked them to be.

$$d(d_1, d_2) = d_1 + d_2$$
$$|a|(d_1, d_2) = \frac{d_1 + d_2}{4 d_1 d_2}$$
$$m(d_1, d_2) = \frac{\pm|d_2 - d_1|}{2\sqrt{d_2 d_1}}$$

How do we get the product of d_1 and d_2 to be perfect squares? One way is for d_1 and d_2 each to be perfect squares themselves. Another way is for d_1 and d_2 to be reciprocals of one another. Then, the product of d_1 and d_2 is the multiplicative identity, 1; but 1 happens to be a perfect square. A third way is for d_1 and d_2 to share the individual factors that make up a perfect square. An example of this is $d_1 = 2$ and $d_2 = 8$. The four twos that make up the perfect square of 16 are distributed unevenly to d_1 and d_2. A fourth way involving fractions is to have one or more factors cancel from the product of d_1 and d_2 leaving a perfect square behind. An example of this $d_1 = 2/3$ and $d_2 = 6$. Here, the threes cancel from the product of d_1 and d_2; and what remains is $2 \cdot 2 = 4$, which is a perfect square. Upon choosing d_1 and d_2, compute m, a, and d to resolve the three variables.

Another situation exists regarding a line going through the focus of a parabola, and it is that the parabola could be a left-right parabola. The diagram labeled Figure 2 illustrates the situation. For variety, we will use p, the directed distance from the focus to the vertex this time instead of the variable a, which measures the bend of the parabola.

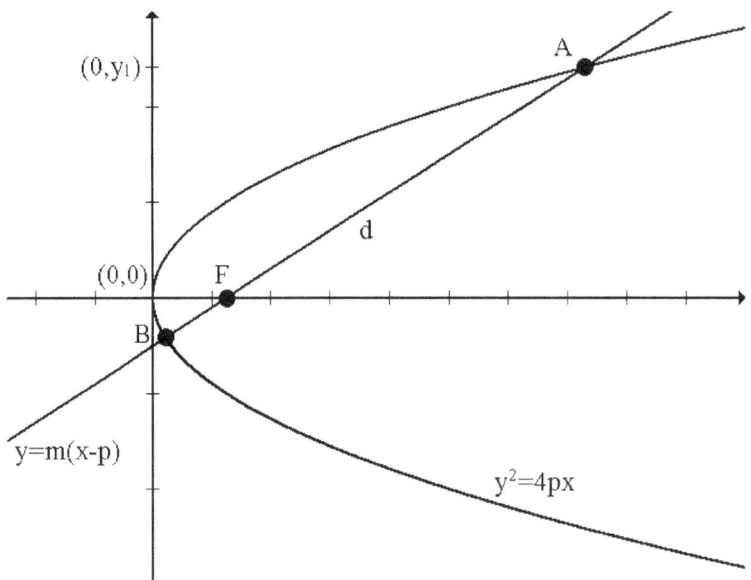

Figure 2: The Setup for L-R Parabola Through the Focus

The vertex of the parabola $4px = y^2$ is at the origin, which is labeled (0, 0). The line $y = m(x-p)$ goes through the point $(p, 0)$, which is the focus of the parabola labeled F. The intersection points between the line and the parabola are labeled A and B, and the distance between these two points is labeled d. Choosing some value of $y_1 > 0$, the coordinates of point A are $\left(\dfrac{y_1^2}{4p}, y_1 \right)$. The projection of point A onto the y-axis is $(0, y_1)$, which is labeled. We split up the distance d between points A and B in the up-down derivation, and we will do so in the left-right derivation as well. We know the coordinates of point A and F in terms of p and y_1, so we can construct the preliminary slope for the line.

$$m = \frac{y_1 - 0}{1/(4p) \cdot y_1^2 - p} = \frac{4py_1}{y_1^2 - 4p^2}$$

Using this slope, we can construct a preliminary equation of the line in terms of p and y_1.

$$y = m(x - p) = \frac{4py_1}{y_1^2 - 4p^2}(x - p) \qquad \text{Solve this equation for } x \text{ to equate functions.}$$

$$(x - p) = y \cdot \frac{y_1^2 - 4p^2}{4py_1}$$

$$\begin{cases} x = y \cdot \dfrac{y_1^2 - 4p^2}{4py_1} + p & \text{Line equation solved for } x. \\[2em] x = \dfrac{1}{4p}y^2 & \text{Parabola equation solved for } x. \end{cases}$$

Solving this system will give us the coordinates of point B. It will also give us the coordinates of point A, but we already know the coordinates of point A are $\left(\dfrac{y_1^2}{4p}, y_1 \right)$. Therefore, the resulting quadratic will factor using the method described earlier.

$$y \cdot \frac{y_1^2 - 4p^2}{4py_1} + p = \frac{1}{4p}y^2$$

$$\frac{1}{4p}y^2 - y \cdot \frac{y_1^2 - 4p^2}{4py_1} - p = 0$$

$$(y - y_1)\left(\frac{1}{4p}y + \frac{p}{y_1} \right) = 0 \qquad \text{As a precaution, the student should check the middle term.}$$

$$y - y_1 = 0, \frac{1}{4p}y + \frac{p}{y_1} = 0$$

$$y = y_1, \quad y = \frac{-p}{y_1} \cdot 4p = \frac{-4p^2}{y_1}$$

$$y = \frac{-4p^2}{y_1} \Rightarrow x = \frac{1}{4p} y^2 = \frac{1}{4p}\left(\frac{-4p^2}{y_1}\right)^2 = \frac{1}{4p} \cdot \frac{16p^4}{y_1^2} = \frac{4p^3}{y_1^2}$$

$$\therefore B(x,y) = \left(\frac{4p^3}{y_1^2}, \frac{-4p^2}{y_1}\right)$$

As before, we have achieved a big moment here because we have the coordinates of point B in terms of p and y_1. Next, we compute the distance between point A and the focus and between point B and the focus and add. This is to avoid a high degree polynomial.

$$d(A,F) = \sqrt{(y_1 - 0)^2 + \left(\frac{1}{4p} y_1^2 - p\right)^2} = \sqrt{y_1^2 + \left(\frac{1}{16p^2} y_1^4 - \frac{y_1^2}{2} + p^2\right)}$$

$$d(A,F) = \frac{1}{4|p|}\sqrt{y_1^4 + 8p^2 y_1^2 + 16p^4}$$

$$d(A,F) = \frac{4p^2 + y_1^2}{4|p|}$$

$$d(B,F) = \sqrt{\left(\frac{-4p^2}{y_1} - 0\right)^2 + \left(\frac{4p^3}{y_1^2} - p\right)^2} = \sqrt{\frac{16p^4}{y_1^2} + p^2\left(\frac{4p^2}{y_1^2} - 1\right)^2}$$

$$d(B,F) = \frac{|p|}{y_1^2}\sqrt{16p^4 + 8p^2 y_1^2 + y_1^4}$$

$$d(B,F) = \frac{|p|(4p^2 + y_1^2)}{y_1^2}$$

$$d = d(A,F) + d(B,F) = \frac{4p^2 + y_1^2}{4|p|} + \frac{|p|(4p^2 + y_1^2)}{y_1^2} = \frac{(4p^2 + y_1^2)^2}{4|p|y_1^2}$$

Recall that $m = \dfrac{4py_1}{y_1^2 - 4p^2}$. Earlier in this chapter, we solved the slope equation for x_1, squared it, and plugged the result x_1^2 into the distance equation. We will do that again here.

$$m = \frac{4py_1}{y_1^2 - 4p^2} \Rightarrow m \cdot y_1^2 - 4py_1 - 4 \cdot m \cdot p^2 = 0$$

$$y_1 = \frac{4p \pm \sqrt{16p^2 - 4m(-4p^2m)}}{2m} = \frac{2p \pm 2p\sqrt{m^2 + 1}}{m}$$

$$\therefore y_1^2 = \frac{4p^2 + 4p^2(m^2 + 1) \pm 8p^2\sqrt{m^2 + 1}}{m^2} = \frac{4p^2}{m^2}(m^2 + 2 \pm 2\sqrt{m^2 + 1})$$

The last thing we need to do is to substitute the expression y_1^2 into the distance equation $d = \dfrac{(4p^2 + y_1^2)^2}{4|p|y_1^2}$. We will also get the dependent $\pm$ signs because y_1^2 is present in two places in the distance equation. However, as before, taking each case in turn yields the same reduction.

$$d = \frac{(4p^2 + y_1^2)^2}{4|p|y_1^2} = \frac{\left(4p^2 + \dfrac{4p^2}{m^2}(m^2 + 2 \pm 2\sqrt{m^2 + 1})\right)^2}{4|p| \cdot \dfrac{4p^2}{m^2}(m^2 + 2 \pm 2\sqrt{m^2 + 1})} = \ldots = \frac{4|p|(m^2 + 1)}{m^2}$$

Most of the algebra has been omitted here, so the student will be asked to supply the missing steps in the exercise set. We need to solve the equation for the other variables, which the student will be asked to do as well. The equations are the following:

$$p = \frac{\pm dm^2}{4(m^2 + 1)} \qquad\qquad m = \pm 2 \cdot \sqrt{\frac{|p|}{d - 4|p|}}$$

Since $p = 1/(4a)$, we can substitute for p in the simplified distance equation. We arrive at the following.

$$d = \frac{m^2 + 1}{|a|m^2}$$

$$a = \frac{\pm(m^2 + 1)}{dm^2} \qquad m = \frac{\pm 1}{\sqrt{|a|d - 1}}$$

The line going through the focus $(k, h + p)$ of the left-right parabola $x = a(y - k)^2 + h$ or $4p(x - h) = (y - k)^2$ with distance d between intersection points is the following:

$$y = \pm 2 \cdot \sqrt{\frac{|p|}{d - 4|p|}} (x - (h + p)) + k .$$

We wish to repeat the three two-variable functions described earlier for the up-down parabola case. We get the same equations for d and a, but m's version is different. We also get the same theorem about the two focal chord pieces as before because the same discriminant is present in the slope equation.

$$d(d_1, d_2) = d_1 + d_2$$

$$|a|(d_1, d_2) = \frac{d_1 + d_2}{4d_1 d_2} \qquad m(d_1, d_2) = \frac{\pm 2\sqrt{d_2 d_1}}{d_2 - d_1}$$

Exercises

In the computational problems only that follow, verify each distance (whether given or found) analytically and graphically.

1. Rationalize the denominator of the equation $d = \dfrac{(m^2 + 1 \pm m\sqrt{m^2 + 1})^2}{|a|(2m^2 + 1 \pm 2m\sqrt{m^2 + 1})}$ to show that $d = \dfrac{m^2 + 1}{|a|}$. Note that the $\pm$ signs in the equation must match, so there are two equations in which to perform this simplification.

2. Find the equation of the positively sloped line going through the focus of the parabola $y = 2x^2 - 3x + 7$ that has a distance of eleven units between the intersection points. Verify the distance analytically and graphically.

3. Find the equation of the downward facing parabola that has the coordinates of the focus at $(1, 1)$ with a line of slope two going through it with a distance between the intersection points of three units.

4. What is the distance between the intersection points of the parabola $y = 0.2x^2 - 3x$ and the line of slope fourteen that passes through its focus.

5. Show a graph of the up-down parabola situation where the two focal chord pieces have lengths of 2/3 and 6. Use the three two-variable equations depicted in the text. Let two of m, a, and d be givens and solve for the third without using the m, a, and d equations in the text. However, $4|a| = 1/d_1 + 1/d_2$ may be used because it was already known before this textbook. Note that this creates three separate algebra problems. Document your work carefully.

6. Redo problem 5, but the student is encouraged to pick a d_1 and d_2 pair that multiply to a perfect square. The instructor, at his or her discretion, may assign a d_1 and d_2 pair to students to work in groups during class time. The students may then present their results to the rest of the class.

7. The parabola $x = 2y^2 - 3y + 2$ has a line of slope six going through the focus. What is the distance between the intersection points?

8. Find the equation of the leftward facing parabola with a focus at $(-3.4, 4.3)$ that has a line of slope 0.61 going through this focus with a distance between intersection points of 1.2 units.

9. Find the equation of the line going through the focus of the parabola $x = -y^2 + 2y + 2$ that has a distance of three units between the intersection points.

10. Using $y_1^2 = \dfrac{4p^2}{m^2}(m^2 + 2 \pm 2\sqrt{m^2 + 1})$, show that the equation $d = \dfrac{(4p^2 + y_1^2)^2}{4py_1^2}$ becomes $d = \dfrac{4|p|(m^2 + 1)}{m^2}$.

11. Solve $d = \dfrac{4|p|(m^2 + 1)}{m^2}$ for p and m.

12. Substitute $p = 1/(4a)$ for all three equations in question 11 to obtain equivalent equations with the variable a instead of the variable p.

13. A leftward parabola has its vertex at (2, 5), and a line goes through its focus. The distance from the vertex to the focus is one-twentieth that of the distance between the intersection points between the parabola and the line. The magnitude of the slope of the line is twice the difference of the two distances mentioned. Find the equations of the parabola and line. Assume $m \geq 0$.

14. Observe the equation $m = \pm\sqrt{|a|d - 1}$. What relationship does this equation tell us about a and d? Derive an inequality, and explain in words what it means.

15. Observe the equation $m = \pm 2 \cdot \sqrt{\dfrac{|p|}{d - 4|p|}}$. What relationship does this equation tell us about p and d?

16. Redo problem 5 but with a left-right parabola.

17. Redo problem 6 but with a left-right parabola.

18. (Theorem) If a circle has a focal chord as a diameter, then this circle is tangent to the directrix. Let's use this theorem in a problem. Let the two focal chord piece lengths be 2/3 and 6 for the parabola $y = -3(x + 5)^2 - 7$. Find the equation of the circle in question, and show that this circle is tangent to the parabola's directrix.

19. (Theorem) If a circle has the focus of a parabola as one endpoint of a diameter and the other endpoint of the diameter as a point P on the parabola, then the circle and parabola share a tangent line through the vertex of the parabola. Use the parabola in question 18 with the point $P(0, -82)$ on the parabola. Find the equation of the circle in question, and find the equation of the mutual tangent line for the parabola and circle.

Some Square Erecting Theorems

Victor Thébault (1882 – 1960), a French mathematician, has three theorems that we call Thébault's theorem. We will present just one of them here. The other two can be looked up on the Internet by the interested student. The motivation for this chapter is a preparation chapter for the *Triangle Geometry* chapter because some of the concepts there rely on erecting shapes on the sides of the triangle. This chapter will only deal with erecting squares, but the Triangle Geometry chapter will also deal with erecting equilateral triangles.

Thébault's Theorem: If four squares are erected outwardly on the sides of a parallelogram, then the quadrilateral formed from the centers of these four squares is a square.

We will do an easier example and a harder example illustrating Thébault's theorem. Of course, we will present the easier example first. The first parallelogram we will provide is the one bound by the four lines $x = 1$, $x = 2$, $y = -3$, and $y = 9$. Because the lines in this parallelogram are vertical and horizontal, we have the special case of a parallelogram known as a rectangle. See Figure 1.

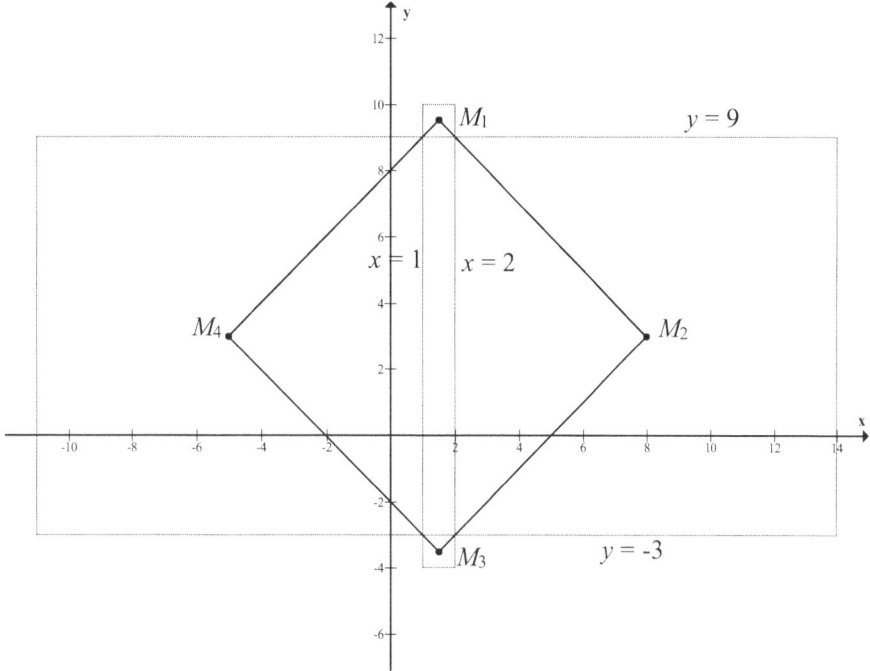

Figure 1: Thébault's Square M_1, M_2, M_3, and M_4

In order to find M_1, it is enough to have two vertices of the square sharing a diagonal. We already have the point (1, 9), so we invest a little bit of effort to find the other vertex sharing this diagonal. The length of a side of the square having M_1 as its center is 1. This is found by computing the distance between (1, 9) and (2, 9). The distance formula can be used, of course, but it is easier just to take the magnitude of the difference of the abscissas (the x-coordinates). Hence, we get $|2 - 1| = |1| = 1$. So we need to go up one unit from the point (2, 9) to get (2, 10). We find that M_1 is the midpoint of the diagonal

between (1, 9) and (2, 10), which is $M_1 = \left(\dfrac{1+2}{2}, \dfrac{9+10}{2}\right) = \left(\dfrac{3}{2}, \dfrac{19}{2}\right)$. We find M_2 through

M_4 similarly to get $M_2 = (8, 3)$, $M_3 = \left(\dfrac{3}{2}, \dfrac{-7}{2}\right)$, and $M_4 = (-5, 3)$.

Thébault's theorem guarantees that M_1, M_2, M_3, and M_4 form a square, but we would like to verify this. The figure above suggests that it is a square because it looks like a square. However, this is not sufficient for us. To demonstrate that this is a square, we would need to show that the four sides are equal in length and that the slopes of the lines are negative reciprocals for adjacent sides and equal for opposite sides. The distance formula handles the lengths of the sides issue, and the slope issue can be dealt with via the slope formula. The student will be asked to demonstrate this in the exercises.

Well, we need another parallelogram for our harder example. Thus, we would like to find the four corners of the square known to exist by way of Thébault's theorem for the parallelogram bound by the four lines $\ell_i: y = m_i x + b_i$, $i = 1, 2, 3, 4$, where $m_1 = m_2 = 1.3$, $m_3 = m_4 = -0.4$, and $b_{j+1} = b_j + j$ with $b_1 = 3$ with j indexed from 1 to 3.

The four lines are not stated directly so that we can plot them. However, we can figure out what the four lines are. The m's are directly given to us, but the b's require a modest amount of effort to find. We are told that $b_1 = 3$. Also, we are told that the b's are recursively defined by the equation $b_{j+1} = b_j + j$. We let $j = 1$ to find that $b_{1+1} = b_2 = b_1 + 1 = 3 + 1 = 4$. Thus, b_2 is 4. Continuing this process, we find that $b_3 = 6$ and $b_4 = 9$. We can now say that the four lines are $\ell_1: y = 1.3x + 3$, $\ell_2: y = 1.3x + 4$, $\ell_3: y = -0.4x + 6$, and $\ell_4: y = -0.4x + 9$. We can plot these four lines with their intersection points and see what we need to do next. See Figure 2.

The coordinates of the four intersection points can be obtained by setting up and solving the appropriate systems of equations involving pairs of lines. Doing this, we get the following four coordinates: $P_1(30/17, 90/17)$, $P_2(60/17, 129/17)$, $P_3(50/17, 133/17)$, $P_4(20/17, 94/17)$.

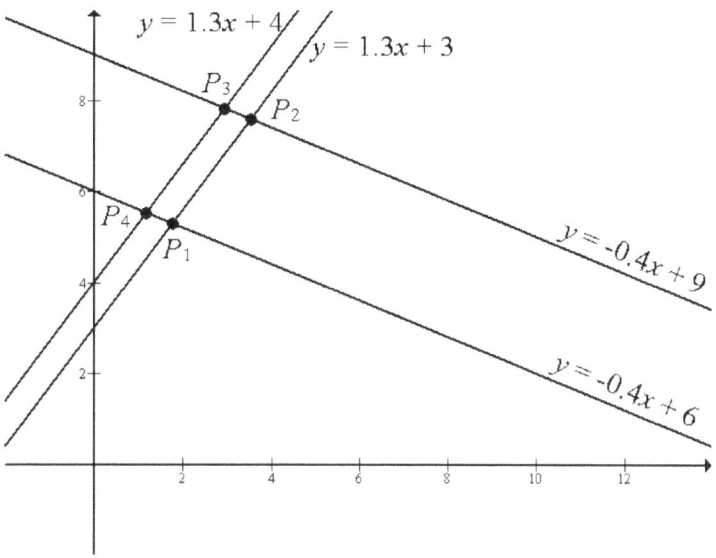

Figure 2: The Four Lines Plotted with Intersection Points $P_1 - P_4$

Next, we need to construct the outward squares on the sides of the parallelogram. The opposite sides of a square are parallel, and the vertices are all right angles. Hence, in order to achieve the right angle, we need the slopes of the lines of adjacent sides of the square to have negative reciprocals (see proof in the *Introduction*). For the side between P_1 and P_2, the slope is 1.3. So then the negative reciprocal would be -1/1.3 = -10/13. Therefore, we need a slope of -10/13 going through P_1 and P_2 to form the two sides of the square adjacent to side P_1P_2. Thus, the equations of the two lines through P_1 and P_2 are

$$y = \frac{-10}{13}x + \frac{1470}{221} \text{ and } y = \frac{-10}{13}x + \frac{2277}{221} \text{, respectively.}$$

The length of P_1P_2 is $d = \sqrt{(x_2 - x_1)^2 + (y_2 - y_1)^2} = \sqrt{\left(\frac{60}{17} - \frac{30}{17}\right)^2 + \left(\frac{129}{17} - \frac{90}{17}\right)^2} =$

$\frac{3\sqrt{269}}{17}$. So we need to go out the length of the side of P_1P_2 along the lines

$y = \frac{-10}{13}x + \frac{1470}{221}$ and $y = \frac{-10}{13}x + \frac{2277}{221}$ to find the other two points of this square. We

will do one of them. To go out a distance of $\frac{3\sqrt{269}}{17}$ from P_1 along the line

$y = \frac{-10}{13}x + \frac{1470}{221}$, we construct a system of equations involving a circle of radius

147

$\dfrac{3\sqrt{269}}{17}$ centered at P_1 and the line $y = \dfrac{-10}{13}x + \dfrac{1470}{221}$. The system is shown below.

$$\left\{\begin{array}{l} \left(x - \dfrac{30}{17}\right)^2 + \left(y - \dfrac{90}{17}\right)^2 = \dfrac{2421}{289} \\[2mm] y = \dfrac{-10}{13}x + \dfrac{1470}{221} \end{array}\right\}$$

Solving this system for (x, y) yields the point (69/17, 60/17). Checking the distance between this point and P_1 yields $\dfrac{3\sqrt{269}}{17}$, which is the needed distance, so this shows that we have the correct point. It isn't necessary to find the fourth point of this square even though we're just one system of equations away from it because we only need the midpoint of the diagonal of this square to find its center. Hence, M_1 is the midpoint of this square, which is $M_1 = \left(\dfrac{1}{2}\left(\dfrac{60}{17} + \dfrac{69}{17}\right), \dfrac{1}{2}\left(\dfrac{129}{17} + \dfrac{60}{17}\right)\right) = \left(\dfrac{129}{34}, \dfrac{189}{34}\right)$.

The distance between P_2 and P_3 works out to be $\dfrac{2\sqrt{29}}{17}$, so this is the radius of the circle centered at P_2. The line through P_2 is $y = \dfrac{5}{2}\left(x - \dfrac{60}{17}\right) + \dfrac{129}{17}$, which could be simplified further before going into the system of equations involving this line and the circle of radius $\dfrac{2\sqrt{29}}{17}$ centered at P_2. We would get the correct point from this system so as to find the midpoint between this point and P_3, which would be M_2. Doing this yields the point $M_2 = \left(\dfrac{57}{17}, 8\right)$.

Continuing on around the parallelogram, we find that $M_3 = \left(\dfrac{31}{34}, \dfrac{257}{34}\right)$ and $M_4 = \left(\dfrac{23}{17}, \dfrac{87}{17}\right)$. We plot these on our graph to see Thébault's square for this parallelogram.

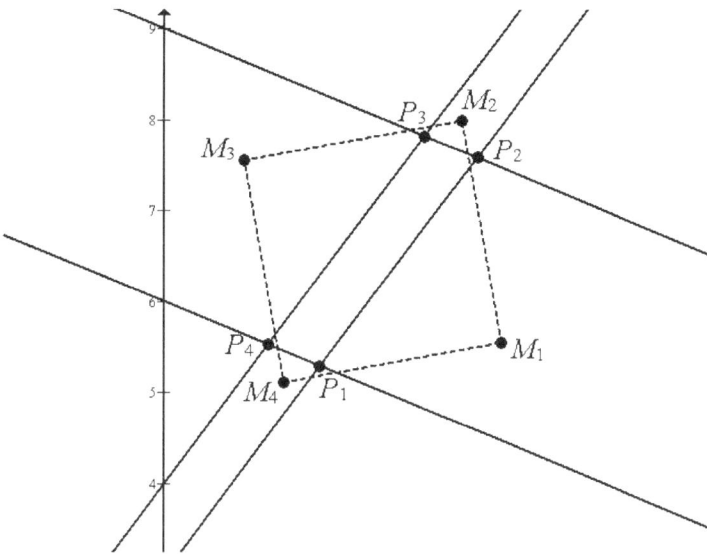

Figure 3: Thébault's Square M_1, M_2, M_3, and M_4

We have a way to get an endless supply of parallelograms. The next theorem shows us how.

Midpoint Quadrilateral Theorem: The midpoints of the sides of any quadrilateral form a parallelogram.

We will work an example using this theorem. Find the parallelogram associated with the sides of the quadrilateral having vertices at $A(-1, -4)$, $B(1, 6)$, $C(6, 7)$, and $D(9, 2)$. Note that when you are dreaming up a quadrilateral, be sure to label the points in such a way so that the sides do not cross. It is good to have a quadrilateral that is convex as well. A convex quadrilateral does not have a side or vertex that pokes back into the body of the shape.

We first need to find the midpoints of the sides. Of course, we will use the midpoint theorem to accomplish this feat. The four midpoints are $M_1 = (0, 1)$, $M_2 = (7/2, 13/2)$, $M_3 = (15/2, 9/2)$, and $M_4 = (4, -1)$. Please see Figure 4 on the next page.

Is this a parallelogram? We would need to show that the slopes of opposite sides are equal. Hence, we compute the slope between M_1 and M_2 to get $m = \dfrac{13/2 - 1}{7/2 - 0} = \dfrac{11}{7}$. We compare that with its opposite side, the side between M_3 and M_4 to get $m = \dfrac{-1 - \frac{9}{2}}{4 - \frac{15}{2}} = \dfrac{11}{7}$.

Hence, we may conclude that these two sides are parallel. The other pair of sides have a slope of $m = -1/2$. Thus, we may conclude that these two sides are also parallel.

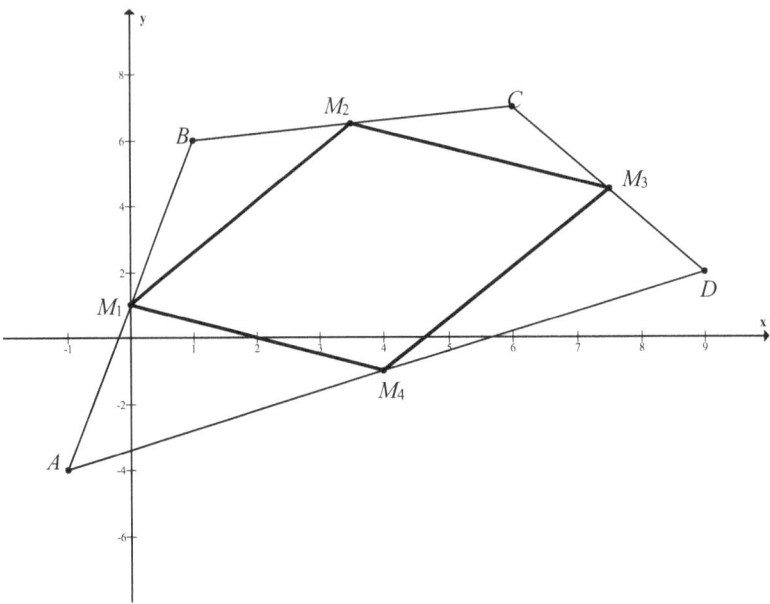

Figure 4: The Parallelogram from the Midpoints of a General Quadrilateral

The main result of this chapter is called van Aubel's Theorem. Thébault's theorem is just a special case of van Aubel's theorem. We will state van Aubel's theorem now.

Van Aubel's Theorem: Given a quadrilateral $ABCD$, construct squares outwardly on each side of the quadrilateral. The centers of these squares can be labeled as M_1, M_2, M_3, and M_4 by going around in order either clockwise or counterclockwise. Then, the line segments $\overline{M_1 M_3}$ and $\overline{M_2 M_4}$ will be perpendicular and of equal length. In the special case that $ABCD$ is a parallelogram, $\overline{M_1 M_3}$ and $\overline{M_2 M_4}$ will also bisect each other forming a square by connecting the points $M_1 M_2 M_3 M_4 M_1$.

Exercises

1. Fill in the missing steps in the second example to find Thébault's square.

2. We used the quadrilateral $A(-1, -4)$, $B(1, 6)$, $C(6, 7)$, and $D(9, 2)$ in the text. Find the centers of the opposite squares, and connect them. Show that these two line segments are perpendicular and of equal length.

3. Find Thébault's square for the rectangle bounded by the lines $x = -4$, $x = 5$, $y = 3$, and $y = 7$. Label all relevant parts in the drawing.

4. Find Thébault's square for the rectangle bounded by the lines $y = x$, $y = x + 6$, $y = -x$, and $y = -x - 6$. Label all relevant parts in the drawing.

5. The four midpoints of the sides of the quadrilateral in exercise 2 are $M_1 = (0, 1)$, $M_2 = (7/2, 13/2)$, $M_3 = (15/2, 9/2)$, and $M_4 = (4, -1)$ as shown in the text. These four points form a parallelogram. Find Thébault's square for this parallelogram.

6. Make up a quadrilateral (preferably convex) that is not regular. Find the two line segments associated with van Aubel's theorem. Show that these two line segments are perpendicular and of equal length. Next, use the midpoint quadrilateral theorem to find the associated parallelogram for your quadrilateral. Finally, find Thébault's square for this parallelogram. Note: Answers will vary. The instructor may assign a quadrilateral for students to work together in groups to solve.

Two Special Quadrilaterals

We will talk about two types of quadrilaterals in this chapter. The first one that we will discuss is a cyclic quadrilateral. The other one is a complete quadrilateral.

Cyclic Quadrilaterals

A cyclic quadrilateral is a quadrilateral that has its four vertices on a circle. Being that it is a quadrilateral, it has four sides. As such, each side has a midpoint, and each side can be represented with an equation of a line. This means that each side has some slope, so we can construct some perpendicular line to it. A term in mathematics is maltitude, and it is a combination of the word midpoint and altitude. A maltitude for a quadrilateral is a line that goes through the midpoint of a side and is perpendicular to the opposite side.

Maltitude Theorem: Let P be the center of the cirumcircle of a cyclic quadrilateral. Let G be the centroid of the cyclic quadrilateral. Then, P' is the reflection of the point P across the point G. In addition, the four maltitudes of a cyclic quadrilateral all pass through a common point, which is P'.

We will work an example illustrating this theorem. Our given cyclic quadrilateral $ABCD$ is $A(-4, -1)$, $B(-4, 7)$, $C(3, 6)$, and $D(4, 3)$. We will proceed on just this information.

We use three of the given points to find the circumcircle of that triangle using any method. The chapter *The Circle* discusses how to do that. Next, we check our fourth point in the equation of the circle to see if we get an identity. Upon seeing the identity, we know that we have a cyclic quadrilateral on our hands. Ptolemy's theorem (discussed soon) can also test a quadrilateral to see if it is cyclic. Then, we find that the circumcircle of the quadrilateral is $(x+1)^2 + (y-3)^2 = 25$. We identify the center of this circumcircle as $P(-1, 3)$.

Next, we need to find the centroid G of the quadrilateral. The centroid $(\overline{X}, \overline{Y})$ of a polygon with n vertices is computed with the equations below. These two formulas are really saying just take the average x-coordinate (we are calling $\overline{X}$) and the average y-coordinate (we are calling $\overline{Y}$). The formula "Avg = sum of the items/number of items" is all that is happening here. Hence, we will add up the x-coordinates and divide by 4 because we have a quadrilateral, which makes $n = 4$. We will do the same for the y-coordinates.

$$\overline{X} = \frac{1}{n} \cdot \sum_{i=1}^{n} x_i \qquad\qquad \overline{Y} = \frac{1}{n} \cdot \sum_{i=1}^{n} y_i$$

$$\overline{X} = \frac{-4 + (-4) + 3 + 4}{4} = \frac{-1}{4} \qquad\qquad \overline{Y} = \frac{-1 + 7 + 6 + 3}{4} = \frac{15}{4}$$

Now that we have P and G, we reflect P across G to find P'. The maltitude theorem states that this point P' will be the point that all four of the maltitudes pass through. The act of reflecting a point across another is explained in *The Deluxe Toolkit* chapter, and another way is presented in *The Circle* chapter. Following the advice in *The Circle* chapter, we find the equation of the circle with center at G with radius $r = d(P, G) =$

$\sqrt{(-1 + \frac{1}{4})^2 + (3 - \frac{15}{4})^2} = \frac{3\sqrt{2}}{4}$. We emphasize our earlier remark in *The Circle* chapter

that because r here is irrational, then P will not correspond to a rational point on the unit circle. Hence, there will not be a Pythagorean triple associated with P. However, this is OK because we want to find P', and this does not require that the corresponding point to P on the unit circle be rational. Thus, the equation of this circle C is the following:

$$\left(x + \frac{1}{4}\right)^2 + \left(y - \frac{15}{4}\right)^2 = \frac{9}{8}.$$

Next, we find the coordinates corresponding to $P(-1, 3)$ on the unit circle:

$$x_u = \frac{x_C - h}{r} = \frac{-1 + \dfrac{1}{4}}{\dfrac{3\sqrt{2}}{4}} = \frac{-\sqrt{2}}{2} \qquad\qquad y_u = \frac{y_C - k}{r} = \frac{3 - \dfrac{15}{4}}{\dfrac{3\sqrt{2}}{4}} = \frac{-\sqrt{2}}{2}$$

This gives us the point $(x_u, y_u) = (-\sqrt{2}/2, -\sqrt{2}/2)$. We find the diametric opposite to this point on the unit circle by switching the signs on x_u and y_u. Hence, we get the point $(x'_u, y'_u) = (\sqrt{2}/2, \sqrt{2}/2)$. We complete the process by finding the corresponding point

P' on C of (x'_u, y'_u). We compute $x'_C = r(x'_u) + h = \dfrac{3\sqrt{2}}{4} \cdot \dfrac{\sqrt{2}}{2} - \dfrac{1}{4} = \dfrac{1}{2}$. We

153

compute $y_C' = r(y_u') + k = \frac{3\sqrt{2}}{4} \cdot \frac{\sqrt{2}}{2} + \frac{15}{4} = \frac{9}{2}$. Then, we form the ordered pair $P'(1/2, 9/2)$. Note that even though P and P' had irrational corresponding points on the unit circle, they were both rational points on C. We remark that we expect that the four maltitudes will go through the point P' that we have just found. Our situation thus far is Figure 1.

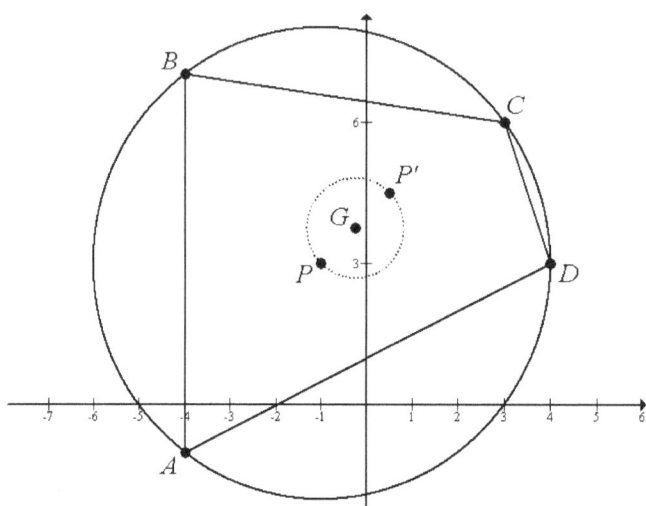

Figure 1: Finding the Expected Point P' for the Maltitudes

We will now find the four maltitude lines. A maltitude line goes through a midpoint of a side of a quadrilateral and is perpendicular to the opposite side. Hence, we need midpoints and perpendicular slopes. We can easily pair the correct midpoint with a perpendicular slope because all four letters must get used. Hence M_{AB} must be paired with the perpendicular slope to side CD, which we will notate as $m_{\perp CD}$. Our four maltitude line problems with solutions after each colon, then, are the following:

Line through $M_{AB}(\text{-}4, 3)$ with slope $m_{\perp CD} = \frac{1}{3}$: $y = \frac{1}{3}x + \frac{13}{3}$

Line through $M_{BC}\left(\frac{-1}{2}, \frac{13}{2}\right)$ with slope $m_{\perp AD} = -2$: $y = -2x + \frac{11}{2}$

154

Line through $M_{CD}\left(\dfrac{7}{2}, \dfrac{9}{2}\right)$ with slope $m_{\perp AB} = 0$: $y = \dfrac{9}{2}$

Line through $M_{AD}(0, 1)$ with slope $m_{\perp BC} = 7$: $y = 7x + 1$

We next form an over determined system of equations with these four maltitude lines. We will solve it to see if a single ordered pair comes out of it. Of course, we are expecting $P'(1/2, 9/2)$, but we must perform the work to see. Having accomplished the feat, we see that they do all intersect at the point P'. Our situation is captured in Figure 2. Note that the midpoints have been relabeled with simpler subscripts.

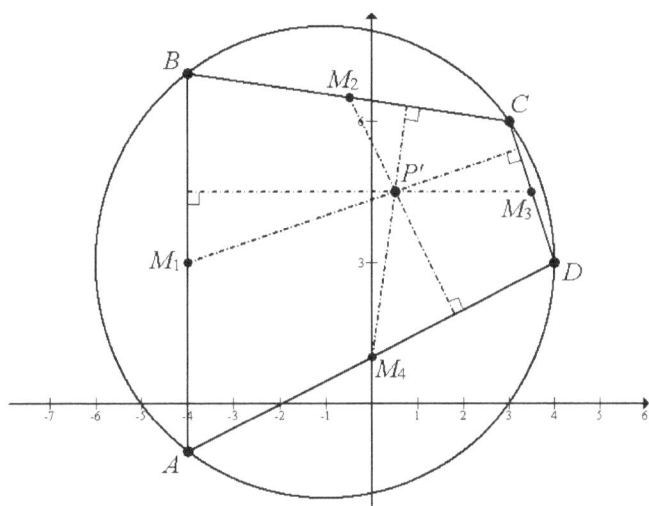

Figure 2: The Four Maltitudes Through the Expected Point P'

Before we move to the complete quadrilateral, we will discuss a property of cyclic quadrilaterals known as Ptolemy's theorem. It states that given a cyclic quadrilateral $ABCD$, the products of the two sets of opposing sides added together is the same as the product of the two diagonals. In symbols, it states that $AB \cdot CD + BC \cdot AD = AC \cdot BD$. This theorem is used in trigonometry to derive what are called the addition formulas. Also, Ptolemy's theorem can be used as a test to determine if a quadrilateral is cyclic.

Complete Quadrilaterals

We will discuss a complete quadrilateral now. What is a complete quadrilateral? It is a quadrilateral formed by extending an arbitrary quadrilateral's four lines such that we get an additional two points. The only way to get these two additional points is that the original quadrilateral must not have any parallel sides. If it does, then extending the

155

parallel lines for that pair of sides will not yield an additional intersection point. A quadrilateral with one pair of parallel sides is a trapezoid, so the parallel sides being extended will not yield an additional intersection point. However, the extended nonparallel sides of a trapezoid will yield one new point on the quadrilateral (the trapezoid), but since we did not get two additional points, the trapezoid cannot be made complete. Squares, rectangles, rhombi, and parallelograms each have two pairs of parallel sides; so these quadrilaterals will not even yield one new additional point when attempting to make them complete. A kite can be made complete as long as it is does not have any parallel sides. Note that a kite with one pair of parallel sides is not possible. However, a kite with two pairs of parallel sides is a rhombus. Thus, as long as a kite is not a rhombus, it can be made complete by extending its sides.

The usual way of defining a complete quadrilateral is to say four general lines that intersect in six points. The boundary of a complete quadrilateral looks like an arrow. The two extra smaller triangles formed make the base of the arrow. The other vertices form the tip of the arrow and the notch. Two of the "vertices" of the complete quadrilateral are 180 degree angles—a straight line. The sides of the arrow each have one of these "vertices."

The distinguishing characteristic of a complete quadrilateral is that it has three diagonals instead of the usual two for other quadrilaterals. It has two interior diagonals—the two usual diagonals that all quadrilaterals exhibit—however, it has a third exterior diagonal between the two vertices that form the base of the arrow as well.

Complete Quadrilateral Theorem: The midpoints of the three diagonals of a complete quadrilateral lie on a straight line. This line is called the complete quadrilateral line. Note that this line is also called the Newton Line, after Sir Isaac Newton (1642 – 1727), an English mathematician and physicist.

A complete quadrilateral has four triangles. Each triangle, then, being a triangle has its own associated orthocenter. The orthocenter of a triangle is the point of concurrence of the three altitudes of the triangle. We may have to extend a side of a triangle to be able to plot its altitude for that side. We have a theorem about these four orthocenters:

Theorem: The four orthocenters belonging to the four triangles of a complete quadrilateral all lie on a line, and this line is perpendicular to the complete quadrilateral line.

We present our next theorem to discuss about complete quadrilaterals. The endpoints of each diagonal of a complete quadrilateral can be made to be diametrically opposed points of a circle. Then, that diagonal becomes a diameter for that circle. Finding the equations of the three circles associated with the three diagonals of a complete quadrilateral, we get a surprising result. Julius Plücker (1801 – 1868), a German mathematician, proved the following theorem.

Plücker's Theorem: The three circles having diameters with the endpoints of the diagonals of a complete quadrilateral have just two common intersection points. Also, the line through these two intersection points is the same line that joins the four orthocenters of the triangles. As such, it is perpendicular to the complete quadrilateral line.

We will call the line through the two common intersection points of the three circles Plücker's line. We will find Plücker's line for the cyclic quadrilateral above, and we will demonstrate that it is perpendicular to its complete quadrilateral line. First, we need to find the complete quadrilateral line. We can compute two of the midpoints of the quadrilateral's diagonals as it is before we complete it because they are already there. Recall that the cyclic quadrilateral's vertices are $A(-4, -1)$, $B(-4, 7)$, $C(3, 6)$, and $D(4, 3)$. Hence, one diagonal is AC, and its midpoint is $M_{AC} = \left(\frac{1}{2}(-4+3), \frac{1}{2}(-1+6)\right) = \left(\frac{-1}{2}, \frac{5}{2}\right)$. We will relabel this as M_1 in the upcoming diagram. The other diagonal is BD, and its midpoint is $M_{BD} = \left(\frac{1}{2}(-4+4), \frac{1}{2}(7+3)\right) = (0, 5)$, which will not be labeled in the upcoming diagram because of lack of room on the diagram.

We extend the sides AB and CD to find the other additional vertex of the complete quadrilateral. We find that it is $V_1(-4, 27)$. We extend the sides BC and AD to find one additional vertex of the complete quadrilateral. We find that vertex to be the point $V_2(76/9, 47/9)$. The line between these two additional vertices is the third diagonal. Its midpoint is $M_3 = \left(\frac{1}{2}\left(\frac{76}{9} + (-4)\right), \frac{1}{2}\left(\frac{47}{9} + 27\right)\right) = (20/9, 145/9)$. We compute the line through $M_{AC}(-1/2, 5/2)$ and $M_{BD}(0, 5)$ as $y = 5x + 5$. This seems like such a simple line for all the effort we put into it. Nevertheless, we find that it checks with $M_3(20/9, 145/9)$. Our situation is captured in Figure 3.

We now proceed to find the equations of the three circles with the endpoints of the three diagonals as diameters. We already have the midpoints of these diagonals. Thus, we already know (h, k) for the three circles. It remains for us to find r for each circle. We find r with the distance formula from (h, k) to either one of the two endpoints.

$$r_1 = d(M_1, A) = \sqrt{\left(-4 - \left(\frac{-1}{2}\right)\right)^2 + \left(-1 - \frac{5}{2}\right)^2} = \frac{7\sqrt{2}}{2}$$

157

$$r_2 = d(M_2, B) = \sqrt{(-4-0)^2 + (7-5)^2} = 2\sqrt{5}$$

$$r_3 = d(M_3, V_1) = \sqrt{\left(-4-\frac{20}{9}\right)^2 + \left(27-\frac{145}{9}\right)^2} = \frac{14\sqrt{65}}{9}$$

The square of these r's is what we put into our three circle equations. Hence, we form the system of three circle equations, which is an over determined system. However, it is OK because it turns out to be consistent. Just use any two of the circles to get the two intersection points, and use the third circle as a check of both of those points. If both of the points yield identities, then the third circle goes through both of those points as the first two circles do. Our system is shown next.

$$\left\{ \begin{array}{l} C_1 : \left(x + \dfrac{1}{2}\right)^2 + \left(y - \dfrac{5}{2}\right)^2 = \dfrac{49}{2} \\[2ex] C_2 : x^2 + (y-5)^2 = 20 \\[2ex] C_3 : \left(x - \dfrac{20}{9}\right)^2 + \left(y - \dfrac{145}{9}\right)^2 = \dfrac{12740}{81} \end{array} \right\}$$

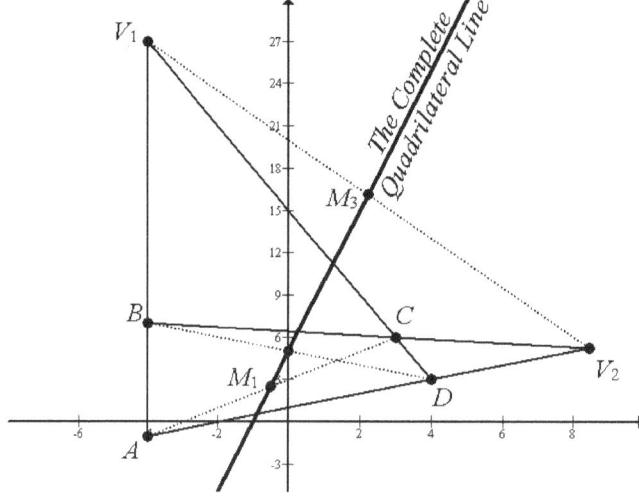

Figure 3: The Complete Quadrilateral Line

See Figure 4. Performing the feat of solving this (and it *is* a feat), we find that the two common points of intersection I_1 and I_2 of these three circles are the following:

158

$$I_1 = \left(\frac{-(1+5\sqrt{129})}{13}, \frac{60+\sqrt{129}}{13} \right) \qquad I_2 = \left(\frac{-1+5\sqrt{129}}{13}, \frac{60-\sqrt{129}}{13} \right)$$

Some remarks about solving the system of three circles are in order. The student needs to be mindful of squaring a square root being the absolute value of the radicand. In other words, $\sqrt{x^2}$ is not equal to x. Rather, it is equal to $|x|$. The path to solving this system generates fourth power terms and third power terms. One gets two polynomial versions by removing the absolute value bars. One version has the fourth and third power terms cancel. It turns out that this is the correct version (thankfully!). Depending on how the student proceeds, nested radicals can show up here or there. Denesting techniques were shown in the *Fractions, Rational Points, and Denesting Radicals* chapter. Each x-value and y-value winds up with two versions. However, since we only get two solutions, it is helpful to graph the four possible ordered pairs to see which two are correct. The other two would be extraneous. Finally, substitute the two correct ordered pairs into the third equation to ensure that both yield identities. Both I_1 and I_2 do check, so these two are the correct points by Plücker's theorem.

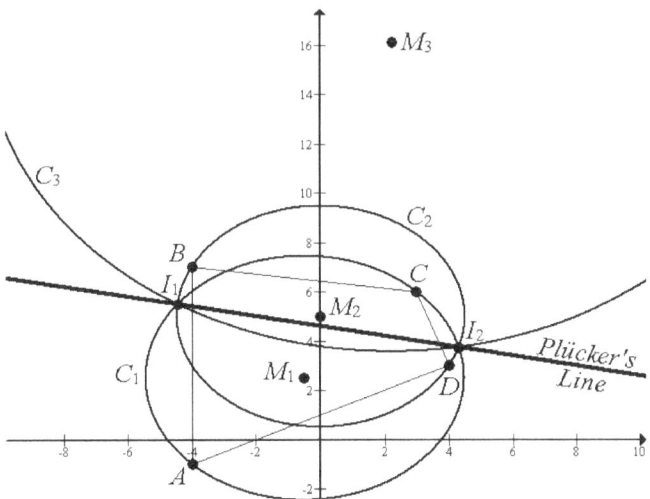

Figure 4: Plücker's Line for the Given Complete Quadrilateral

We find the slope between I_1 and I_2 to be -1/5. The square roots just disappear. Square roots do seem to have some sort of magic about them. Sending a line with slope $m = -1/5$ through either I_1 or I_2 produces the result $y = \frac{-1}{5}x + \frac{23}{5}$. Again, the square roots just disappear. The complete quadrilateral line was found to be $y = 5x + 5$. These slopes are negative reciprocals, so we have confirmed that Plücker's line is perpendicular to the

159

complete quadrilateral line. Is this the same line that all four of the orthocenters pass through? This will be left as an exercise for the student to show. Incidentally, we will discuss orthocenters again in the *Triangle Geometry* chapter.

Harvey gives us our final theorem in this chapter. His theorem is about a complete quadrilateral. It says that the four triangles in a complete quadrilateral each have an orthocenter and circumcenter. The four perpendicular bisectors of each orthocenter and circumcenter pair are concurrent. That is to say that we bisect the line segment between an orthocenter and circumcenter pair with a perpendicular line. All four such occurrences will yield a single intersection point. We can call this point Harvey's Point for a complete quadrilateral.

Exercises

1. For parts *a*, *b*, and *c* below, the vertices of a quadrilateral are given. Check to see if the quadrilateral is cyclic and/or complete. If it is not cyclic, write, "This quadrilateral is not cyclic." If it cannot be made complete, write, "This quadrilateral cannot be made complete." For a cyclic quadrilateral, find the expected and actual meeting point of the maltitudes and compare. In addition, if it is a cyclic quadrilateral, verify Ptolemy's theorem. For a quadrilateral that can be made complete, find the two new vertices that complete it. Then, find the complete quadrilateral line, the line through the orthocenters of the four triangles associated with the complete quadrilateral, the vertices I_1 and I_2, and Plücker's line through them. Show all work, and document your algebra.

> a) $A(-3, -4)$ $B(-1, 5)$, $C(3, 9)$, $D(2, -1)$
> b) $A(-2, 8)$, $B(-1, 9)$, $C(7, 5)$, $D(2, 0)$
> c) $A(-15, 2)$, $B(-2, 15)$, $C(10, 7)$, $D(3, -10)$
> d) $A(0, 0)$, $B(1, 0)$, $C(1, 3)$, $D(0, 2)$

2. Identify the four triangles in the given complete quadrilateral $A(-4, -1)$, $B(-4, 7)$, $C(3, 6)$, and $D(4, 3)$ used in this chapter. Compute the orthocenters of each triangle to obtain four points. Find the equation of the line between two of the orthocenters, and ensure that the other two orthocenters are on this line. Compare this result with Plücker's line in the example.

3. Find the point of concurrency for Harvey's theorem for complete quadrilaterals in the complete quadrilateral in the previous exercise. Show that each of the four perpendicular bisector equations yield an identity when this point is inserted in the equation.

Linear and Parabolic Asymptotes

This chapter is about asymptotes. Linear asymptotes can be horizontal, vertical, or oblique, which are also called slant. Parabolic asymptotes can be upward or downward. In this book, we are interested in lines and conic sections; linear asymptotes are lines, and parabolic asymptotes are one of the conic sections. Ellipses and hyperbolas are not asymptotes, but hyperbolas do have linear asymptotes. Furthermore, they both have a pair of linear directrices, which are lines. The main motivation for this chapter is to discuss parabolic asymptotes because the topic is given too little attention in the college algebra texts. Some books will discuss it at some length, but most that I have seen do not cover the topic. We will begin our discussion with horizontal asymptotes.

Rational functions, like many functions we study, have end behavior that can be determined. A rational function is usually defined as $f(x) = \dfrac{g(x)}{h(x)}$, where $g(x)$ and $h(x)$ are polynomials and $h(x) \neq 0$. The technique of choice is long division of polynomials, which produces the following:

$$f(x) = q(x) + \frac{r(x)}{h(x)}$$

The polynomial $q(x)$ is called the quotient polynomial, and the polynomial $r(x)$ is called the remainder polynomial. The polynomial $h(x)$ is called the divisor, and the polynomial $g(x)$ is called the dividend. The end behavior is determined by the quotient polynomial. If $q(x) = c$, $c \in \Re$, then the end behavior is the horizontal asymptote $y = c$.

Perform long division on the rational function $f(x) = \dfrac{3x^2 - 4}{2x^2 + 7x + 8}$ to determine the end behavior asymptote and the remainder polynomial.

$$
\begin{array}{r}
\dfrac{3}{2} - \dfrac{(21/2)x + 16}{2x^2 + 7x + 8} \\[2ex]
2x^2 + 7x + 8 \overline{\smash{\big)}\, 3x^2 + 0x - 4} \\[1ex]
-3x^2 - \dfrac{21}{2}x - 12 \\[1ex]
\hline
\dfrac{-21}{2}x - 16
\end{array}
$$

Note the sign change on this line to enable adding.

We see that $q(x) = 3/2$, which is a constant, and $r(x) = -(21/2)x - 16$. Thus, the end behavior asymptote is the horizontal line $y = 3/2$. The following graph, which is labeled Figure 1, confirms this result graphically. The point A is the intersection of the graph with its asymptote.

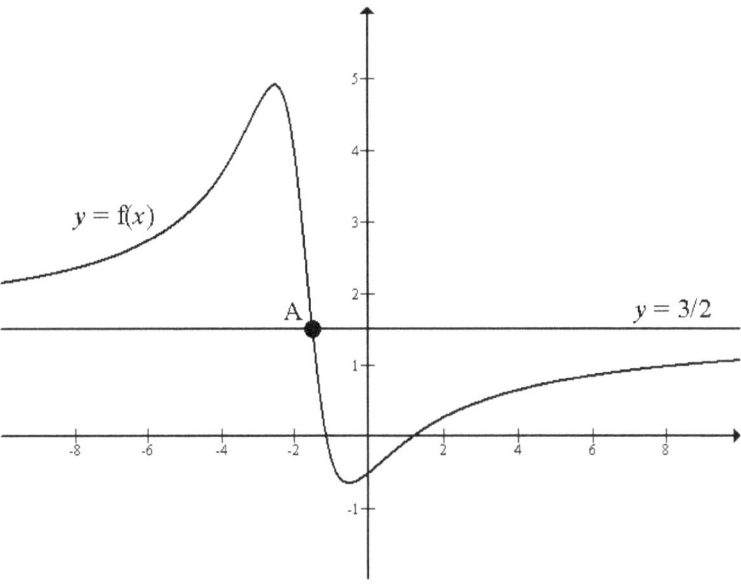

Figure 1: A Rational Function Crossing its Horizontal Asymptote

If the degree of $g(x)$ is less than the degree of $h(x)$, then long division of the polynomials is not going to generate anything other than $g(x) / h(x)$. Thus, $q(x) = 0$, and $r(x) = g(x)$. Therefore, the end behavior asymptote is the line $y = 0$. The following, then, is true if $\deg(g(x)) < \deg(h(x))$.

$$f(x) = \frac{g(x)}{h(x)} = q(x) + \frac{r(x)}{h(x)} = 0 + \frac{r(x)}{h(x)}$$

Thus, for example, if $h(x)$ is of degree 2, which is a quadratic function, then the degree of $r(x)$ must be 1 or 0. So, $r(x)$ is some line $y = mx + b$ where $m \neq 0$ when $\deg(r(x)) = 1$, and $m = 0$ when $\deg(r(x)) = 0$, which makes $r(x) = b$.

Where does a rational function cross its end behavior asymptote? We set the rational function equal to the end behavior asymptote and solve for x. Thus, $f(x) = q(x)$. However, $f(x) = q(x) + \frac{r(x)}{h(x)}$. So, $q(x) = q(x) + \frac{r(x)}{h(x)}$, which means that $\frac{r(x)}{h(x)} = 0$. However, when we multiply both sides by $h(x)$, we get just $r(x) = 0$. This analysis shows that we can just solve for the roots of $r(x)$ to find where a function crosses its asymptote. Thus, for example, in the previous example, we set $r(x) = \dfrac{-21}{2}x - 16 = 0$ to find $x = -32/21$. Then, we plug this x-value into either f or q to find the y-coordinate. Since $q(x) = 3/2$, we do not have to compute anything. Thus, $(-32/21, 3/2)$ is the place in the previous example where the function crossed its end-behavior asymptote, which was labeled as point A.

To find the vertical asymptotes of a rational function, we find where the denominator is zero. Thus, we set $h(x) = 0$ and solve for x. Only the real solutions for x produce vertical asymptotes of the rational function at that point. Solutions for x that are imaginary do not generate vertical asymptotes on the graph of $f(x)$. We assume that the polynomials $g(x)$ and $h(x)$ do not have any common factors because if they did, then the solutions to $h(x) = 0$ will only find holes in the graph of $f(x)$, not vertical asymptotes. Thus, when looking for vertical asymptotes, the student would do well to factor the numerator and denominator to cancel any common factors; but be sure to label those points as holes on the graph of f. Find the vertical asymptotes of the rational function $f(x) = 5/(4x^2 - 13x + 9)$. We note that the numerator does not have any factors with a variable to create any potential holes. We identify $h(x)$ as $4x^2 - 13x + 9$. We set this equal to zero and solve.

$$4x^2 - 13x + 9 = 0 \Rightarrow = \frac{13 \pm \sqrt{13^2 - 4 \cdot 4 \cdot 9}}{2 \cdot 4} = \frac{13 \pm 5}{8} = \frac{9}{4}, 1.$$

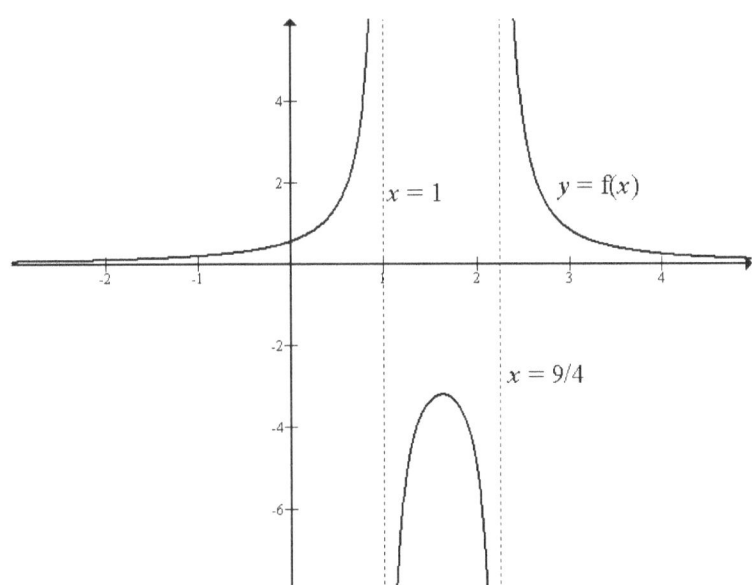

Figure 2: Two Vertical Asymptotes and a Horizontal Asymptote

163

Thus, we expect to see vertical asymptotes at the x-values of 9/4 and 1. We also note that $\deg(g(x)) < \deg(h(x))$, so we expect a horizontal asymptote at the line $y = 0$. The following diagram, which is labeled Figure 2, confirms the horizontal and two vertical asymptotes.

We now move to the slant asymptote. This is the situation when $q(x)$ is a first degree polynomial. Let us investigate the following rational function.

$$f(x) = \frac{10x^3 - 41x^2 + 41x - 9}{2x^2 - 7x + 4}$$

We perform the long division like so.

$$
\begin{array}{r}
5x - 3 + \dfrac{3}{2x^2 - 7x + 4} \\[2ex]
2x^2 - 7x + 4 \overline{\smash{\big)}\, 10x^3 - 41x^2 + 41x - 9} \\
\underline{-10x^3 + 35x^2 - 20x} \qquad \text{Note the sign changes to add.} \\
-6x^2 + 21x - 9 \\
\underline{6x^2 - 21x + 12} \quad \text{Note the sign changes again.} \\
3
\end{array}
$$

We can now identify $q(x)$ as $5x - 3$, which is a first degree polynomial, so we expect $y = 5x - 3$ to be the slant asymptote for the function. We observe that $r(x)$ is just the constant 3, which means that the function $f(x)$ will not cross its asymptote $q(x)$. The denominator $h(x)$ when solved for its zeros yields the x-values of $\dfrac{7 \pm \sqrt{17}}{4}$, which are approximately 0.719 and 2.781, so we expect to see vertical asymptotes at those domain values. The following diagram, which is labeled Figure 3, confirms all of these observations.

We now move to the main motivation for this chapter—the parabolic asymptotes. When we perform the long division for a rational function and find that the polynomial $q(x)$ is quadratic, then we know that the end behavior asymptote is a parabola. It seems that a parabolic asymptote strikes fear into students, but it is no different from a linear asymptote in the sense that the function will approach this asymptote as the magnitude of x increases without bound. Yes, the shape of a parabolic asymptote is not a line, and students are more comfortable with lines as asymptotes. However, asymptotic behavior of functions can take many forms, and the parabolic asymptote is a fine first exposure to nonlinear asymptotes.

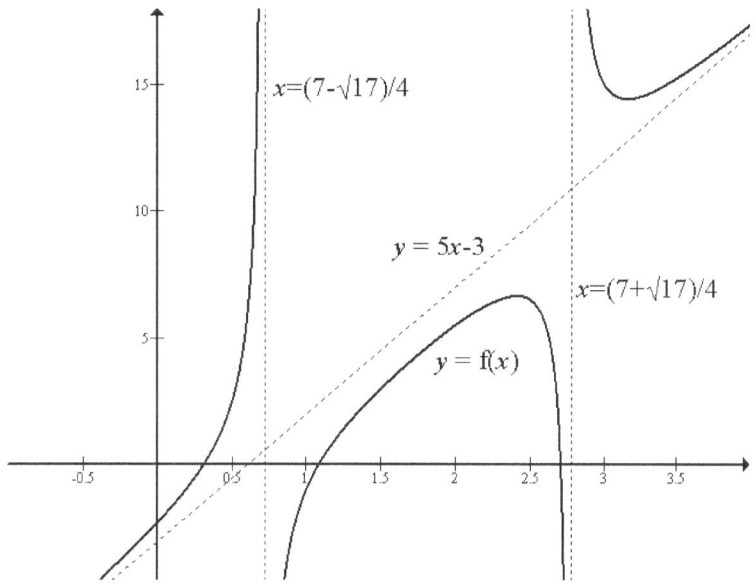

Figure 3: A Slant or Oblique Asymptote

For our first parabolic asymptote, we will study the function $y = f(x) = \dfrac{x^3 + 1}{x}$. We will do the long division and identify $q(x)$ and $r(x)$. We also note that x is not a factor of the numerator, so the graph will not have any holes. Of course, $h(x)$ is the denominator x, so the zero of $h(x)$ occurs when $x = 0$. Therefore we expect a vertical asymptote at the line $x = 0$. Note the supplied missing powers of x in the long division.

$$
\begin{array}{r}
x^2 + \dfrac{1}{x} \\
x\,\overline{\smash{)}\,x^3 + 0x^2 + 0x + 1} \\
\underline{-x^3} \\
1
\end{array}
$$

Note the sign change to add.
Bring down the straggling 1.

We observe that $q(x)$ is the polynomial x^2, and $r(x)$ is the constant 1. Thus, the function $f(x)$ has a parabolic asymptote of $y = x^2$. Solving $r(x) = 0$ yields the contradiction $1 = 0$, which shows that the function f does not cross its parabolic asymptote. The following diagram, which is labeled Figure 4, confirms all of these observations.

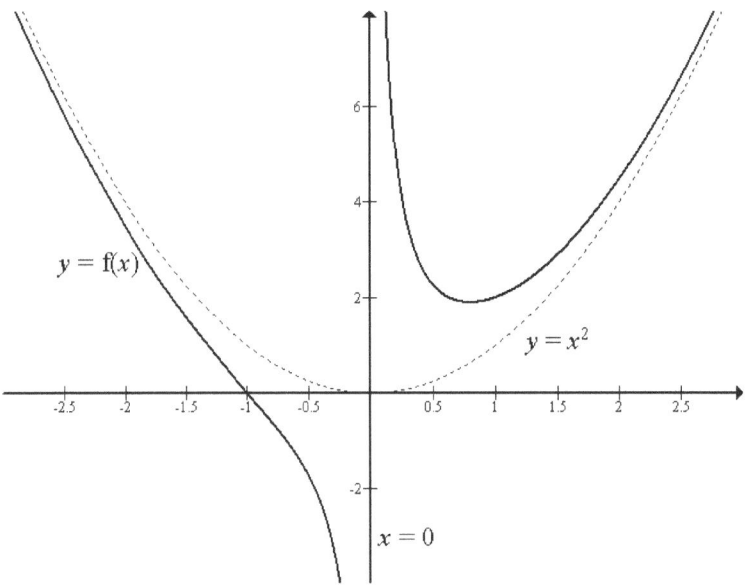

Figure 4: A Parabolic Asymptote

For our next parabolic asymptote, we will study the following rational function:

$$f(x) = \frac{6x^4 + 2x^3 + 23x^2 + 6x + 21}{2x^2 + 5}$$

Long division yields: $\qquad f(x) = 3x^2 + x + 4 + \frac{x+1}{2x^2 + 5}$

We identify that $q(x) = 3x^2 + x + 4$, which means that the function f has a parabolic asymptote. We solve $r(x) = 0$, which means that we solve $x + 1 = 0$. Thus, we find that the function f crosses its end behavior parabolic asymptote at $x = -1$. The denominator $h(x)$ is an irreducible quadratic, so the rational function f will not have any vertical asymptotes. These observations are borne out on the following graph, which is labeled Figure 5. Point A is where the function crosses its end-behavior asymptote. Note that this rational function semi-closely follows its parabolic asymptote over the whole domain. Usually, a rational function has wild and erratic behavior prior to succumbing to the end behavior; however, this is an exception to that tendency.

166

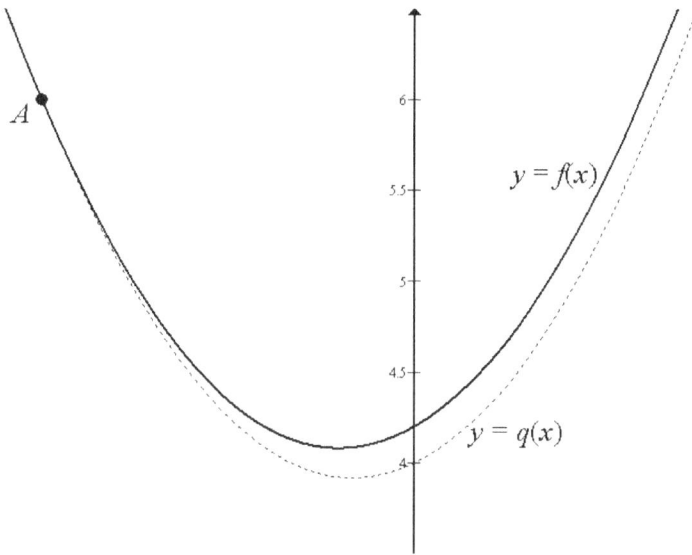

Figure 5: The Graph of $f(x)$ and Its End-Behavior Asymptote $q(x)$

Exercises

The following exercises are for this chapter. Analyze the following rational functions. In each case provide a complete graph with all of the relevant details.

1. $f(x) = \dfrac{2x-1}{2x+1}$

2. $f(x) = \dfrac{14x^2 - 11x + 2}{14x + 2}$

3. $f(x) = \dfrac{-35x^4 - 6}{84x^3 - 59x^2 - 43x + 30}$

4. $f(x) = \dfrac{21x^3 - 4x^2}{3x + 10}$

5. $f(x) = \dfrac{(5x-2)(1-x)}{(4x^2 - 9)}$

6. $f(x) = \dfrac{32x^5 - 1}{8x^3 + 1}$

7. $f(x) = 4x^2 + \dfrac{x}{1-x}$

8. $f(x) = 1 + x + x^2 - \dfrac{4x^2}{2-x}$

9. Writing in mathematics: Analyze several rational functions where $\deg(g(x)) = \deg(h(x)) + 1 = 3$. Thus, in general, $h(x) = ax^2 + bx + c$ for this problem. Make up your own examples. Your examples should push the boundaries of not quite having vertical asymptotes but appearing to have them on the graphing screen, which is known as hidden behavior. This can be done in several ways, so try to portray each way in your examples. Use inequalities to show why the observed behavior occurs. [Hint: Think about the discriminant of $h(x)$.]

167

10. Repeat problem #9, except $\deg(g(x)) = \deg(h(x)) + 2 = 4$.

11. Construct and analyze a rational function with a cubic asymptote.

12. Construct and analyze a rational function with a quartic asymptote.

Some Tangency Problems

This chapter deals with the concept of tangency. We discuss tangency in the calculus sequence; however, we do it there in terms of the calculus. The problems discussed in the present chapter are through algebraic methods. Thus, we do not require calculus in the pursuit of answers for these problems. The student saw the slope function and the perpendicular slope function in *The Deluxe Toolkit* chapter, which gives the slope of the tangent line and normal line. However, these functions will not help much with the problems discussed in this chapter.

Our first topic of discussion is an arbelos. Arbelos is a Greek word that means shoemaker's knife. When we put two semicircles of the same or differing radii side by side in the upward orientation on the x-axis (or any horizontal line $y = y_0$), with them just touching one another, and stretch another semicircle over them having the same baseline, the area enclosed by the three semicircles is called an arbelos because it looks like a shoemaker's knife from ancient Greece. See Figure 1.

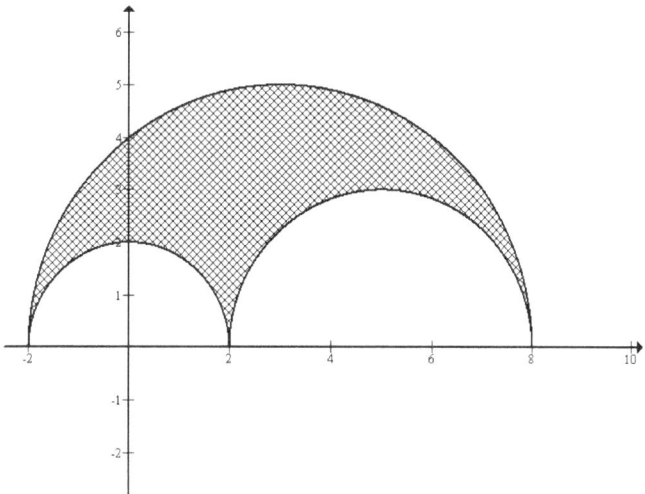

Figure 1: An Arbelos

Let us fill in some details for this picture. Point A is the left intersection point of the large semicircle with the x-axis. Point B is the right intersection point of the large semicircle with the x-axis. Point C is the common intersection point of the two smaller semicircles with the x-axis; point C is an important reference point that we will denote as (x_0, y_0) if the base of the outer semicircle is not on the line $y = 0$ (the x-axis). Drop a vertical line through point C, which we can denote symbolically as $x = x_0$. In the picture for this particular example, $x_0 = 2$, so the line through point C is $x = 2$. The intersection point of this vertical line with the large semicircle is point D. Now, the line AD intersects the left smaller circle at point E, and the line BD intersects the right semicircle at point F. The line EF is tangent to the two smaller circles. Figure 2 depicts what we have said thus far.

Does it look like a parallel line to the tangent line ran through point D would make a tangent line to the big circle? It turns out that it is, in fact, a tangent line as well. Hence, finding the slope of the tangent line to the two smaller circles nets the benefit of finding the slope of the tangent line to the big circle as well. Thus, if we want to find the equation of the tangent line to a circle at some point (x, y) on it, all we have to do is to set up the arbelos situation to get it. It may be prudent to center the circle $(x - h)^2 + (y - k)^2 = r^2$ at the origin by changing the equation to $x^2 + y^2 = r^2$ to make the algebra nicer, and then translate it back to its original place with its original h and k values.

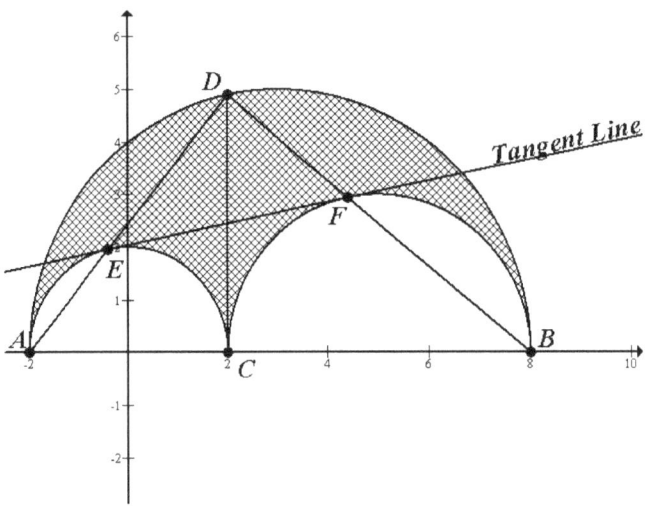

Figure 2: Tangent Line to the Two Smaller Circles

Can we say anything more about the arbelos? Yes, we can. The next thing we will discuss is Archimedes' Rectangle, named after Archimedes (c. 287 BC – 212 BC), an ancient Greek mathematician. The quadrilateral $CEDF$ is a rectangle. An angle in a semicircle is a right angle. This was proven by the ancient Greek mathematician Thales of Miletus (c. 625 BC – 547 BC). Hence, angle AEC is a right angle, and the angles CFB and ADB are also right angles. One of these right angles is an angle of the quadrilateral, and the other two right angles have supplementary angles (i.e. right angles) as angles in the quadrilateral. Hence, the quadrilateral $CEDF$ has three right angles. The sum of the angles in an n sided polygon is $180(n - 2)$. Thus, a quadrilateral must have $180(4 - 2) = 360$ total degrees. The three right angles take up $3(90) = 270$ degrees. We subtract to get

$360 - 270 = 90$ degrees. Hence, angle ECF measures 90 degrees. Thus, the quadrilateral $CEDF$ is a rectangle.

Another way to demonstrate that this particular quadrilateral is a rectangle is to compute the slopes of the lines CE and CF. If the three conditions $CE \parallel DF$, $CF \parallel DE$, and $DF \perp DE$ hold, then these conditions show that this quadrilateral is a rectangle as well. This is because if opposite sides are parallel and two adjacent sides are perpendicular, then any pair of adjacent sides are perpendicular by the parallel lines cut by a transversal arguments. Thus, all the angles in the quadrilateral are right angles, which makes the quadrilateral a rectangle. Figure 3 shows Archimedes' Rectangle.

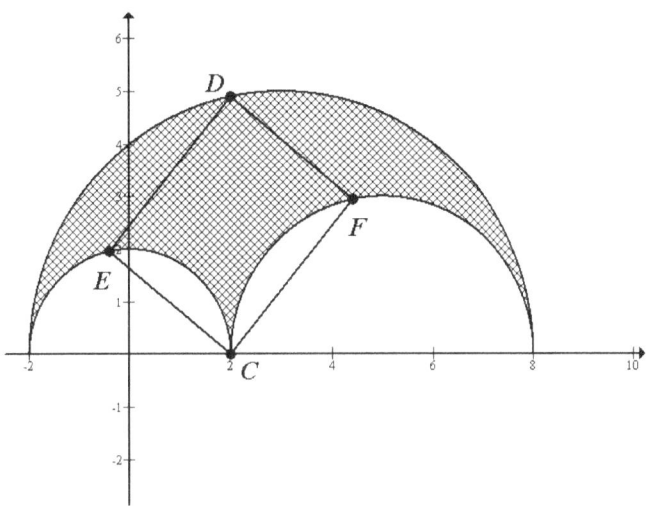

Figure 3: Archimedes' Rectangle

What is the area of the arbelos? It is the area of the big semicircle minus the combined area of the two smaller semicircles. The area of a circle is πr^2, so the area of a semicircle is $\frac{1}{2}\pi r^2$. Thus, the area of the big semicircle is $\frac{1}{2}\pi(5)^2 = \frac{25}{2}\pi$. The area of the left smaller semicircle is 2π, and the area of the right smaller semicircle is $\frac{9}{2}\pi$. Thus, the area of the arbelos is $\frac{25}{2}\pi - 2\pi - \frac{9}{2}\pi = \frac{12}{2}\pi = 6\pi$. The area of the circle having the vertical line in the arbelos as a diameter has the same area as the arbelos. The student should be able to demonstrate this. Similarly, the length of the top of the arbelos can be computed with the circumference formula for a circle modified for a semicircle and compared to the bottom of the arbelos.

The next item to discuss is Archimedes' Twin Circles. Let us start by showing the picture of what we mean. See Figure 4.

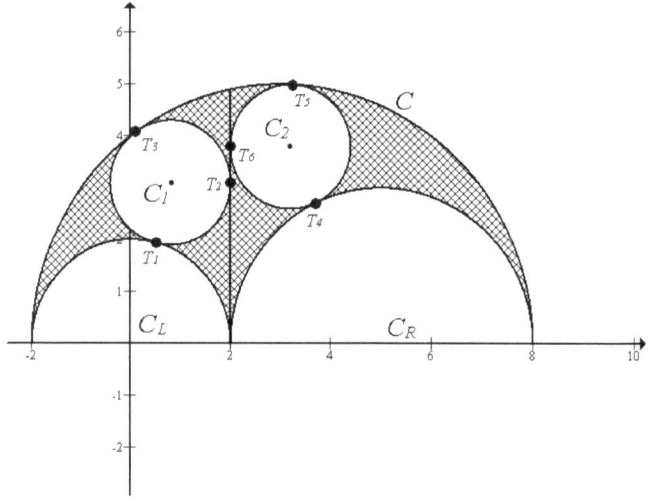

Figure 4: Archimedes' Twin Circles with Six Tangency Points

We see the big semicircle labeled C along with the two smaller semicircles labeled C_L and C_R, where L and R stand for left and right, respectively. The bases of these semicircles are on the line $y = y_0$, which for this particular example is $y_0 = 0$. These three semicircles create the arbelos as we have previously seen. The vertical line $x = x_0$ in the arbelos is shown. It is the line $x = 2$ with the range restrictions $0 \le y \le 2\sqrt{6}$. The two Archimedes' twin circles are labeled C_1 and C_2 for the left twin and the right twin, respectively. We also see the six points of tangency that are just labeled arbitrarily as T_1, T_2, T_3, T_4, T_5, and T_6.

In order to find the equations of the two twin circles, we need to know their centers and radii. Being that they are twins, they have the same radius. The Archimedean radius of each twin circle can be shown to be $r_A = \dfrac{r_L \cdot r_R}{r_L + r_R}$, where r_L and r_R are the radii of the left and right smaller circles in the arbelos, respectively; and r_A represents the Archimedean radius of the twin circles. It can be shown that the coordinates of the centers of the Archimedes' twins are the following: $C_1(x, y) = (x_0 - r_A, y_0 + 2\sqrt{r_L \cdot r_A})$ and $C_2(x, y) = (x_0 + r_A, y_0 + 2\sqrt{r_R \cdot r_A})$. Note that these formulas are good for finding the centers of Archimedes' twins for any arbelos. Armed with the equations of the twin circles, we set up the six systems of equations involving one of the twin circles paired in turn with the big semicircle, the appropriate smaller circle, and the vertical line. Solving each of these six systems yields the six tangency points T_1 through T_6. For example, in symbols we write $T_1(x, y) = C_1 \cap C_L$. This is the intersection point of circle C_1 with circle C_L (as a system of two circle equations) to yield the tangent point labeled as T_1.

The next item we will discuss is the Archimedes' quadruplets. These were discovered by Frank Power in 1998. Thus, students in this course get to explore a very recent find in mathematics. We start with the arbelos, but we find and label a few different points than before. See Figure 5.

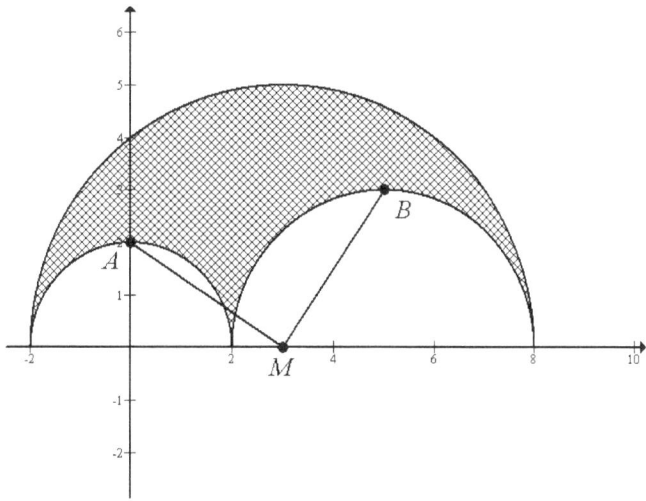

Figure 5: The First Preparation Diagram for Archimedes' Quadruplets

Points A and B are the highest points on the two smaller semicircles in the arbelos. Would we be understood if we said the top vertices of these circles in the manner of how we discuss ellipses? I think it is fine vocabulary to use. Point M is the midpoint of the base of the big semicircle of the arbelos. We can also think of point M as the center of the big circle if we included its lower half. Next, we connect point A to M to form the line segment AM, and we connect point B to M to form the line segment BM.

After this, we construct a perpendicular line segment to AM through the point A. The center of this line segment is at point A. The length it protrudes out on either side of A is the Archimedean radius r_A. The endpoints of this line segment are the centers of two of the Archimedes' quadruplets. We repeat this process for point B. Construct a perpendicular line segment to BM through point B. The center of this line segment is at

173

point B. The length it protrudes out on either side of B is the Archimedean radius r_A. The endpoints of this line segment are the centers of the other two Archimedes' quadruplets. As a computational note, it may be easiest to find the four endpoints of the two perpendicular segments by solving the two systems of two equations involving the line segment as a line and the circle of radius r_A centered at A and B, respectively. The two systems below are constructed with the perpendicular slope formula, the point-slope equation of a line, and the equation of a circle and use the notation A_x, A_y, B_x, B_y, M_x, and M_y, for the x and y coordinates of $A(x, y)$, $B(x, y)$, and $M(x, y)$ respectively. See Figure 6 for an illustration of what we have thus far.

$$\begin{cases} y = \dfrac{A_x - M_x}{M_y - A_y}(x - A_x) + A_y \\ (x - A_x)^2 + (y - A_y)^2 = r_A^2 \end{cases} \qquad \begin{cases} y = \dfrac{B_x - M_x}{M_y - B_y}(x - B_x) + B_y \\ (x - B_x)^2 + (y - B_y)^2 = r_A^2 \end{cases}$$

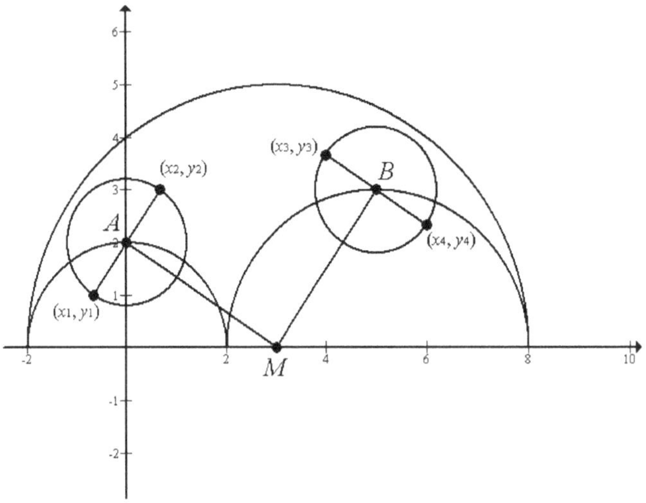

Figure 6: The Second Preparation Diagram for Archimedes' Quadruplets

Let us say that the four solutions to these systems are (x_1, y_1), (x_2, y_2), (x_3, y_3), and (x_4, y_4), which are labeled as such in Figure 6. Then, we can construct the four equations of the four Archimedes' quadruplets circles C_i, $i = 1, 2, 3,$ and 4. The indexing variable i lets us write the four equations compactly. Note that C_1 and C_2 here have nothing whatsoever to do with the Archimedes' twins discussed earlier.

$$C_i : (x - x_i)^2 + (y - y_i)^2 = r_A^2, \ i = 1, 2, 3, \text{ and } 4$$

174

We now have the equations of the four Archimedes' quadruplets. Observe in the next figure that two of the quadruplets are tangent to each other at point A, and the other two quadruplets are tangent to each other at point B. It remains to find the four points of tangency of these quadruplets with the big semicircle C. We can set up and solve the four systems of two circle equations involving an Archimedes' quadruplet with the big semicircle C to yield the four points of tangency. However, we also have at our disposal the four lines from M through the centers of each of the Archimedes' quadruplets circles. These four lines intersect the big semicircle C at precisely the points of tangency of the four Archimedes' quadruplets. The points of tangency are labeled in order around the circle from left to right as T_1, T_2, T_3, and T_4. Figure 7 illustrates these facts.

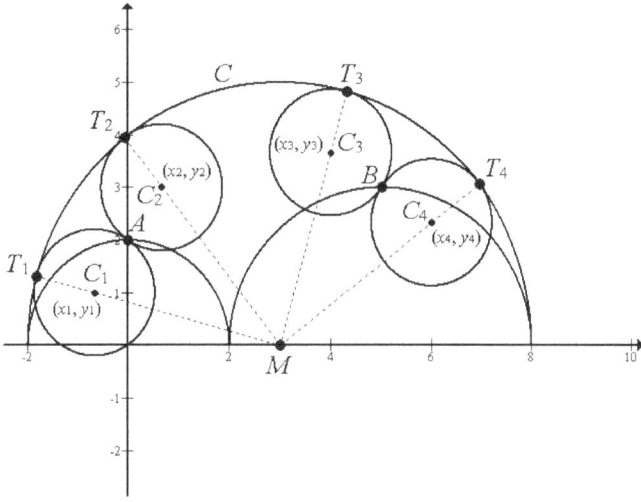

Figure 7: The Archimedes' Quadruplets

As a note of precaution, the CAS (Computer Algebra System) you use to find the exact numerical values or algebraic expressions may yield some messy computer output that probably can be simplified further. In particular, some output may contain nested radicals that certainly can be denested. See the chapter on fractions and denesting radicals for information on this. Also, some of the algebraic expressions may not be in a form to your liking. Learn how to utilize the CAS commands to get it to produce the clean results you need. I had trouble with my CAS when trying to find the coordinates of T_1, T_2, T_3, and T_4—the tangent points of the Archimedes' quadruplets with the big semicircle C.

More information is available about the arbelos on the Internet. Just Google the word arbelos, and look at what is returned. We turn our attention now to analyzing the three pair-wise tangent circles that are centered at the three vertices of a non-degenerate triangle. Figure 8 shows a picture of what we mean.

175

In order to find the equations of the three circles, we must first compute the lengths of the three sides of the triangle a, b, and c, which are opposite the vertices A, B, and C, respectively. Then, we compute the semiperimeter $s = p/2$, which is half of the perimeter of the triangle. When writing the semiperimeter as s, be careful as to how you write it so as not to confuse it with the number 5. The coordinates of the centers of the three circles are the vertices of the triangle. The radii of the three circles are the following:

$$r_A = p/2 - a = \frac{-a+b+c}{2}, \; r_B = p/2 - b = \frac{a-b+c}{2}, \; r_C = p/2 - c = \frac{a+b-c}{2}$$

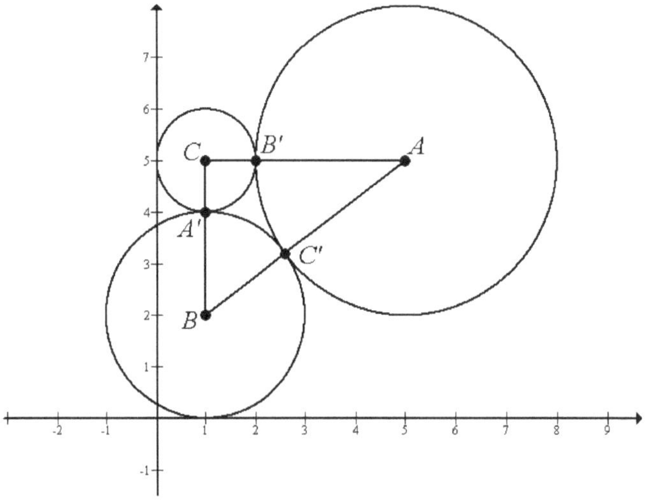

Figure 8: Three pair-wise Tangent Circles

Observe that r_A in the first formula above has nothing whatsoever to do with the Archimedean radius described earlier. The points of tangency of the circles can be computed by solving the three systems of equations involving the possible pairs of circles. In addition, the points of tangency can be found by using the extended midpoint formula given in *The Deluxe Toolkit* chapter. For example, use the weights $w_1 = r_A/b$ and $w_2 = 1 - r_A/b$ with the points A and C to find B'. However, only the system of equations of the two circles demonstrates that it is a tangent point. Thus, be mindful of whether you are just wanting the location of the point, which can be obtained more easily with the extended midpoint formula, or whether you want to demonstrate if the two circles in question are tangent or not, which must be handled with the system of equations yielding a single intersection point.

The points of tangency are labeled with primes such that A', B', and C' are on the sides opposite the vertices A, B, and C, respectively. A surprising fact is that the lines

AA', *BB'*, and *CC'* intersect at a single point. In addition, the perpendicular lines through *A'*, *B'*, and *C'* also intersect or concur at a common point. These two examples provide an insight as to how some of the triangle concurrencies in the *Triangle Geometry* chapter are found. They usually involve some manipulation or other of just a few basic concepts. Obviously, most of the triangle concurrencies at the website cited in that chapter get exotic. But, here is a beginning taste of the process. The *Triangle Geometry* chapter gets into quite a number of those concurrencies and a host of other related things. For the record, the mutual intersection point of the three lines *AA'*, *BB'*, and *CC'* is called the Gergonne point, *Ge*; and the mutual intersection point of the three perpendicular lines through the points *A'*, *B'*, and *C'* is called the incenter, *I*. However, the construction in Figure 8 is a different way of finding the Gergonne point and incenter than the methodology presented in the *Triangle Geometry* chapter. In mathematics we often have more than one way of getting to a result.

We have something to say about triangle *A'B'C'*. It is in perspective with triangle *ABC*. The perspector is the Gergonne point, *Ge*. Thus, according to Desargues' theorem, there must be a perspectrix. This perspectrix is called the Gergonne line. We name the triangle *A'B'C'*, the Gergonne triangle after Joseph Diez Gergonne (1771 – 1859), a French mathematician and geometer. See Figure 9.

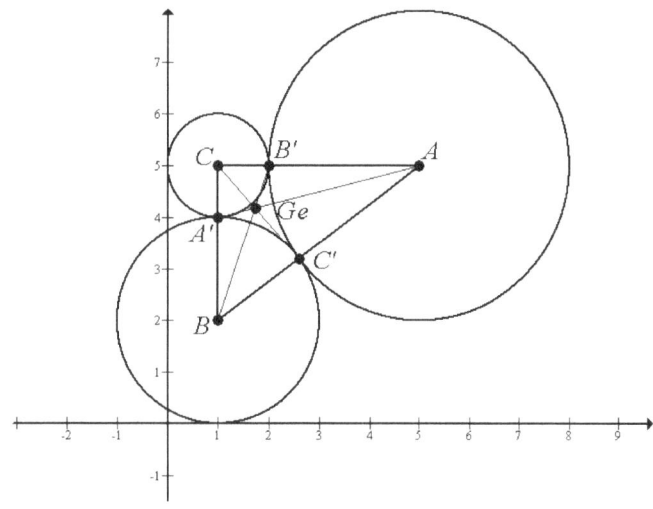

Figure 9: The Gergonne Point

One property of the incenter, *I*, is that it is the same distance from each side of the triangle's sides. Thus, the distance to each side is the radius of the circle that is tangent to the three sides of the triangle. To find this radius, just compute the distance between the points *I* and *A'* using the distance formula. If you need to show that the distance to the sides is the same, then compute all three distances and show that $d(I, A') = d(I, B') =$

177

d(*I*, *C'*). The circle tangent to the three sides of the triangle is called the incircle, and its radius is called the inradius, *r*. With the incenter, *I*, and the inradius, *r*, we can construct the equation of the incircle. See Figure 10. One further comment about this diagram is that the incenter, along with the perpendicular line segments from it to the sides of the triangle, partitions the triangle into three kites. Google the word kite along with the word math to find out some interesting properties of kites.

Our next discussion will be on finding the equation of the tangent line to a parabola. This topic is covered in the calculus class, but it uses the concept of the derivative, which is a calculus concept. We seek a way to find the equation of the tangent line through algebraic methods. Fortunately, we have two such ways that we will present.

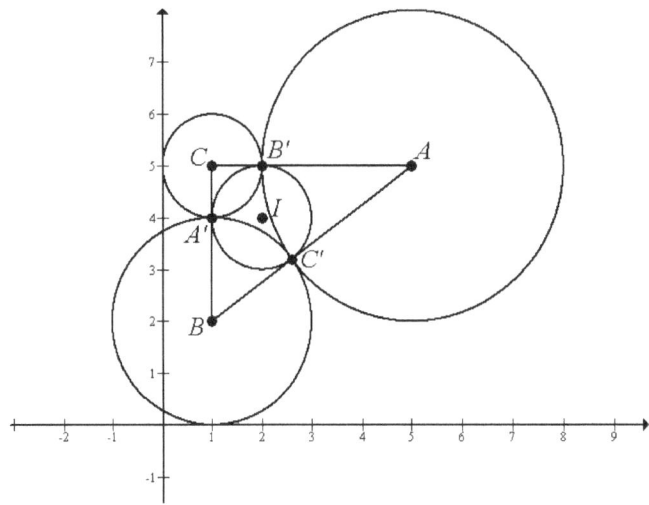

Figure 10: The Incircle and Incenter

The first way was discovered by one of my fellow tutors at Oakland Community College when I challenged him to find the equation of the tangent line to a specific parabola at a specific point *without using calculus*. He accepted the challenge after asking me a few questions about the boundaries of the ground rules. In short order, he found an algebraic method that yielded the tangent line. He told me when he was finished that he knew it was correct because he checked it with calculus and got the same line. This is how he did it.

Example: Find the equation of the tangent line to the parabola $y = 3x^2 + 5x - 4$ at the point (1, 4).

Assume the tangent line is $y = mx + b$ with m and b to be determined. We have the point of tangency given by (1, 4), so we can plug this into the tangent line equation.

$$y = mx + b \Rightarrow 4 = m \cdot 1 + b \Rightarrow b = 4 - m$$

Thus, b is represented algebraically in terms of m. The equation of the tangent line is now $y = mx + 4 - m$. Construct and begin to solve the system of equations involving the tangent line and the parabola.

$$\begin{cases} y = 3x^2 + 5x - 4 \\ y = mx + 4 - m \end{cases}$$

$mx + 4 - m = 3x^2 + 5x - 4$

$3x^2 + 5x - 4 - mx - 4 + m = 0$

$3x^2 + (5 - m)x + m - 8 = 0$ Stop at this point.

We have a quadratic equation in x. However, this equation involves another variable m, so what is this equation saying? The set of all lines going through the point (1, 4) yields two intersection points with the parabola whose x-coordinates are the roots of this equation. In other words, let us for the moment just pick a slope of two for the line. Thus, $m = 2$ for this example. With m as 2, the solutions to this equation yield the two x-coordinates where this line intersects the parabola. However, m can be any real number. We would like this equation to tell us the particular value of m so that we get only one x-coordinate solution. This particular value of m is the slope of the tangent line through that point. How do we get the equation to tell us this one value of m? We use the discriminant of the quadratic equation, and set it equal to zero. When the discriminant is zero, we get one real solution. This concept was covered in the theory of quadratic equations, and we need it here. So even though we set up the equation as a quadratic for x, we are really after another variable m, which we can get through an understanding of the nature of x. This is a first rate insight. Thus, we have the following:

$3x^2 + (5 - m)x + m - 8 = 0$ We need to identify the coefficients A, B, and C.

$A = 3$, $B = 5 - m$, and $C = m - 8$

$B^2 - 4AC = (5 - m)^2 - 4(3)(m - 8)$ Set this equal to zero and solve for m.

$25 - 10m + m^2 - 12m + 96 = 0$

$m^2 - 22m + 121 = 0$

$(m - 11)^2 = 0$

$\therefore m = 11$

We have found the unique slope yielding just one intersection point with the parabola—the slope of the tangent line. It remains to compute $b = 4 - m = 4 - 11 = -7$. We can now write the equation of the tangent line $y = 11x - 7$. Solving the system of equations involving this line with the parabola yields the single point $(1, 4)$, which demonstrates that this line is the tangent line to the parabola at that point. See Figure 11 for an illustration of this situation.

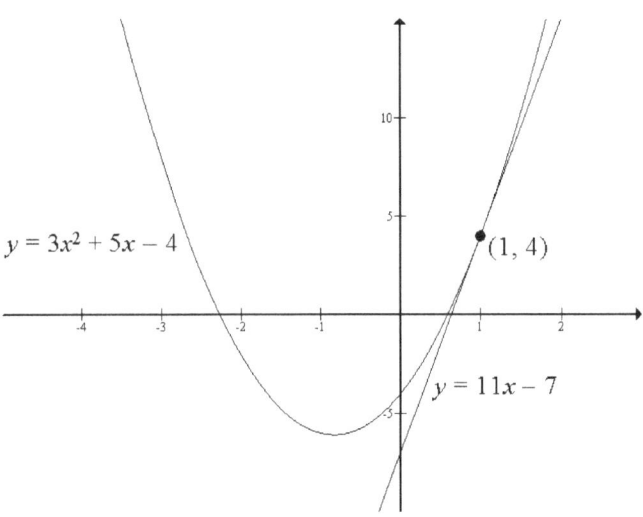

Figure 11: A Parabola with the Tangent Line at a Point

This tutor performed wonderfully under the challenge I issued him. My motive for challenging him was for him to say, "I cannot think of a way without calculus." Then, I was going to show him the way to do it based on a distance property of the parabola I had recently found when I read about Apollonius, one of the ancient Greek mathematicians. This property would yield the tangent line. I didn't think he would come up with a way without calculus, but I humored him and let him work on it without interrupting him. His results speak for themselves, and I am so impressed with them that I included them in this book. After he showed me his solution, I showed him Apollonius' discovery, which is described next.

Another way exists to find the tangent line to a point on a parabola, which was discovered by Apollonius of Perga (c. 225 BC – c. 175 BC), an ancient Greek mathematician. This next method, coupled with the previous method just described, gives the student two distinct algebraic methods for finding the tangent line equation to a parabola at a specific point. Frégier's theorem discussed in the next chapter (a theorem highlight chapter) gives the student a third method. If the student has already had calculus, then that makes four methods that are available to the student to accomplish the feat. See Figure 12 for Apollonius' observation on the distance relationships for a tangent line to a parabola. Later in this chapter, we will see the general tangent line equation and how to spin it off to the various forms it can take for specific conic section equations.

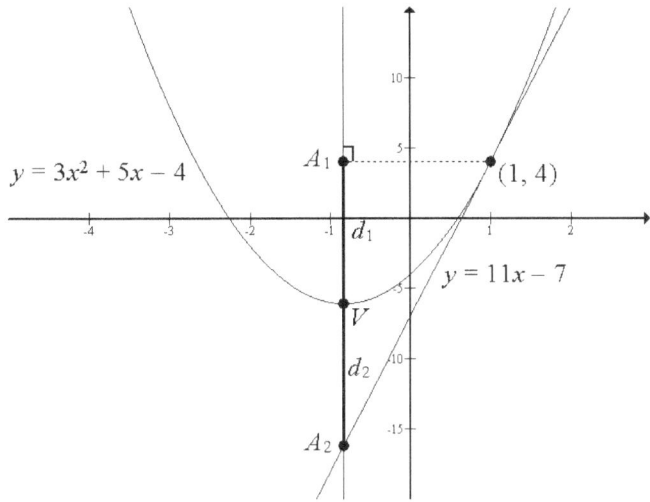

Figure 12: Apollonius' Distance Relationships

Apollonius' observation in Figure 12 is that if $d_1 = d_2$ on the axis of symmetry, then the line through A_2 and the given point on the parabola (in this example $(1, 4)$) will be a tangent line. We will develop a procedure for finding a tangent line with this observation by Apollonius. A terminology note is in order here. If we know that a line is a tangent line to a parabola, then the line segment on the axis of symmetry between A_1 and A_2 is called the *subtangent*. The subtangent is bisected at the vertex of the parabola.

Given a point P on the parabola for which to find the tangent line, we perform the following steps. Drop a perpendicular from P (in this example, P is $(1, 4)$) to the axis of symmetry. The point where it hits the axis of symmetry can be called A_1. In this example, A_1 is the point $(-5/6, 4)$. Next, compute the distance d_1 (the Apollonian distance for this parabola with this tangent line) between the vertex and A_1. In this example, $d_1 = |A_{1y} - V_y| = |4 - (-73/12)| = 121/12$. We use absolute value bars with subtraction of y-coordinates to get distance. We can do this because we know the x-coordinate is fixed at -5/6 because these points are on the axis of symmetry. With left-right parabolas (and square root functions), we would modify this procedure to hold the y-coordinates fixed. The next step is that we subtract this distance from the corresponding coordinate of the vertex (i.e. establish the Apollonian $d_1 = d_2$ relationship in the diagram to yield the tangent line). The computation is $A_{2y} = V_y - d_1 = -73/12 - 121/12 = -194/12 = -97/6$. Note that with a downward parabola, we would add at this point, not subtract. Just look at the picture to

181

guide your efforts. The previous remarks about all these points being on the axis of symmetry allow us to write $A_2 = (-5/6, -97/6)$. The line through A_2 and P will be tangent to the given parabola. This line can be found through standard methods.

Our next topic in this chapter is from Johannes Regiomontanus (1436 – 1476), a German mathematician. He posed a question about where a person should stand so that a statue on a pedestal would appear largest. The answer to this question comes through understanding circles. The top of the statue is one point. The bottom of the statue is a second point. The location of the person's eye is the third point. We know that three noncollinear points determine a circle, so we understand that depending on where the person stands, it determines the circle through those three points. Well, what of that? Does this bring us any closer to the answer? It does. Consider Figure 13.

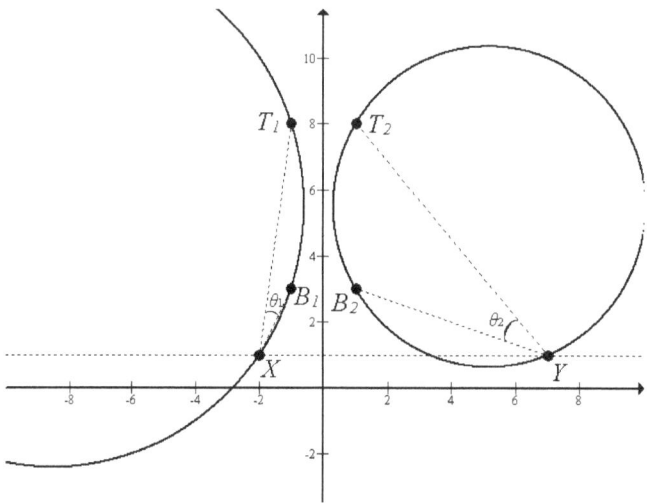

Figure 13: Two Viewing Distances for Statues of Equal Height

Let us explain this picture. A person's eye is at point $X(-2, 1)$ viewing a statue on a pedestal with base at $B_1(-1, 3)$ and top at $T_1(-1, 8)$. These three coordinates make a rather large circle with a correspondingly small angle of view θ_1 for the statue. The person moves her eye by walking over to point $Y(7, 1)$ to view the backside of the statue with the base at $B_2(1, 3)$ and top at $T_2(1, 8)$. These three coordinates make a considerably smaller circle with a correspondingly bigger angle of view θ_2 for the statue. Hence, when she is at point Y, the bigger angle of view makes the statue appear larger to her. We are going to assume that she just walks along a level floor, so the y-coordinate for her eye will not change. How can she position her eye so that she makes the smallest possible circle yielding the largest possible angle of viewing? It turns out that the smallest possible circle is when the circle is tangent to her eye. Don't be alarmed though because it won't hurt her.

So if we find the equation of the circle tangent to her eye's movement horizon, we can find where she should position her eye.

To solve this, we let the line $y = y_0$ represent all of the positions that her eye can be. In this example, the line is $y = 1$. Then, we need to know the coordinates of the three points. These are (1, 3), (1, 8), and $(x_0, 1)$ where x_0 is the x-coordinate of the best place for her eye to be to view the statue. Observe the third point has a y-coordinate of one. This is equivalent to having a system of equations with a circle and the line $y = 1$. Next, we would like to find the equation of the circle. Thus, we construct the system of three equations involving our three points.

$$\begin{cases} (1-h)^2 + (3-k)^2 = r^2 \\ (1-h)^2 + (8-k)^2 = r^2 \\ (x_0 - h)^2 + (1-k)^2 = r^2 \end{cases}$$

Pairing the first two yields $k = 11/2$. Pairing the first and the third equations and simplifying yields the following:

$$x_0^2 - (2h)x_0 + 2h + 13 = 0$$

We have seen the power of the discriminant being zero with a two-variable equation with the tangent to the parabola. Let's do that again here.

$$4h^2 - 4(1)(2h + 13) = 0$$
$$h^2 - 2h - 13 = 0$$
$$h = 1 \pm \sqrt{14}$$

We have two values of h now. The plus version represents being on the right hand side of the vertical line $x = 1$, which is where the statue is on the graph. The minus version represents being on the left hand side of it. Let us continue if we can by carrying both of these h values along for the ride. We will solve the x_0 equation now with these h values.

$$x_0^2 - (2h)x_0 + 2h + 13 = 0 \quad \text{Replace } h \text{ with } 1 \pm \sqrt{14}.$$
$$x_0^2 - 2(1 \pm \sqrt{14})x_0 + 2(1 \pm \sqrt{14}) + 13 = 0$$
$$x_0^2 - 2(1 \pm \sqrt{14})x_0 + 15 \pm 2\sqrt{14} = 0 \quad \text{Note the coefficients } A, B, \text{ and } C.$$

Note the dependent $\pm$ signs. Both in the upper or both in the lower position means we have only two equations here, not four. These are dependent because they came from h. The variable h is a variable and can be any real number, but it can only be one value once it has chosen. Thus, h cannot be $1+\sqrt{14}$ in the first position and $1-\sqrt{14}$ in the second position in the x_0 equation because then h has chosen to be two values at the same time (i.e. in the same equation). This is beyond its capacity. Now, we apply the quadratic formula.

$$x_0 = \frac{2(1\pm_h\sqrt{14})\pm\sqrt{4(1\pm_h\sqrt{14})^2 - 4(1)(15\pm_h 2\sqrt{14})}}{2}$$

We have more to say here about these $\pm$ signs. We got the first one from the h equation earlier. Now, the x_0 equation has provided another. We need to keep track of which ones are which. The one from the h variable has spread out and multiplied during the solution process. These have been designated with the subscript h to denote their origin. Hence, all of these must be in the upper position, or all of these must be in the lower position. The remaining $\pm$ sign has no subscript. This one was obtained because of the quadratic nature of x_0. We thus only have two independent $\pm$ signs present in this expression for x_0. Thus, 2 independent occurrences of $\pm$ times 2 positions each yields 4 possible values for x_0 as it is present in the form we see it now. We continue.

Observe that a four can be factored out of the big radical in the numerator as a two. The rest of the terms have a two as a factor. Thus, a big cancellation can occur right away.

$$x_0 = 1\pm_h\sqrt{14}\pm\sqrt{(1\pm_h\sqrt{14})^2 - (15\pm_h 2\sqrt{14})}$$

$$x_0 = 1\pm_h\sqrt{14}\pm\sqrt{1\pm_h 2\sqrt{14}+14-15\mp_h 2\sqrt{14}}$$

$$x_0 = 1\pm\sqrt{14}$$

The value under the big radical was zero. However, we knew that because we had found the h values that made it so. But these values for x_0 are the same as those for h. Is there a mistake here somewhere? No, everything is correct. Remember what we are trying to accomplish here. We want the circle that is tangent to a line (kind of the opposite goal with the arbelos theory we learned earlier in this chapter). Since this line is horizontal, the center of the circle would be vertically above the point of tangency. Hence, the point of tangency and the center of the circle are on the vertical line $x = h$, which for this example is $x = 1\pm\sqrt{14}$.

We can now compute r^2 with the h and k values that we have obtained. We can use any of the three equations above, but look at that last equation. It shows $(x_0 - h)^2$. Since these are the same, we can save time by using the third equation. Shortcuts like these can make tough jobs easier. We just have to notice that the shortcut is there in the first place. Thus, $r^2 = (1 - k)^2 = (1 - 11/2)^2 = 81/4$. We can now construct the equations for the two circles that we have found.

$$C_1: (x - 1 - \sqrt{14})^2 + (y - 11/2)^2 = \frac{81}{4}$$

$$C_2: (x - 1 + \sqrt{14})^2 + (y - 11/2)^2 = \frac{81}{4}$$

Observe that we have two solutions here. This means that the two values of x_0 represent the two best places for the lady to place her eye to view the statue. Because the statue is vertical, she gets the same viewing angle for both positions of x_0. However, what if the statue is not vertical? That is to say, suppose the statue has an overhang. What would happen then? All of the same calculations would happen, except we would get two nonsymmetrical solutions on each side of the base of the statue. The x_0's would still have to be under the h's because these are still tangent circles to the horizontal line. One of the angles would be bigger than the other (meaning the two tangent circles have different radii), but these would be the biggest angles possible on that side of the statue. We label the variables with subscripts L and R to denote their left and right counterparts to the statue. See Figure 14.

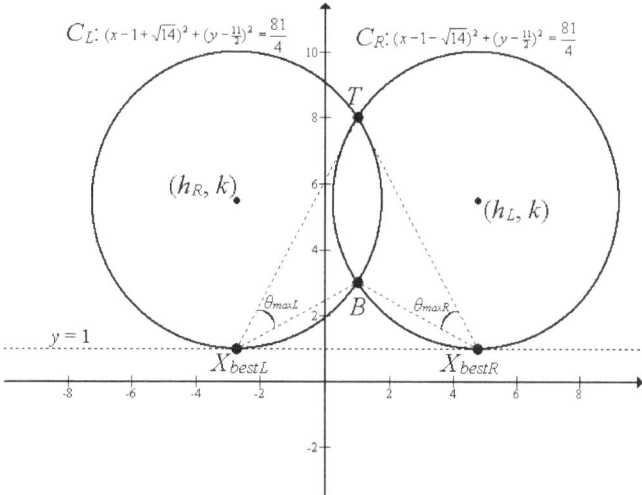

Figure 14: The Largest Viewing Angle Comes from the Tangent Circles

185

One final comment about these statue problems is that we are not attempting to find what this maximum angle is. That is a job that trigonometry can solve for us, and the qualified student is welcome to find that angle. We will be content just to know that we have the maximum angle possible, whatever that angle may be.

Apollonius of Perga showed another property of a conic section—in this case a hyperbola. Let us say the asymptotes to the hyperbola meet at C, the center of the hyperbola. Given a tangent line to a hyperbola that meets the hyperbola at T and that the tangent line intersects the asymptotes at the points P and Q, then the following two distance relationships hold: $d(C,P) \cdot d(C,Q)$ is a constant (the constant is $a^2 + b^2$) and $d(T,P) = d(T,Q)$. These two facts allow us to find the tangent line to a hyperbola under various circumstances.

We will note next an additional way to get the tangent line to an ellipse or hyperbola, but we will not provide any exercises for it in this section until we have covered the incenter in the *Triangle Geometry* chapter.

The normal line to a point $P(x_0, y_0)$ on the ellipse bisects the angle between the two focal radii. The tangent line to the ellipse at that point is perpendicular to the normal line at that point. We can use the concept of the incenter of the three points in question to get this tangent line because a line from a vertex to the incenter bisects the angle of that vertex. For the hyperbola, the tangent line to a point (x_0, y_0) on the hyperbola bisects the angle between the two focal radii at that point. We can use the incenter of the three points in question to get this tangent line.

Our next item to discuss is the general tangent line equation. Given a general non-rotated conic section $ax^2 + by^2 + 2cx + 2dy + e = 0$ and a point $P(x_0, y_0)$ on the conic section, the equation of the tangent line to the conic section at the point $P(x_0, y_0)$ is $ax_0 x + by_0 y + c(x + x_0) + d(y + y_0) + e = 0$. This equation is easy to remember if we see that a variable squared is the variable sub-naught times the variable itself and that twice a variable is the variable sub-naught plus the variable itself. Then, the tangent line follows easily from the general non-rotated conic section equation.

We can input a point of tangency into the usual conic section equation for just one of the occurrences of the squared variable present in the equation. For example, the tangent line to the tall ellipse $\dfrac{x^2}{b^2} + \dfrac{y^2}{a^2} = 1$ at the point $P(x_0, y_0)$ is $\dfrac{x_0 \cdot x}{b^2} + \dfrac{y_0 \cdot y}{a^2} = 1$. If h and k are present, which for the tall ellipse is $\dfrac{(x - h)^2}{b^2} + \dfrac{(y - k)^2}{a^2} = 1$, then we expand the numerators out and replace a variable squared with that variable times its sub-naught version and twice a variable with the sum of the variable and its sub-naught version. Thus, the expanded tall ellipse equation is $\dfrac{x^2 - 2hx + h^2}{b^2} + \dfrac{y^2 - 2ky + k^2}{a^2} = 1$. We rewrite this as $\dfrac{x^2 - h(2x) + h^2}{b^2} + \dfrac{y^2 - k(2y) + k^2}{a^2} = 1$. We make the replacements as we have stated to get $\dfrac{xx_0 - h(x + x_0) + h^2}{b^2} + \dfrac{yy_0 - k(y + y_0) + k^2}{a^2} = 1$. Thus, this would be the equation of the tangent line to a tall ellipse at the point $P(x_0, y_0)$.

We present a theorem involving tangents and conic sections. If the vertices of a triangle lie on a conic section, then the tangent lines to the conic section at the vertices meet the opposite sides of the triangle (making extensions to the sides of the triangle as necessary) in three collinear points.

Our next theorem about tangents involves an ellipse. We will state this theorem, but we will not ask a question about this theorem. The ambitious student can explore this theorem at will, however. Let AB and CD be two chords of an ellipse that intersect at the point H. Let the tangents at A and C meet at the point P, and let the tangents at B and D meet at the point Q. Then, the points P, H, and Q are collinear. Furthermore, let the tangents at A and D meet at the point R, and let the tangents at B and C meet at the point S. Then, the points R, H, and S are collinear.

Another ellipse theorem involving tangents is presented next. Let P, Q, and R be three points on an ellipse such that the three tangent lines form a triangle ABC. Let P lie opposite vertex A; Q lies opposite vertex B; R lies opposite vertex C. Then, the following distance relationship holds: $d(B, P) \cdot d(C, Q) \cdot d(A, R) = d(P, C) \cdot d(Q, A) \cdot d(R, B)$.

The next chapter is a theorem highlight chapter about Frégier's theorem. Frégier's theorem will let us find the equation of the tangent line to a specific point on any conic section without the use of calculus. However, it is a bit cumbersome compared to the relative ease of the general tangent line equation given in this chapter.

Exercises

1. Compute the arc length of the top part of the arbelos in Figure 1. Compute the arc length of the bottom part of the arbelos in Figure 1. Draw another arbelos with the same enclosing outer semicircle but with a different sized pair of smaller semicircles. Compute the arc length of this new pair of smaller semicircles, and compare it to the other arc lengths already computed. Conjecture what would happen with any two sizes of the

smaller semicircles enclosed by the same outer semicircle. Conjecture what would happen with any outer semicircle with any pair of enclosed semicircles.

2. Regarding Figure 2, find the equation of the circle having the line segment CD as a diameter. Compute the area of this circle, and compare it to the area of the arbelos, which was 6π in the text.

3. Find the four coordinates of Archimedes' Rectangle in Figure 3.

4. Use the points E and F from problem 3 to find the equation of the tangent line labeled as such in Figure 2.

5. Use the slope of the tangent line in problem 4 to find the equation of the tangent line through point D in Figure 2. Solve the system of equations involving this tangent line and the big semicircle C to demonstrate that there is only one intersection point. In other words, the resulting quadratic equation yields one solution of multiplicity 2. Even multiplicity from the theory of polynomial equations means that the quadratic *bounces* off the line. This means that it is tangent to the line.

6. Use the arbelos theory we have discussed to find the equation of the tangent line to the circle $x^2 + y^2 = 1$ at the point (3/5, 4/5). This circle is called the unit circle, and this point is one of the eight rational points on this circle from the Pythagorean triplet (3, 4, 5). Solve the system of equations involving the tangent line you found with the unit circle to demonstrate that the point (3/5, 4/5) is the only intersection point. See the remarks in question 5.

7. Regarding the Archimedes' Twins diagram, the slope for the line through AD is $\dfrac{\sqrt{6}}{2}$, and the slope for the line through BD is $\dfrac{-\sqrt{6}}{3}$. The angle in a semicircle is a right angle, according to Thales' theorem. Thus, these two slopes are perpendicular, which means that they are supposed to be negative reciprocals of one another. Explain why we do not have any contradictions in this collection of statements.

8. Use Figure 4 to complete the parts to this problem.

    a) Find the equations of the three semicircles labeled C, C_L, and C_R.
    b) Compute the Archimedean radius r_A.
    c) Identify x_0 and y_0 for the diagram so as to use them in the upcoming computations.
    d) Find the two centers C_1 and C_2 of the Archimedes' twin circles in the diagram.
    e) Construct the equations for C_1 and C_2.

f) Construct and solve the relevant six systems of equations involving the vertical line, C_L, and C_R with each Archimedes' twin circle to find the six labeled points of tangency T_1 through T_6.

9. Follow the procedure outlined in question 8 to analyze the arbelos in question 6.

10. This problem deals with finding the equations of the four Archimedes' quadruplets discussed in the text.

a) Find the points A, B, and M in Figure 5.
b) Set up and solve the two systems that yield the four centers of the Archimedes' quadruplets.
c) Construct the equations for the four Archimedes' quadruplets.
d) Find the equations of the four lines through M and the center of an Archimedes' quadruplet circle.
e) Set up and solve the appropriate systems of equations involving an equation from part c and an equation from part d to yield the coordinates of the four tangency points T_1 through T_4 as pictured in Figure 7.

11. Follow the procedure outlined in question 10 to analyze the arbelos in question 6.

12. This problem asks the student to write a paper about the arbelos. Consult the text and the earlier problems in this exercise set to help fill in all the details.

a) The outer semicircle is the top half of the circle $(x - 2)^2 + (y + 3)^2 = 25$.
b) The point on the circle with which to find the tangent line through is $(6, 0)$.
c) Draw the arbelos, and find all the things discussed in the text related to the arbelos.
d) For this arbelos, find the Archimedes' twins discussed in the text along with the six points of tangency.
e) For this arbelos, find the Archimedes' quadruplets and the four points of tangency.
f) Include all relevant diagrams fully labeled. Provide the supporting algebra fully documented. Organization is important. Show your professionalism. Diagrams created with software look more professional than hand sketches. Word-processed papers look more professional than hand written ones.

13. A statue has a base at $(2, 3)$ and its top is at $(2, 12)$. A spectator's eye moves along the line $y = 1$. Find the best position for the spectator to place her eye so that the statue appears largest.

14. A statue with an overhang has a base at (2, 3) and its top at (4, 12). A spectator's eye moves along the line $y = 1$. Find the best position for the spectator to place her eye so that the statue appears largest. What is the "other" supposedly extraneous solution telling us?

15. Find the equation of the tangent line to the ellipse $\dfrac{(x-1)^2}{25} + \dfrac{(y+2)^2}{100} = 1$ at the point $P(x_0, y_0) = (-2, 6)$ using the idea of the general tangent line equation modified for this specific ellipse.

16. The parabola $y = x^2$ has a triangle inscribed in it at the $A(1, 1)$, $B(0, 0)$, and $C(3, 9)$.

   a) Find the equations of the three tangent lines to the parabola at the vertices of the triangle.
   b) Find the coordinates of the three intersection points of a tangent line with the opposite side of the triangle.
   c) Show that the coordinates obtained in part b are collinear.

17. Find the equation of the tangent line to the hyperbola $\dfrac{x^2}{9} - \dfrac{y^2}{16} = 1$ at the point $T(3.75, 3)$ using the Apollonius relationship discussed in the chapter. [Hint: The student will have to derive a method for this. Let $P = (P_x, P_y)$ and $Q = (Q_x, Q_y)$ so that the appropriate equations can be established. I boiled it down to a polynomial in P_x that was the following: $50P_x^4 - 279P_x^3 - 81P_x - 4050 = 0$, which I solved using the Rational Roots theorem. There is just one rational root, and the other three are not rational.]

18. Find the equation of the circle passing through the point (-3, 1) that is tangent to the line $y = 2x + 1$ at the point (4, 9). [Hint: Use the fact that the center of a circle lies on the perpendicular bisector of any chord. Also, the center lies on the normal line to a point of tangency.]

19. Find the equations of the two circles containing the points (2, 3) and (4, 7) that are both tangent to the line $y = x - 4$. [Hint: Use the fact that all radii are equal in a circle. Also, use the distance from a point (i.e. the center of the circle) to a line formula.]

Frégier's Theorem

Frégier's work was published in 1816 in Gergonne's *Annales de Mathematique*, a French mathematics magazine in the early 1800's. Frégier was a French mathematician. Frégier's theorem is important because it will allow us to find the equation of the tangent line to any point on a conic section. The slope of the tangent line to a curve in general is important because it represents the rate of change of the original curve. If the curve represents where an object is in time, (time as x, position as y), then its change in position in time is velocity. If the curve represents velocity, then this rate of change is acceleration.

Many physical problems in the world require the related rate of change curve to the original curve in the course of solving the problem. We are part way there without calculus (which gives us the entire related rate of change curve) because we can get any specific points on the related rate of change curve that we desire with Frégier's theorem. More specifically, the slope of the tangent line to a curve at some value x is the instantaneous rate of change of the curve at that point. If the original curve depicts a particle's velocity, for example, then the instantaneous rate of change at some value x is that particle's acceleration. Having said this, using Frégier's theorem is not the preferred way to get the particle's acceleration. We would prefer calculus because it is less laborious, and it gives us the entire instantaneous rate of change curve. However, Frégier's theorem does provide an algebraic way (i.e. without calculus) to get the instantaneous rate of change at any point we choose on a conic section, which boosts the algebraic skill level of the student. Thus, we have presented the pros and cons of Frégier's theorem. We should mention that the slope function and perpendicular slope function depicted in *The Deluxe Toolkit* chapter were obtained using calculus. Thus, the student benefited from calculus without having to do the calculus part of the computation to get those functions.

Frégier's Theorem (1816): Given a conic section, a point P on it, and a right angle with its vertex located at P rotating freely. If the lines comprising the right angle intersect the conic in two points, the line determined by those two intersection points will pass through a fixed point Q. The line PQ is a normal line to the conic at P.

Reading over this theorem, we might have a couple of questions. The first may be, "What does it mean to rotate the right angle freely?" A right angle is comprised of two perpendicular lines. If one of the lines has a slope of m, then the other line making up the right angle has a slope of $m_\perp = \dfrac{-1}{m}$. As long as the choices for m and $m_\perp$ provide the required pair of intersection points to the conic, then the line through those two intersection points will pass through the point Q.

The second question may be, "What is a normal line through P?" A normal line is perpendicular to the tangent line at the point P. If we use the subscripts *tan* and *norm* when we wish to reference the tangent line and normal line, respectively, we have the relationship $m_{tan} = \dfrac{-1}{m_{norm}}$. This is just the perpendicular slope equation, but it applies because these two slopes are perpendicular.

We will start with the parabola for an example. Suppose we want the find the equation of the tangent line to the parabola $y = -4x^2 + 3x - 5$ at the point (2, -15). The situation is captured in Figure 1.

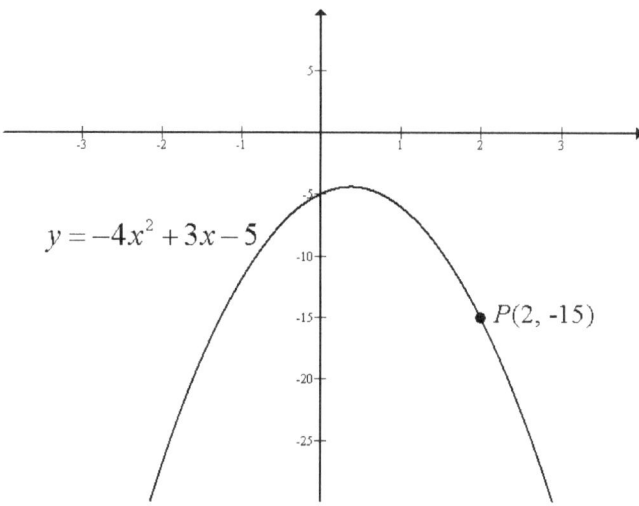

Figure 1: The Given Parabola and Point P

Frégier's theorem makes no demands about the orientation of the right angle at P. As long as we find a right angle that intersects the conic section in two places, we're good. The only line that a parabola cannot beat is a vertical line. Any other slope chosen for a line through P, no matter how ridiculously large it is, will eventually be overtaken by a parabola producing a second intersection point (with P being the first intersection point between the line and the parabola). Having said that, we need to perform the algebra with our choices of the pair of slopes comprising the right angle. Thus, we should choose the simplest slopes that come to our imaginations. I am thinking of ± 1 for the first pair of perpendicular slopes making the right angle. The other pair of perpendicular slopes can be 2 and -1/2.

192

Let's find the line ℓ_1 going through the first pair of slope choices for the right angle. We need to find the equation of the line through the point (2, -15) with a slope of 1. Performing this feat yields the line $y = x - 17$. The other slope in this pair is -1. Thus, we need to find the equation of the line through the point (2, -15) with a slope of -1. This situation yields the line $y = -x - 13$. Next, we need to find the two intersection points that this right angle makes with the conic section. Once we accomplish this, we can find the equation of the line through those two points to yield ℓ_1. To that end we set up the system of equations with the conic and the line with a slope of 1. We also set up the system of equations with the conic and the line with a slope of -1.

$$\begin{cases} y = -4x^2 + 3x - 5 \\ y = x - 17 \end{cases} \qquad \begin{cases} y = -4x^2 + 3x - 5 \\ y = -x - 13 \end{cases}$$

The left system yields the solutions (2, -15), (-3/2, -37/2). Of course, we understand that the second point is the one we will use to find the line ℓ_1. Solving the other system yields the solutions (2, -15), (-1, -12). Thus, the second point is the other point needed to find the line ℓ_1.

Our next problem, then, is to find the equation of the line through the two points (-3/2, -37/2) and (-1, -12). We first find the slope through these two points, m_1. Then, we finish performing this feat to yield the line $\ell_1 : y = 13x + 1$. Going through a similar process for the other right angle with slopes 2 and $-1/2$ yields the line $\ell_2 : \dfrac{29}{2}x + \dfrac{23}{8}$. However, I would like to comment on the algebra to get to this second line.

When we choose $m = 2$, and send the line with this slope through $P(2, -15)$, we get the line $y = 2x - 19$. Generating the system of equations involving this line and the parabola yields an equation in the x variable that after simplification is $4x^2 - x - 14 = 0$. This equation factors. We do not have to use the AC method (also called the Factor by Grouping method) to find how this equation factors. The reason why we do not have to do this is because we know that the line goes through $P(2, -15)$. Hence, we know one of the roots automatically, namely $x = 2$. If $x = 2$ is a root of $4x^2 - x - 14 = 0$, then $(x - 2)$ is a factor. We construct the other binomial by dividing $4x^2/x = 4x$ and $-14/-2 = 7$. Thus, the other binomial is $(4x + 7)$. It is always prudent to check the middle term when doing this shortcut just to make sure everything is OK.

Similarly, when we do this for the $m = -1/2$ slope, we wind up with the equation $8x^2 - 7x - 18 = 0$. We benefit because we already know one root is $x = 2$. Thus, $(x - 2)$ is one of the two binomial factors. The other one is obtained by dividing $8x^2/x = 8x$ and

-18/-2 = 9 yielding (8*x* + 9). Furthermore, because we knew one root, we also know that these equations would factor in the first place. Thus, because we knew that these lines passed through the point *P*(2, -15), it made a few nice reductions in the amount of work we had to do to find the two intersection points that yielded the line ℓ_2.

Now that we have ℓ_1 and ℓ_2, we need to find the intersection point between them. This will yield the point *Q*, which is on the normal line to the parabola at *P*. As we discussed, the line *PQ* is perpendicular to the tangent line to the parabola at *P*. Thus, we will find $\ell_1 \cap \ell_2$. This notation says to find the intersection point between the objects ℓ_1 and ℓ_2. Since these objects are both lines, this translates to a system of two linear equations. The solution to this system is *Q*(-5/4, -61/4), which can be found by standard methods.

Next, we find the equation of the line between *P* and *Q*. This line is the normal line to the point *P* on the parabola. We compute the slope as follows:

$$m_{norm} = \frac{P_y - Q_y}{P_x - Q_x}$$

This slope turns out to be 1/13. We can now run this slope through either *P* or *Q* to get the equation of the normal line. We will depict both versions of the equation of the normal line through *P*.

$$y - P_y = m_{norm}(x - P_x)$$
$$y - Q_y = m_{norm}(x - Q_x)$$

Placing both of these equations into *y* = *mx* + *b* form will show that both of these versions of the line are actually the same line. This line is $y = \frac{1}{13}x - \frac{197}{13}$.

Once we have the point *Q*, we do not have to find the equation of the normal line. We do not even have to find the slope of the normal line. If we need the normal line or its slope, we know how to find it. Given that we have found the point *Q*, we can find the slope of the tangent line with the following equation.

$$m_{tan} = \frac{Q_x - P_x}{P_y - Q_y}$$

This works out to be -13. Observe that this is the negative reciprocal of 1/13, the slope of the normal line. Contrary to having two options to find the normal line, we can only get the equation of the tangent line through P by sending a line with the m_{tan} slope through P, not Q. If we do send a line with the m_{tan} slope through Q, we have made a mistake. Plotting this equation with graphing software would show the mistake pictorially. Hence, we get only one version of the equation for the tangent line.

$$y - P_y = m_{tan}(x - P_x)$$

This equation works out to be $y = -13x + 11$. We do, however, have a way of checking that this is the equation of the tangent line to the parabola at P. We set up a system of equations with the parabola and the tangent line equation. Solving this system yields a single solution of multiplicity 2. The graph of this scenario is Figure 2. The lines do not look perpendicular because the graph had to be scaled to get a nice picture. Also, it looks like Q is on the parabola, but do not be deceived by this. It is not on it. Most of the slopes have been labeled, but the one dotted line not labeled has a slope of -1/2.

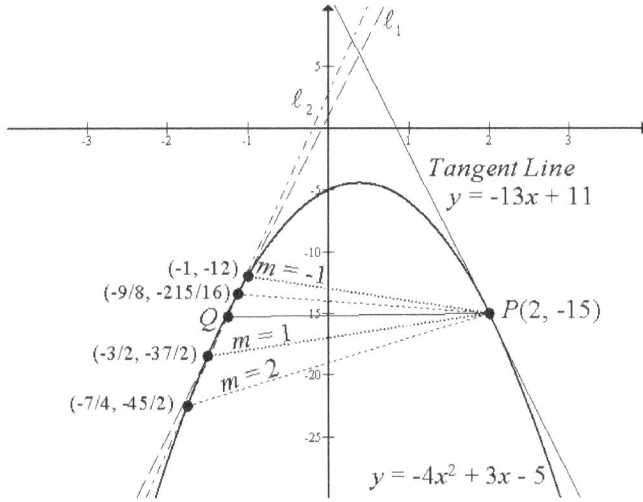

Figure 2: The Parabola and the Tangent Line at P.

We turn our attention now to the remaining conic sections: the circle, ellipse, and hyperbola. We will develop a method to handle all three of these cases. If we have a parabola, we will deal with it as shown above. However, if we are confronted with a circle, ellipse, or hyperbola, we will deal with any of these with the same method described next.

A parabola has an axis of symmetry. This is because if we fold the parabola over its axis of symmetry, we get the two halves directly on top of one another. Thus, the axis of symmetry is like a mirror with one half of the parabola being the mirror image of the other half. It is true that a circle has an infinite number of axes of symmetry. Any line going through the center of the circle is an axis of symmetry for that circle. However, we will just concern ourselves with horizontal and vertical axes of symmetry here. Thus, the circle has a horizontal and a vertical axis symmetry. An ellipse only has two axes of symmetry, and a hyperbola also has only two axes of symmetry.

We review the equations of the circle, ellipse, and hyperbola.

Circle: $(x-h)^2 + (y-k)^2 = r^2$

Ellipse: $\dfrac{(x-h)^2}{a^2} + \dfrac{(y-k)^2}{b^2} = 1, a \geq b$ or $\dfrac{(x-h)^2}{b^2} + \dfrac{(y-k)^2}{a^2} = 1, a \geq b$

Hyperbola: $\dfrac{(x-h)^2}{a^2} - \dfrac{(y-k)^2}{b^2} = 1$ or $\dfrac{(x-h)^2}{b^2} - \dfrac{(y-k)^2}{a^2} = 1$

Observe that the centers of the circle, ellipse, and hyperbola are all (h, k). Thus, the lines $x_{VA} = h$ and $y_{HA} = k$ are the equations of the vertical and horizontal axes of symmetry, respectively. The subscripts VA and HA stand for the vertical axis (of symmetry) and the horizontal axis (of symmetry), respectively.

Given the equation of the circle, ellipse, or hyperbola, and the point P for which we seek the equation of the tangent line, we would like to find the points P_{VA}, P_{HA}, and P'. These are the points obtained by reflecting the point P across the vertical axis of symmetry, the horizontal axis of symmetry, and both axes of symmetry, respectively. The point P' can also be obtained by reflecting the point P across the center.

The rationale for locating these points is that the points P_{VA} and P_{HA} create a right angle with P as its vertex. Thus, the line through P_{VA} and P_{HA} is ℓ_1. We have it that one of the lines formed from the intersection with the conic of the lines from a right angle at P is basically handed to us with little effort. The other line, ℓ_2, will take a bit more work; however, we can still take advantage of the twofold symmetry. We can use the line PP' as one of the lines making a right angle with its "other" intersection point with the conic at P'. The other line, then, making the right angle has to be perpendicular to PP' and runs through P. This is where we must invest some effort. The slope of this perpendicular line to PP' is the following:

$$m_\perp = \frac{P'_x - P_x}{P_y - P'_y}$$

We can find the equation of this line by running this slope through the point P as follows:

$$y - P_y = m_\perp (x - P_x)$$

We construct the system of equations involving this line with the conic to find the "other" intersection point. However, we already know one of the roots, namely, P_x. We employ the shortcut as described earlier to lighten our work to get this intersection point. We use this point along with P' to find the equation of ℓ_2. The intersection point of ℓ_1 and ℓ_2 is the point Q. Now that we have Q, we can employ the equations used in the parabolic version described earlier in this chapter to obtain the equation of the tangent line through P (and the normal line through P as well, if needed). Let's work through an example of this type. Find the equation of the tangent line to the circle $(x-3)^2 + (y+2)^2 = 16$ at the point $P(59/13, 22/13)$. The situation is shown in Figure 3.

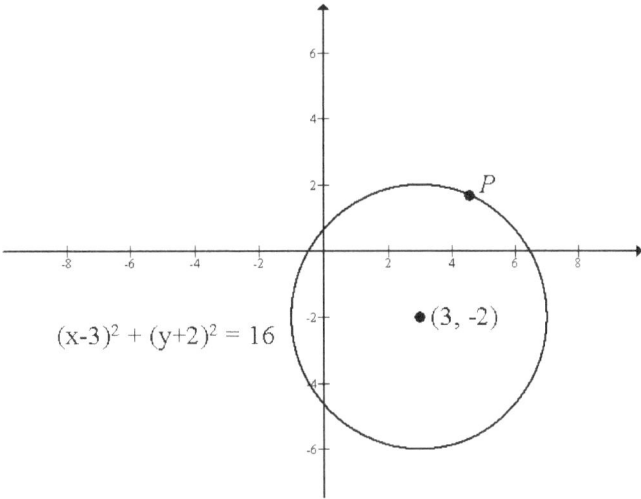

Figure 3: A Conic (Circle) with Center (3, -2) and Given Point P

The first thing we need to do is to find the two axes of symmetry $x_{VA} = h$ and $y_{HA} = k$. This is evident from the form of the equation as we see it. The two axes are $x_{VA} = 3$ and $y_{HA} = -2$. However, if the equation were not in this standard form, we would have to complete the square on both variables first to get it into this form.

197

Next, we find the points P_{VA}, P_{HA}, and P'. We accomplish this by inspecting the diagram. We see that P is to the right of the vertical axis of symmetry. Hence, we compute the distance that P is from the vertical axis as $d(P_x, h) = |P_x - h| = |59/13 - 3| = 20/13$. Then, we subtract this distance from h because we want to go that distance to the left of the vertical axis. Thus, we obtain $P_{VAx} = h - d(P_x, h) = 3 - 20/13 = 19/13$. Of course, P_{VAx} is the x-coordinate of the point P_{VA}. If P were on the left of the vertical axis, then we would add the distance to h to get P_{VAx}. Look at the picture to guide your efforts.

We only need to do one more distance computation. We need to find P_{HAy}. The y-coordinate of P_{VA} is the same as the y-coordinate of P. Thus, by finding P_{VAx}, we have P_{VA}. Similarly, once we have P_{HAy}, we will have P_{HA} because the x-coordinate of P_{HAy} is the same as the x-coordinate of P. Our last computation is, therefore, to compute the distance that P is from the horizontal axis. It is $d(P_y, k) = |P_y - k| = |22/13 - (-2)| = |22/13 + 2| = 48/13$. Since P is above the horizontal axis $y_{HA} = -2$, we subtract from k to get the y-coordinate of P_{HA}, which we notate as P_{HAy}. Thus, $P_{HAy} = k - d(P_y, k) = -2 - 48/13 = -74/13$. Again, if P were below the horizontal axis, we would have added to k; just look at the picture to guide your efforts. To find P', we just plug in our newly found x and y-coordinates. Thus, $P'(x, y) = P'(19/13, -74/13)$.

We now have the counterparts to $P(59/13, 22/13)$. They are $P_{VA}(19/13, 22/13)$, $P_{HA}(59/13, -74/13)$, and $P'(19/13, -74/13)$. We will update our picture to reflect the events thus far. See Figure 4.

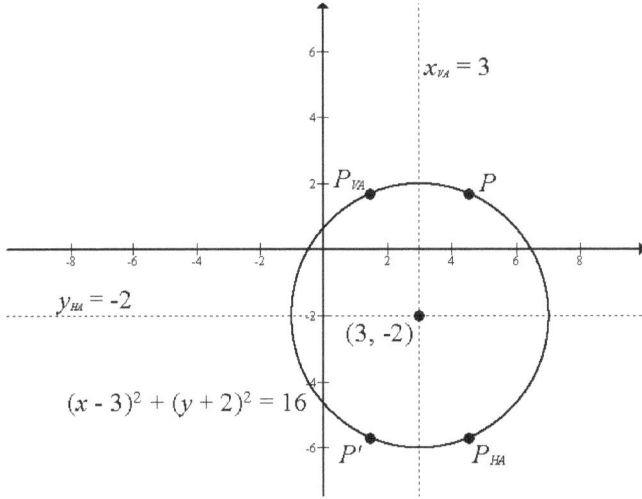

Figure 4: Given Point P Reflected Across Two Axes of Symmetry

198

Now, we can find the equation of ℓ_1. This line goes through the points P_{VA} and P_{HA}. We compute the slope, m_1; and then, we send it through either point or through the center. Doing this yields the equation $\ell_1 : y = \dfrac{-12}{5}x + \dfrac{26}{5}$.

Next, we will work on finding ℓ_2. One line forming the second right angle through P is the line PP'. Since the slope for the line ℓ_1 was $m_1 = -12/5$, the slope for the line PP' is $12/5$, its opposite. We already know the other intersection point with the conic, namely P'. We just need to find the equation of the perpendicular line to PP'. Thus, we know its slope is the negative reciprocal of the slope of the line PP'. This slope was the opposite of the slope for ℓ_1. Thus, understanding these symmetries saves a lot of computational effort. We see by inspection, then, that the slope of $m_{PP'\perp} = -5/12$. We send a line with this slope through P to yield the equation $y = \dfrac{-5}{12}x + \dfrac{43}{12}$. We construct a system of equations with this line and the conic section (our original circle in this example) to yield the other intersection point (59/13, 22/13). When doing this, remember that we already know one of the roots, namely $x = P_x$, which is $x = 59/13$ for this example. Thus, the quadratic equation in the x variable will factor yielding the other desired root. When showing your work, it is suggested also that a CAS be employed from time to time to get passed problematic points en route to an answer. Notice that this "other" intersection point is the original point P. This suggests a property of circles that we will get to momentarily. First, we will finish illustrating this method.

We now have the two intersection points of the conic with the second right angle with vertex at P, namely P'(19/13, -74/13) and (59/13, 22/13), which also happens to be P; however, we will not label it as P because this is, technically, the "other" intersection point. These two points will be used to find the equation of ℓ_2. We use the slope formula to find $m_2 = 12/5$. Then, we use the point-slope form of the equation of a line to yield the line $\ell_2 : y = \dfrac{12}{5}x - \dfrac{46}{5}$. Next, we can find the coordinates of the point Q by finding the intersection point of the two lines ℓ_1 and ℓ_2. In symbols, we write this as $\ell_1 \cap \ell_2$. The system of equations that we have is the following:

$$\begin{cases} \ell_1 : y = \dfrac{-12}{5}x + \dfrac{26}{5} \\ \ell_2 : y = \dfrac{12}{5}x + \dfrac{46}{5} \end{cases}$$

The solution to this system is $Q(3, -2)$. Notice that the point Q is the center of the circle. This suggests another property of circles that we will get to momentarily. Now that we have the point Q, we can finish the problem off with the methodology depicted for the parabola. This means that PQ is the normal line to the conic at P, and m_{tan} is the negative reciprocal of m_{norm}. Those equations and example computations were already depicted there. Thus, the method is complete for circles, ellipses, and hyperbolas. The equation of the normal line to this conic at P is $y = \dfrac{12}{5}x - \dfrac{46}{5}$ (which happened to be ℓ_2 because this was a circle, but this won't happen with hyperbolas and ellipses). The equation of the tangent line to this conic at P is $y = \dfrac{-5}{12}x + \dfrac{43}{12}$ (which happened to be the perpendicular line to PP' because this was a circle, but this won't happen with hyperbolas and ellipses). Our final diagram for this situation is Figure 5.

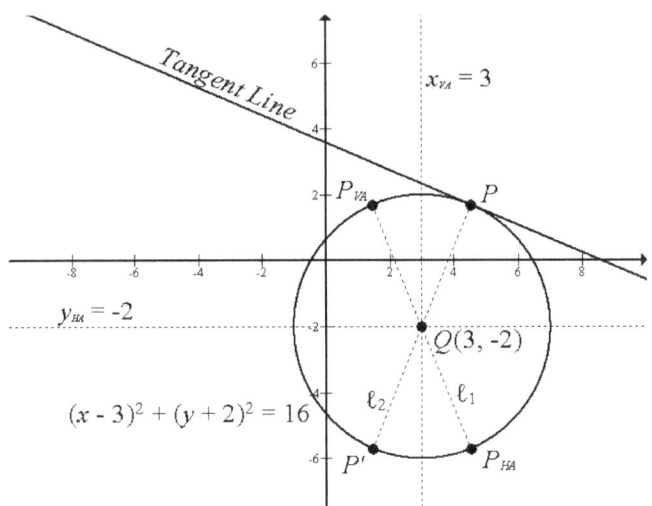

Figure 5: The Tangent Line to the Conic at Given Point P

We noted a couple of properties of circles that this problem demonstrated. The first property of the circle demonstrated was that the tangent line to a circle is perpendicular to the radius at that point. Hence, we could have found the negative reciprocal of the slope between the center of the circle and point P. Thus, we have the following for a circle:

$$m_{norm} = \frac{P_y - k}{P_x - h} \qquad\qquad m_{tan} = \frac{P_x - h}{k - P_y}$$

We can send these slopes through P to get the normal line and tangent line, respectively. Thus, finding the equation of the tangent line to a circle at a given point P is really easy with this understanding. It is definitely a heck of a lot better than using Frégier's theorem if we had forgotten that relationship. However, the above shortcut for circles does not apply to ellipses and hyperbolas. Thus, Frégier's theorem will have to be used for those situations.

The second property of circles suggested by this problem was that point Q was the center of the circle. This was no accident. Thales' theorem states that an angle in a semicircle is a right angle. This was proven by the ancient Greek mathematician Thales of Miletus (c. 625 BC – 547 BC). We learned this in *The Deluxe Toolkit* chapter as well as in the *Some Tangency Problems* chapter. A semicircle can be formed by cutting the circle in half with any line that goes through the center of the circle. Our first right angle to find ℓ_1 went through the center of the circle. This result is due to Thales' theorem. Our second right angle, which turned out to be composed of just one line, ℓ_2, also went through the center of the circle. Of course, if both of these lines went through the center of the circle, then their intersection point must be the center of the circle. That is why we saw that Q was the center of the circle.

Having stated the above two properties of the circle, we do not discount Frégier's theorem for finding the equation of the tangent line to a circle at a given point P. It did work just as any theorem works when it is properly applied. Frégier's theorem is our only option with a hyperbola and ellipse, however, as was already mentioned, unless we have a special circumstance for P. If the given point P is a vertex of the ellipse or hyperbola, then the slope of the tangent line through P will either be zero (i.e. $y = P_y$) or undefined (i.e. $x = P_x$). If P is not a vertex, then we must defer to Frégier's theorem.

Exercises

1. Use Frégier's theorem as outlined in this chapter to find the equation of the normal and tangent lines to the given conic at the indicated point P. In each case identify the point Q. Label the algebra and diagrams. Don't forget to use a CAS to help you. Some of these problems are real challenges.

a) $y = x^2$, $P(1, 1)$
b) $y = 2x^2 - 2x + 8$, $P(1, 8)$
c) $x^2 + y^2 = 25$, $P(3, -4)$
c) $x^2 + 4y^2 = 16$, $P(0, 2)$
d) $x^2 - 4y^2 = 16$, $P(4, 0)$

e) $\dfrac{x^2}{25} + \dfrac{y^2}{9} = 1$, P is the point in quadrant IV below the focus.

f) $\dfrac{x^2}{25} + \dfrac{y^2}{16} = 1$, $P(3, 16/5)$

2. Use any shortcuts to obtain the equation of the normal and tangent lines to the conic at the given point P. State clearly what shortcut you are employing and why that shortcut works.

a) $x^2 + y^2 = 81$, $P(-9, 0)$

b) $y^2 = 4px$, $P(0, 0)$

c) $(x - h)^2 = 4p(y - k)$, $P(h, k)$

d) $x^2 + y^2 = 1$, $P(3/5, -4/5)$

e) $(x - 4)^2 + (y - 3)^2 = 16$, $P(6.4, 0.2)$

The Polar Line and Conic Section Diameters

The polar line and conic section diameters each relate to the concept of tangency, but they do so in their own way. We will present the polar line first in this highlight chapter. Then, we will discuss the conic section diameters.

A polar line has associated with it a pole, which is a point. The pole can be anywhere in the plane of the conic section. However, we will restrict our discussion to poles that are on the conic section or outside the conic section. Higher-level courses than this one will plumb poles and polars deeper than this course.

A tangent line can also be a polar line. In this case, the pole associated with the polar line is the point of tangency. Thus, the pole in a case like this is the point of tangency. We have covered this case rather extensively elsewhere. Thus, we move to the case of the pole P being outside of the conic section. Note that the polar line will intersect the conic in two special places, which we will call points T_1 and T_2. The two lines PT_1 and PT_2 are the two tangent lines from the pole P to the conic section.

The device we learned in the *Some Tangency Problems* chapter to obtain the equation of the tangent line to a conic at a given point $P(x_0, y_0)$ is valid here for finding the polar line. We replace occurrences of x^2 with xx_0 and occurrences of y^2 with yy_0. We also replace the first power terms as well. Occurrences of $2x$ get replaced with $x + x_0$ and occurrences of $2y$ get replaced with $y + y_0$. Alternatively, we may replace an x with $(x + x_0)/2$ and a y with $(y + y_0)/2$. We then simplify the equation of the polar line.

Next, we construct a system of equations involving the polar line with the conic section. The solutions to this system are the points T_1 and T_2. Thus, if we start with a pole P and a conic section, then we perform the substitutions in the above paragraph to find the equation of the polar line. Then, we solve the system of equations involving the polar line and the conic section to find the points of tangency T_1 and T_2.

Two circles of unequal radii, as long as one is not completely inside the other, have a common pole. This common pole is on the line that goes between the centers of the two circles. This common pole is also on any line going through corresponding points on the two circles. Thus, the line going through the tops of the two circles travels through the common pole. As long as the two circles do not overlap (but they can be tangent to each other), the line that goes through the top of one circle and the bottom of the other circle will yield a polar line in the two circles such that we get the internal tangents to the two circles.

We will work an example. Find the common poles for finding the external tangent lines to the circles C_1: $x^2 + y^2 = 1$ and C_2: $(x-4)^2 + y^2 = 4$.

We will first find the common poles of C_1 and C_2. We need to find the line between the centers of the two circles. Thus, $(h_1, k_1) = (0, 0)$ and $(h_2, k_2) = (4, 0)$. The line through these two points is $y = 0$. Next, we find the top points on the two circles. For C_1, this is the point $(0, 1)$; and for C_2, this is the point $(4, 2)$. The line through these two points is $y = \frac{1}{4}x + 1$. The common pole for the external tangents for these two circles is the intersection point of these two lines, which is the point $P_E(x_0, y_0) = P(-4, 0)$. We would use the point $(4, -2)$ on C_2 to find the common pole for the internal tangents, which works out to be $P_I(4/3, 0)$. Now that we have the common poles, we use it to find the polar lines for C_1 and C_2 with respect to these poles. For the rest of this example, we will focus on just the external tangents. The internal tangents will be left as an exercise.

We transform the equation of C_1 to $xx_0 + yy_0 = 1$. Then, we plug in the coordinates of the common pole P_E to get $x(-4) + y(0) = 1$. We lose the y-variable, so we are left with the vertical line $x = -1/4$ as the polar line for C_1. We then expand the equation of C_2 to $x^2 - 8x + y^2 = -12$. Then, we transform this to the polar line equation $xx_0 - 4(x + x_0) + yy_0 = -12$. We make the replacements with the coordinates of P to get $x(-4) - 4(x + (-4)) + y(0) = -12$. We lose the y-variable again to get the vertical line $x = 7/2$ as the polar line for C_2. We now have the polar lines for C_1 and C_2, so we can now construct and solve the system of equations involving each circle with its polar line. The systems are depicted below.

$$\begin{cases} x^2 + y^2 = 1 \\ x = \dfrac{-1}{4} \end{cases} \qquad \begin{cases} (x-4)^2 + y^2 = 4 \\ x = \dfrac{7}{2} \end{cases}$$

The solution to the first system is $\left(\dfrac{-1}{4}, \dfrac{\pm\sqrt{15}}{4} \right)$, and the solution to the second system is $\left(\dfrac{7}{2}, \dfrac{\pm\sqrt{15}}{2} \right)$. These are the points of tangency, which we combine with the common pole P_E to produce the equations of the tangent lines through these two circles.

Next, for C_1 we find the equation of the tangent line through $T_1\left(\dfrac{-1}{4}, \dfrac{\sqrt{15}}{4} \right)$ and $P(-4, 0)$.

This is the line $y = \dfrac{\sqrt{15}}{15}x + \dfrac{4\sqrt{15}}{15}$. The equation of the tangent line through $P(-4, 0)$

$T_2\left(\dfrac{-1}{4}, \dfrac{-\sqrt{15}}{4}\right)$ and is $y = \dfrac{-\sqrt{15}}{15}x - \dfrac{4\sqrt{15}}{15}$. These two tangent lines that go through the

pole $P(-4, 0)$ also are tangent to C_2. The solution points of the second system above are the points of tangency on C_2, which we can label as T_3 and T_4. These two points will check in the two tangent line equations for C_1. We exhibit our situation in Figure 1. We will discuss the Monge line next—a theorem relating common poles of three circles.

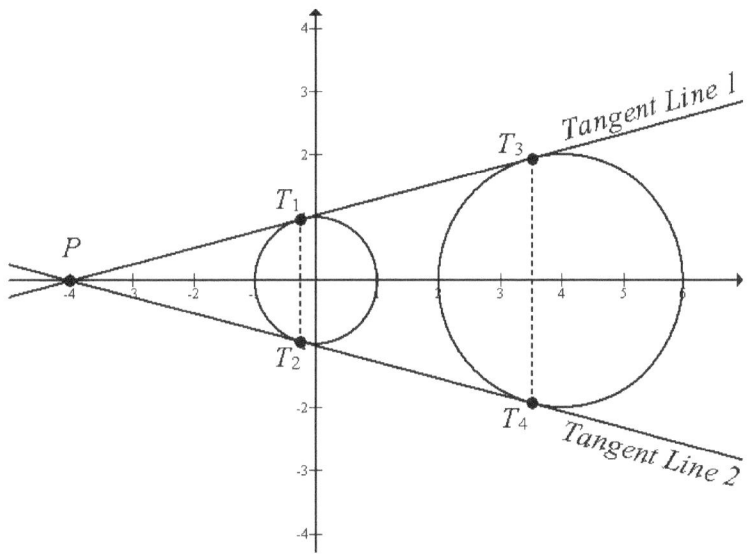

Figure 1: The Common Pole P, Polar Lines (Dotted), and Tangent Lines

Theorem (The Monge Line): If three circles are such that no two have the same radius and no circle is completely inside another, then the three common poles of three such circles taken pair wise are collinear. The line so formed is called the Monge line, after Gustav Monge (1746 – 1818), a French mathematician.

We will work an example of the Monge line. Find the Monge line for the three circles C_1: $x^2 + y^2 = 1$, C_2: $(x-4)^2 + y^2 = 4$, and C_3: $x^2 + (y-7)^2 = 9$.

We do not need to find the equations of the polar lines and tangent lines to find the pair wise common poles of the three circles. All that we have to do is to find the three common poles. The common pole for the circles C_1 and C_2 has already been found above.

It is $P_{1,2}(-4, 0)$. We proceed to find the other two common poles $P_{1,3}$ and $P_{2,3}$ in the same way as outline above.

Let us find the common pole $P_{1,3}$. The line through the centers of C_1 and C_3 is the line $x = 0$. The line through the top points of C_1 and C_3 will not help us here. Let us choose a different set of corresponding points on C_1 and C_3—the rightmost points. These points would be $(1, 0)$ on C_1 and $(3, 7)$ on C_3. We find that the line through these two corresponding points is $y = \dfrac{7}{2}x - \dfrac{7}{2}$. When $x = 0$ from the line through the two centers, we find that the common pole is $P_{1,3}(0, -7/2)$.

We will find the common pole $P_{2,3}$ next. The line through the center of C_2 and $C3$ is $y = \dfrac{-7}{4}x + 7$. Choosing the top point on the circles C_2 and C_3, we find that the line through those two points is $y = -2x + 10$. Finding the intersection point of these two lines yields the common pole $P_{2,3}(12, -14)$.

Next, we will find the equation of the Monge line for this situation. We now have the three pair wise common poles $P_{1,2}(-4, 0)$, $P_{1,3}(0, -1/2)$, and $P_{2,3}(12, -14)$. We use two of them to find the Monge line, and the third can be used as a check point to ensure that we did our work correctly. Using the first two, we find that the equation of the Monge line is $y = \dfrac{-7}{8}x - \dfrac{7}{2}$, which checks with the third point. We exhibit this situation in Figure 2. Note that a zoom square window was not able to be used.

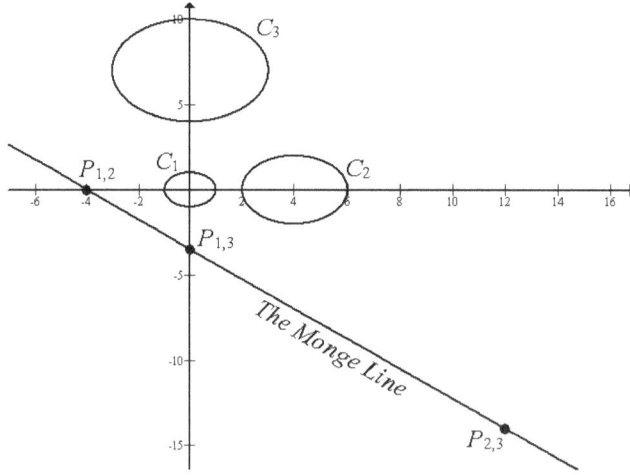

Figure 2: The Monge Line for Circles C_1, C_2, and C_3

We have discussed the polar line for conic sections in general, and we have given a circle specific application of common poles. We now turn our attention to conic section diameters. Yes, all four of the conic sections have diameters. We present the definition of a diameter.

Definition (Diameter): The line running through the midpoints of system of parallel chords of slope m for any conic section is called a diameter of the conic.

A reduced form of the general second degree equation that has no xy term is the following (and please note the 2's): $ax^2 + by^2 + 2cx + 2dy + e = 0$. We present an equation of the diameter of a conic section in terms of the constants a, b, c, and d for this reduced second degree equation. If the slope of the system of parallel chords is m, then the equation of the diameter of the conic section is $(ax + c) + m(by + d) = 0$.

We will work an example using the diameter of a conic section. Find the diameter of the ellipse $\dfrac{x^2}{9} + \dfrac{y^2}{4} = 1$ if the system of parallel chords has slope $m = 1/3$.

We do not have the constants in the reduced second-degree equation identified. Thus, we multiply both sides of the ellipse equation by 36. This changes the equation of the ellipse to $4x^2 + 9y^2 = 36$. The constant e is not carried over to the diameter equation, so we have gone far enough with this transformation. We can now make the necessary identifications: $a = 4$, $b = 9$, $m = 1/3$, and $c = d = 0$. We plug these into the diameter equation $(ax + c) + m(by + d) = 0$ to get $(4x + 0) + (1/3)(9y + 0) = 0$. After a slight amount of further work, we find that the equation of the diameter is $4x + 3y = 0$.

In order to obtain a tangent line using a diameter, we solve the system of equations involving the conic section and the equation of the diameter (for the given slope m). Depending on the conic section, we will get at least one point of intersection. Using the slope m of the slope of the system of parallel chords with the point or points of intersection of the diameter line with the conic section, we construct the equation or equations of the tangent lines at those intersection points. In these diameter of a conic section situations, the line we are finding will be the tangent line or tangent lines for the conic section at that point. Thus, we solve the below system of equations.

$$\begin{cases} \dfrac{x^2}{9} + \dfrac{y^2}{4} = 1 \\ 4x + 3y = 0 \end{cases}$$

We find that the solutions to this system are the points $\left(\dfrac{-3\sqrt{5}}{5}, \dfrac{4\sqrt{5}}{5} \right)$ and

$\left(\dfrac{3\sqrt{5}}{5}, \dfrac{-4\sqrt{5}}{5} \right)$. We then use these points with the given slope of $m = 1/3$ to find that the

equations of the two tangent lines are $y = \dfrac{1}{3}x + \sqrt{5}$ and $y = \dfrac{1}{3}x - \sqrt{5}$. We exhibit our

findings with Figure 3.

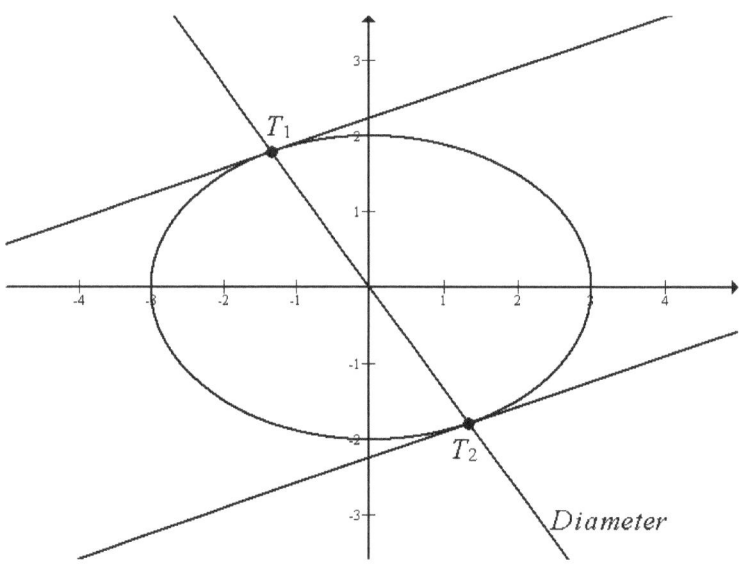

Figure 3: The Diameter of the Ellipse when $m = 1/3$.

As a further note, we could take a particular chord of the given slope m from the system of parallel chords and put it in a system of equations involving the diameter of the conic section for that slope. This would give us the intersection point of the diameter with that chord. We could find the midpoint of that chord to compare. Of course, the diameter is supposed to bisect any such chord, so the intersection point and the midpoint should be one and the same point.

We will discuss the concept of conjugate diameters for an ellipse. If a diameter of an ellipse bisects all chords that are parallel to another diameter of the ellipse, then the other diameter will bisect all chords that are parallel to the first diameter. We call two such diameters *conjugate diameters* of the ellipse. Let m_1 be the slope of one conjugate diameter and m_2 be the slope of the other conjugate diameter. Then, $m_1 m_2 = \dfrac{-b^2}{a^2}$ for a

208

wide ellipse and $m_1 m_2 = \dfrac{-a^2}{b^2}$ for a tall ellipse. Note that this is not really negative reciprocals because conjugate diameters are most of the time not perpendicular, but we may be tempted to think that these slopes are perpendicular because of the negative sign.

Exercises

1. Find the polar line of $y = 3(x-1)^2 + 5$ for the pole $P(1, 1)$. Then, find the equations of the two tangent lines to the parabola through P.

2. Find the polar line of $x^2 + \dfrac{y^2}{5} = 1$ for the pole $P(-2, 1)$. Then, find the equations of the two tangent lines to the ellipse through P.

3. Find the polar line of $x^2 + y^2 = 1$ for the pole $P(1, 2)$. Then, find the equations of the two tangent lines to the circle through P.

4. Find the polar line of $\dfrac{(x-2)^2}{9} - \dfrac{(y+5)^2}{25} = 1$ for the pole $P(3, 0)$. Then, find the equations of the two tangent lines to the hyperbola through P.

5. For the circles C_1: $x^2 + y^2 = 1$ and C_2: $(x+4)^2 + (y+5)^2 = 16$, find the common poles for their internal and external tangents. With those poles, find the associated polar lines (two for each circle). Then, find the equations of the two external tangents and two internal tangents. Provide a complete diagram and label all four points of tangency.

6. To problem 5, add circle C_3: $(x-1)^2 + (y-3)^2 = 4$. Find the Monge line, and label all three common poles for the external tangents. Is it true for these three circles that the three common poles for internal tangents lie on a line? Investigate this. [Hint: The three internal tangent poles do not lie on a line, but some groups of three points in the figure are collinear.]

7. For each of the conic sections in problems 1 – 4, if the system of parallel chords has slope $m = 2$, then find the diameters for these conic sections.

8. Find the equations of the tangent lines to the conic sections in problem 7 at the point or points of contact of the diameter with the conic section.

9. For the ellipse $\dfrac{x^2}{25} + \dfrac{y^2}{144} = 1$, find the diameter for the system of parallel chords of slope $m = 4$. Next, find the conjugate diameter. Pick a chord in each system, and show that those diameters bisect the chord.

Triangle Geometry

Clark Kimberling has a website called the *Encyclopedia of Triangle Centers* at the web address http://faculty.evansville.edu/ck6/encyclopedia/ETC.html. It is a deep website cataloging some 3000 plus important points of a triangle. Each triangle center is assigned a number in the format X(number). For example, the Prasolov point is X(68). We will not discuss triangle centers by their number assignment. Rather, we will discuss each by its name. Obviously, we are only going to scratch the surface here in our treatment. In addition, some of these points are collinear, which means we have some associated lines with the triangle to find as well. Finally, we have some associated triangles to the original triangle; and since three points determine a circle (as well as another triangle), we will get some other associated circles (triangles) to the original triangle. Other circles will arise for other reasons as well. Two sets of points on the triangle are called splitters and cleavers. We will learn what those are, how to find them, and the special points they match up with. Our task in this chapter is to find the locations of some of these important points, lines, triangles, and circles. All of this stuff is a branch of mathematics all its own, or at the very least a sub-branch of geometry. However, it has reached a level of maturity and depth that is very impressive. I have selected those things that I thought were within reach of the college algebra graduate. Nevertheless, this chapter is very rich mathematically in spite of the sparse selection from what we know in general about the triangle centers (i.e. the 3000 plus points).

Suppose we wish to place a triangle ABC onto the Cartesian plane, and we know the lengths of the three sides a, b, and c. One way to do this is to plot one end of side c (vertex A) at the origin, the other end of side c (vertex B) at $(c, 0)$, and the third point (vertex C) at (x, y) to be found by solving the system of equations below.

$$\begin{cases} x^2 + y^2 = b^2 \\ (x-c)^2 + y^2 = a^2 \end{cases}$$

Solving the above system yields a pair of points $(x, \pm y)$ where $x = \dfrac{-a^2 + b^2 + c^2}{2c}$, $y = \dfrac{2A}{c}$, where A is the area of the triangle with sides a, b, and c, which can be computed with Heron's formula (described in *The Deluxe Toolkit* chapter). We can modify our results by reflecting the triangle across the x-axis, y-axis, or both. In addition, we can move the vertices $A(x, y)$, $B(x, y)$, and $C(x, y)$ by the constants h and k to yield the new vertices $A(x + h, y + k)$, $B(x + h, y + k)$, and $C(x + h, y + k)$. Also, we can switch the places of x and y in the three points to yield the inverse triangle. These maneuvers give us tremendous latitude in getting a triangle with those side lengths onto the coordinate axes. Of course, one of the triangle's sides will be horizontal or vertical with this outlined protocol. If it is desired to have a different orientation than the eight possible presented here, the reader can consult *The Circle* chapter for information on how to do this.

When we are given three noncollinear points (or plot them ourselves), we may find the equations of the three lines that go through those points in pairs. These lines represent the sides of the triangle. Of course, the lines go on forever, so just the parts of the lines between the vertices are the sides of the triangle. We can compute the lengths of the sides using the distance formula. Typically, we label the vertices of the triangle with capital letters going around the triangle in a counterclockwise direction. The sides of the triangle will typically get the lowercase version of the letter of the vertex opposite to the line for that side.

Since we know that three noncollinear points determine a circle, we can find the equation of the circle going through them. The name for this circle is called the circumcircle. The radius of the circumcircle is usually denoted with an R. The center of the circumcircle is known as the circumcenter. Sometimes, the circumcenter is denoted with an O, but other times it may be denoted with a D.

Given a point P on the circumcircle of triangle ABC, the three reflections of P across the sides of the triangle all lie on a straight line. This line is called the Steiner line of the point P with respect to triangle ABC. We will hear of Steiner again shortly.

Next, we can find the midpoints of each side of the triangle. We can connect a vertex with a line to the midpoint of the opposite side. This line is called a median. Since a triangle has three medians, we can perform the feat three times. A wonderful thing happens when we do this, namely, that the three medians meet at a single point. We say the medians of a triangle concur. It is also a theorem that the distance from the vertex to the meeting point of the three medians is two-thirds of the length of that median. In addition, a median bisects the area of a triangle.

In general, when we connect a vertex to a point on the opposite side of a triangle, that line is called a cevian, named after Giovanni Ceva (1647 – 1734), an Italian mathematician. The median is a special type of cevian. Ceva proved the following theorem in 1678 about cevians now named after him; however, we also know that an eleventh century king in Saragossa, named Yusuf Al-Mu'taman ibn Hűd, also discovered and proved this same theorem, which was certainly well before Ceva.

Ceva's Theorem: Given: Triangle ABC and points on the sides of the triangle A', B', C' such that A' is opposite vertex A, B' is opposite vertex B, and C' is opposite vertex C. If the product of the ratios of corresponding parts of the sides of triangle ABC is one, then the cevians concur. In symbols, if $\dfrac{\overline{AC'}}{\overline{C'B}} \cdot \dfrac{\overline{BA'}}{\overline{A'C}} \cdot \dfrac{\overline{CB'}}{\overline{B'A}} = 1$, then the three cevians $\overline{AA'}, \overline{BB'}, \overline{CC'}$ concur. Furthermore, if the three cevians concur, then the ratios of corresponding parts of the sides of triangle ABC is one.

In the case of the three medians of the triangle, since each cevian splits that side of the triangle in half, both of the segments on a side are equal. Hence, the ratio of the two segments on each side is one. The product of three ones is, of course, one. Thus, the hypotheses (the prerequisite conditions needed to use the conclusion (the "then" part) of a theorem) of Ceva's theorem are met. Therefore, we conclude that the three medians of a triangle concur by Ceva's theorem. The place where the three medians meet is called the centroid of the triangle. We usually denote the centroid with a G; however, the notation centroid(x, y) is another way to notate it. The centroid has a physical meaning. It is the balancing point. If we cut out a similar triangle out of a sturdy material of the same thickness across the triangle, and we support the triangle at its centroid with a sharp point, it will balance at that point and not fall off of the support there.

We will discuss the Steiner circles and the Steiner point next. Given a triangle ABC, reflect the vertices A, B, and C across the centroid G yielding the new points A', B', and C', respectively. Then, the three circles $AB'C'$, $A'BC'$, and $A'B'C$ are called the Steiner circles. The three Steiner circles concur at the Steiner point St, the first two letters of his last name. Furthermore, the Steiner point lies on the circumcircle. If we reflect the Steiner point across the circumcenter, we get its diametrically opposed point. This point is called the Tarry point T.

The three midpoints of a triangle also make a triangle. This triangle is called the medial triangle. The medians of the medial triangle are the same as those of the original triangle. The medial triangle also shares the same centroid as the original triangle. The medial triangle is a similar triangle to the original triangle.

Something that may not be so obvious is finding three cevians that trisect the perimeter of the triangle. Let A' be opposite A, and B' be opposite B, and C' be opposite C such that the distances between the cut points is the same. Then, the cevians AA', BB', and CC' concur at the point known as the trisected perimeter point. We can show that the perimeter is trisected by computing the six distances involved and showing that the following relationship holds.

$$\overline{AB'} + \overline{AC'} = \overline{BA'} + \overline{BC'} = \overline{CA'} + \overline{CB'} = \tfrac{1}{3}P, \text{ where } P \text{ is the perimeter}$$

The next thing we might reasonably notice is that when we draw the altitudes of the triangle, the three lines always seem to intersect. Of course, this is no accident. The intersection of the three altitudes of a triangle is called the orthocenter. We usually denote it with an H, but we can also use orthocenter(x, y). The orthocentroidal circle of a triangle is the circle that has, as endpoints of a diameter, the orthocenter H and the centroid G.

The three feet of the altitudes also make a triangle. This triangle is called the orthic triangle. The incenter (discussed soon) of the orthic triangle is the same point as the orthocenter of the original triangle. In 1775, a problem known as Fagnano's problem asked for the triangle of minimum perimeter that could be inscribed in a given acute triangle, named after Giovanni Fagnano (1715 – 1797). Fagnano proposed the problem, and then he used calculus to solve the problem. It turns out that the orthic triangle is the solution to that problem. The orthic triangle is in perspective with the original triangle. The orthocenter is the center of perspectivity. Using Desargues' theorem, we find the axis of perspectivity, which we call the orthic axis. We saw an exercise on this in the *Desargues' Theorem* highlight chapter.

When we look at six of the points we have so far (i.e. the three midpoints of the sides and the three feet of the altitudes), we observe that they seem to all lie on a circle. This has been proven to be the case. Also on this circle are the midpoints of the line segments from the orthocenter to their associated vertices, which are called Euler points and are usually denoted E_A, E_B, and E_C. This is because Euler showed in 1765 that a certain special circle bisects any line segment from the orthocenter to the circumcircle. Thus, we have nine points that lie on this special circle. What shall we call this circle having nine points?

The mathematician Olry Terquen was the first to name the nine-point circle, and he proved the existence of the circle containing the nine points including the midpoints of the sides of the triangle, the feet of the altitudes of the triangle, and the midpoints of the line segments from the orthocenter to the vertex. However, Karl Wilhelm Feuerbach (1800 – 1834), a German geometer, is recognized as discovering six of the nine points on the nine-point circle. The center of the nine-point circle is usually denoted with an N; however, we can also use nine-point(x, y). The nine-point center is the midpoint between the orthocenter and the circumcenter. The radius of the nine-point circle is half that of the radius of the circumcircle.

Leonhard Euler (1707 – 1783) discovered a line that bears his name—the Euler line (1765). The circumcenter (D), orthocenter (H), and centroid (G) are collinear. Later, it was discovered that the nine-point center (N) also lies on the Euler line. In addition, the following relationships hold:

$$d(D, H) = 3d(D, G)$$
$$d(D, H) = 6d(G, N)$$
$$d(D, H) = 2d(N, H)$$

The next item on our list is the incircle. Just as there is a unique circle going around the triangle, we have a unique circle that fits inside the triangle. This circle just touches the sides of the triangle. When something touches in mathematics, we really mean that it is tangent. That is a specialized meaning of the word tangent because tangent has a different meaning in trigonometry. We are not doing trigonometry in this course, so tangent has only the one meaning for us here—touching. We use the symbol I to denote the incircle. The radius of the incircle is denoted with an r.

The upcoming formulas use the conventions that the coordinates of the vertices A, B, and C are (x_a, y_a), (x_b, y_b), and (x_c, y_c), respectively. Furthermore, the lengths of the sides opposite to vertices A, B, and C are a, b, and c, respectively. What do we need to find a circle? We need its center and its radius. The center-radius form of the equation of a circle is $(x - h)^2 + (y - k)^2 = r^2$. For the incircle we have the following:

$$h = \frac{ax_a + bx_b + cx_c}{a + b + c}, \quad k = \frac{ay_a + by_b + cy_c}{a + b + c}, \quad r = \frac{2A}{a + b + c}, \text{ where } A = \text{area of } \triangle ABC.$$

The center of the incircle lies on the angle bisectors for each of the three vertices. The median lines have already been discussed. A very important triangle center is called the symmedian or Lemoine point, named after Emile Lemoine (1840 – 1912), a French geometer. The symmedian point, usually denoted K, involving the angle bisectors and median lines will be treated in a theorem highlight chapter immediately following this chapter. Another point we will mention relating to the Euler line and incenter concepts is the Schiffler point of a triangle ABC, and this is the point of concurrence of the Euler lines of the four triangles ABC, ABI, AIC, and IBC, where I is the incenter of triangle ABC.

If we extend the sides of the triangle, we get three circles that are externally tangent to the triangle. We shorten the vocabulary used to say this a bit and just say excircles. We use the symbols I_a, I_b, and I_c to denote the excircles opposite to the vertices A, B, and C, respectively. Similarly, h, k, and r get appropriate subscripts to identify the specific excircles they belong to. For example, h_a is the x-coordinate of I_a. We get the following formulas for the three excircles.

$$h_a = \frac{-ax_a + bx_b + cx_c}{-a + b + c}, \quad k_a = \frac{-ay_a + by_b + cy_c}{-a + b + c}, \quad r_a = \frac{2A}{-a + b + c}, \text{ where } A = \text{area of } \triangle ABC.$$

$$h_b = \frac{ax_a - bx_b + cx_c}{a - b + c}, \quad k_b = \frac{ay_a - by_b + cy_c}{a - b + c}, \quad r_b = \frac{2A}{a - b + c}, \quad \text{where } A = \text{area of } \triangle ABC.$$

$$h_c = \frac{ax_a + bx_b - cx_c}{a + b - c}, \quad k_c = \frac{ay_a + by_b - cy_c}{a + b - c}, \quad r_c = \frac{2A}{a + b - c}, \quad \text{where } A = \text{area of } \triangle ABC.$$

Let D be the circumcenter of $\triangle ABC$, and let R be the circumradius. Let a, b, and c be the lengths of the $\triangle ABC$. Let d be the distance between the inradius and circumradius. Also, let I, I_a, I_b, and I_c represent the centers of the incircle and three excircles. It is known that the following relationships exist:

$$R = \frac{abc}{\sqrt{(a + b + c)(-a + b + c)(a - b + c)(a + b - c)}}$$

$$(\overline{DI})^2 + (\overline{DI_a})^2 + (\overline{DI_b})^2 + (\overline{DI_c})^2 = 12R^2$$

$$r_a + r_b + r_c - r = 4R$$

$$r r_a r_b r_c = A^2, \text{ where } A = \text{area } \triangle ABC.$$

$$r = \frac{abc}{4sR}, \text{ where } s = \frac{a + b + c}{2} \text{ is the semiperimeter.}$$

$$A = rs$$

$$d^2 = R(R - 2r) \text{ This is Euler's theorem.}$$

$$R \geq 2r \text{ This follows from Euler's theorem.}$$

Feuerbach added four more significant points to the nine-point circle when he discovered that the nine-point circle is tangent to the incircle and the three excircles. In fact, we call the point of tangency of the incircle with the nine-point circle the Feuerbach point. We call the other three points of tangency with the nine-point circle the Feuerbach triangle.

Now that we have the incircle and three excircles, we get a whole new world of combinations of intersection points with these four circles and the original triangle.

The first one we will discuss is the Yff point. Let I be the incenter of triangle ABC. Bisect angle AIB, and intersect this line with side AB. This yields the point C'. Similarly, bisect angle BIC, and intersect this line with side BC. This yields the point A'. Finally, bisect angle AIC, and intersect this line with side AC. This yields the point B'. *The DeluxeToolkit* chapter has advice on how to bisect an angle. Also, the incenter is on all three angle bisectors. Thus, we have two methods available to us to bisect the three angles. The lines AA', BB', and CC' concur at the Yff point.

The next triangle center we will discuss is the Gergonne point, named after Joseph Diez Gergonne (1771 – 1859), a French mathematician and geometer. The Gergonne point is usually denoted with the symbol Ge; however, we can also use Gergonne(x, y).

Gergonne Point and Triangle Theorem: Given $\triangle ABC$. Let the incircle of $\triangle ABC$ be tangent to the sides of $\triangle ABC$ at the points A', B', and C' opposite the vertices A, B, and C, respectively. Then, the cevians AA', BB', and CC' concur at the Gergonne point of $\triangle ABC$. The points A', B', and C' form a new triangle called the Gergonne triangle or contact triangle.

Next, we will talk about the three Nobb's points and the associated Gergonne line. We can pair up a side of a triangle with the side of the Gergonne triangle not having a vertex on that side. For example, for side AB of $\triangle ABC$, we pair that with side $A'B'$ of the Gergonne triangle. Observe that the letters match up. Each pairing like this results in an intersection known as a Nobb point. It is a fact that the three Nobb's points are collinear. In symbols, we say that AB and $A'B'$ meet at C'', AC and $A'C'$ meet at B'', and BC and $B'C'$ meet at A''. Observe that a Nobb point gets the "missing" letter. The Nobb's points A'', B'', and C'' are collinear; and we call this line the Gergonne line. The original triangle and Geronne's triangle are in perspective with the perspector at the Gergonne point and the perspectrix being the Gergonne line. The intersection point of Gergonne's line and Euler's line is Evans point.

Lazare Carnot (1753 – 1823), a French mathematician, proved the following relationship between the signed distances from the circumcenter (called D in this equation) to the sides of the triangle. Let A' be on the perpendicular from D to the side opposite A. Let B' be similar for vertex B, and let C' be similar for vertex C. Of course, R is the circumradius, and r is the inradius. Let sign A be negative only when the entire perpendicular segment from D to A' lies outside the triangle, except the point A'. Let this convention hold similarly for sign B and sign C. Then, the following equation is Carnot's theorem.

$$(\text{sign } A)\,\overline{DA'} + (\text{sign } B)\,\overline{DB'} + (\text{sign } C)\,\overline{DC'} = R + r$$

It is appropriate here to discuss the pedal triangle of a point because Carnot's theorem discusses a particular pedal triangle. If we choose a point on the plane that is not any of the three vertices of the original triangle and draw perpendiculars to the sides of the triangle extending the sides when appropriate, we get three new points. These three points make a new triangle, and we call this triangle the pedal triangle of that point with respect to that triangle. If the point chosen is the orthocenter, then the resulting pedal triangle is the orthic triangle. If the point chosen is the incenter, then the resulting pedal triangle is Gergonne's triangle. Does it ever occur that those three points do not form a triangle? In other words, is it possible that the three new points are collinear—a degenerate pedal triangle? Yes, it is possible. When the point chosen is on the circumcircle of the triangle, then the three points form a line, called the Wallace-Simson line, which we discussed in a highlight chapter earlier in this book.

In a similar vein to Carnot's theorem, we have Viviani's theorem, named after Vincenzo Viviani (1622 – 1703), an Italian mathematician and engineer. Viviani's theorem states that given an equilateral triangle $\triangle ABC$ and a point P, construct the three points A', B', and C'—the feet of the perpendiculars from P to the sides of the triangle, extending the sides when necessary. The sign convention is the same as it was for Carnot's theorem. The relationship of the signed distances from P to the triangle's sides and the altitude h of the equilateral triangle is the following equation.

$$(\text{sign } A) \ \overline{PA'} + (\text{sign } B) \ \overline{PB'} + (\text{sign } C) \ \overline{PC'} = h,$$

Another triangle, although not a pedal triangle, is the tangential triangle. The tangential triangle of triangle ABC is the triangle created by the tangent lines to the circumcircle at its three vertices. The medial triangle and the tangential triangle are in perspective with the perspectrix being the orthic axis. The Gob point is the perspector of the orthic triangle and the tangential triangle.

The Prasolov point is obtained by the following procedure. Let a', b', and c' be the vertices of the orthic triangle opposite the vertices A, B, and C of the original triangle ABC. Reflect the points of the orthic triangle across the nine-point center to obtain the vertices a'', b'', c'' of a new triangle. This new triangle is in perspective with triangle ABC, and its perspector is called the Prasalov point. Note that the three vertices of the orthic triangle are on the nine-point circle. When we reflect those vertices across the nine-point center, we are finding the diametrically opposed points to those vertices on the nine-point circle. Hence, we may use the procedure outlined in *The Circle* chapter to accomplish this. We studied the Prasolov point in the *Desargues' Theorem* chapter.

Next, we will discuss the Nagel point (1836), named after Christian Heinrich von Nagel (1803 – 1882), a German clergyman, geometer, and educator. Nagel's point deals with the three excircles I_a, I_b, and I_c. Let the point of tangency of $\triangle ABC$ with I_a be A'. Let the point of tangency of $\triangle ABC$ with I_b be B'. Let the point of tangency of $\triangle ABC$ with I_c be C'. Then, the cevians AA', BB', and CC' concur at the Nagel point of $\triangle ABC$. Another way exists to find the Nagel point, which is through the three splitters of a triangle. A splitter has a vertex at one end, and it bisects the perimeter at its other end. Let A' be the other endpoint of the splitter for vertex A. Let B' be the other endpoint of the splitter for vertex B. Let C' be the other endpoint of the splitter for vertex C. Then, the cevians AA', BB', and CC' concur at the Nagel point of $\triangle ABC$. Sometimes, the Nagel point is denoted with an Na, the first two letters of his last name.

The points A', B', and C' from the cevians for Nagel's point form another triangle called the extouch triangle because that's where the excircles touch the original triangle. An equation exists that computes the area of the extouch triangle, denoted A_T. Also, note that the area of the extouch triangle is the same area as the Gergonne triangle (also called the intouch triangle) above. $A_T = \dfrac{2Ar^2s}{abc}$, where s is the semiperimeter, r is the inradius, and A is the area of $\triangle ABC$. We have an equation that computes the area of Gergonne's triangle, denoted by A_{Ge}, is $A_{Ge} = \dfrac{rA}{2R}$, where A is the area of $\triangle ABC$.

Another triangle center is called the Weill point, W. It is the centroid of its intouch triangle. The line through the circumcenter O and the incenter I contains the Weill point. Furthermore, the relations $\dfrac{d(W,I)}{d(I,O)} = \dfrac{r}{3R} = \dfrac{(a+b-c)(a-b+c)(-a+b+c)}{6abc}$ hold.

The next point we will discuss is called the Mittenpunkt. The three lines going from the excenter through the midpoint of its associated side are concurrent in the Mittenpunkt. We will notate this point as M or as Mittenpunkt(x, y). This is another point that Nagel studied in 1836.

Another point we are interested in is the cleavance center of the triangle. This point is the intersection of the three cleavers of the triangle. A cleaver has one point at a midpoint of the triangle. The other point bisects the perimeter. A couple of notable properties of a cleaver is that it is parallel to one of the angle bisectors of the triangle, and the cleavance center is the same as the center of the Spieker circle. The Spieker circle is the incircle of the medial triangle of $\triangle ABC$. The Spieker center is the incenter of its medial triangle. We will notate this point as Spieker(x, y) or cleavance(x, y). We also use the symbol Sp, the first two letters of his last name.

The Spieker center, incenter, centroid, and Nagel points are collinear. A fixed set of proportions exists between these four points as well. The name for this line is the Nagel line. Using G for centroid, the fixed proportions on the Nagel line are the following:

$$d(Na, Sp) : d(Na, G) : d(Na, I) = 3 : 4 : 6.$$

We now know about the orthocenter and the Nagel point. A circle exists with these two points as endpoints on a diameter. This circle is called the Fuhrmann circle. The center of the Fuhrmann circle is usually denoted Fu, the first two letters of his last name. The radius of the Fuhrmann circle is the same as the distance between the circumcenter and the incenter. We also know these distance relationships: $d(I, Fu) = 2d(N, I) = R - 2r$, where N is the nine-point center. With the conventions that r_F is the radius of the Fuhrmann circle, R is the circumradius, and a, b, and c are the lengths of the sides of the original triangle ABC, W. Fuhrmann, a German mathematician, in 1890 showed the following:

$$r_F = R\sqrt{\frac{a^3 - a^2b - ab^2 + b^3 - a^2c + 3abc - b^2c - ac^2 + c^3}{abc}}$$

One of the triangle centers is called the de Longchamps point. Consider the circle with the circumcenter as its center and having the orthocenter as a point on the circle. Then, the de Longchamps point is the point diametrically opposite the orthocenter on this circle. Another way exists to find the de Longchamps point. Construct the anticomplimenary triangle (also called the antimedial triangle), which is the triangle that has the original triangle as its medial triangle. Then, the de Longchamps point is the orthocenter of this new triangle. We have a number of known distance relationships involving the de Longchamps point. We will represent these points with the following conventions. The de Longchamps point is L. The orthocenter is H. The circumcenter is D. The Gergonne point is Ge. The incenter is I. The nine-point center is N. The inradius is r. The distance relationships are the following:

$$LG = \frac{4}{3}DH, \ LH = 2DH, \ LN = \frac{3}{2}DH, \ LD = DH,$$

$$LGe = \frac{2(a^3 + b^3 + c^3 - a(b^2 + c^2) - (a^2 + bc)(b + c) - 2abc) \cdot IL}{(a + b + c)(a^2 + b^2 + c^2 - 2(ab + ac + bc))}$$

$$LI = \frac{1}{2r}\sqrt{(a - b)^2[a^2 + ab + b^2 - (a + b)c] + c^2(a - c)(b - c)}$$

In 1804, Benjamin Bevan wanted it proven that the circumcenter of $\triangle ABC$ was the midpoint of the incenter and the center of the circle going through the three excenters. Additionally, he wanted it proven that the radius of this circle (i.e. the radius of the circumcircle of the three excenters—and note that the triangle through the three excenters is called the excentral triangle) was twice the radius of the circumcircle of $\triangle ABC$. John Butterworth proved Bevan's problem that same year. However, we call the center of the circumcircle of the excentral triangle the Bevan point, not the Butterworth point. The Bevan point can be found another way. Run a line through each excenter perpendicular to its associated side. The point of concurrence of these lines is the Bevan point. We call the circle through the three excenters the Bevan circle.

We have an inradius theorem that relates the three altitudes of $\triangle ABC$ to the inradius, r. It is the following: $\dfrac{1}{r} = \dfrac{1}{\alpha} + \dfrac{1}{\beta} + \dfrac{1}{\chi}$, where α, β, χ are the lengths of the altitudes through vertices A, B, and C, respectively.

We have a few inequalities related to triangles to discuss. The first is $R \geq 2r$, which follows from Euler's theorem discussed earlier. The next is Pedoe's inequality (1943). This states that if in one triangle with sides a, b, and c with area f and if in a second triangle with sides A, B, and C with area F, then the following is true. If the two triangles are similar, then the inequality becomes equality; so Pedoe's inequality can be a test for similar triangles other than computing the ratios of corresponding sides to see if they are all equal.

$$A^2(-a^2 + b^2 + c^2) + B^2(a^2 - b^2 + c^2) + C^2(a^2 + b^2 - c^2) \geq 16Ff$$

The next inequality we will discuss is the Hadwiger-Finsler inequality (1937). This states that with a triangle with sides a, b, and c with area f, the following holds.

$$a^2 + b^2 + c^2 \geq (a-b)^2 + (b-c)^2 + (c-a)^2 + 4\sqrt{3} \cdot f$$

The last inequality we will discuss is Weitzenböck's inequality. This states that with a triangle with sides a, b, and c with area f, the following holds. If the triangle is an equilateral triangle, then the inequality becomes equality; so Weitzenböck's inequality can be a test for an equilateral triangle. Of course, if $a = b = c$, then the triangle is equilateral, which is the easiest test for an equilateral triangle.

$$a^2 + b^2 + c^2 \geq 4\sqrt{3} \cdot f$$

The next few special points are obtained differently than our approach has been thus far. We will put a shape outside the triangle on each of its sides, adjusting the size of the shape according to the length of that side. We did this sort of thing already in the *Some Square Erecting Theorems* chapter.

The first shape we will put on each side (facing outward, not inward) of the triangle is an equilateral triangle. Doing this results in three new points (which are the new vertices of the equilateral triangles called the external Fermat triangles), which we will call A', B', and C' being opposite to A, B, and C, respectively. The lines AA', BB', and CC' concur at the point known as the Fermat point or the external Fermat point, named after Pierre de Fermat (1601 – 1665), a French lawyer and amateur mathematician. We denote this point as F or as F_1 if we are going to discuss his other point.

Do not think that because he was an amateur mathematician that his stuff is simple. Proving Fermat's last theorem took the best mathematicians over three centuries. Fermat proposed his last theorem in 1637, and Andrew Wiles solved it in 1994. Thus, the world had to wait 357 years. That's a long time, and a number of excellent mathematicians had tried during those centuries to prove Fermat's Last Theorem and came up short. We will just deal with the particular point of a triangle bearing his name for now. The interested student can find an abundant amount of material on Fermat by just searching for his name on the Internet.

The significance of the Fermat point F is that the total distance from this point to the vertices is a minimum compared to other points chosen inside or on the boundary of the triangle. Thus, suppose three cities located at the vertices of a triangle want to build roads connecting their cities. The minimum expenditure would be to build roads from each city to the Fermat point. Choosing a different road plan results in more road having to be built. This ignores such things as lakes or mountains being in the paths of any of the roads, but barring this sort of geographical problem, the Fermat point is the best.

If we erect equilateral triangles inwardly, we get three new points A'', B'', and C'', which are the new vertices of the triangles (called the internal Fermat triangles) opposite to vertices A, B, and C, respectively. The three lines AA'', BB'', and CC'' concur at the second Fermat point or internal Fermat point. We denote this point as F_2. The line between the two Fermat points is called the Fermat axis. The midpoint of the centroid and orthocenter lies on the Fermat axis. Furthermore, the midpoint of the two Fermat points lies on the nine-point circle. The two Fermat points lie on a special circle called Lester's circle, after June Lester. Lester's circle is discussed in the *Concyclic Points* chapter (following the symmedian point highlight chapter), which goes into detail (with pictures) about the outward erected and inward erected equilateral triangles.

Two points, called the Wernau points $W+$ and $W-$, are formed in the following way. We get $W+$ with the external Fermat triangle's new points A', B', and C'. We form the circles $AB'C'$, $A'BC'$, and $A'B'C$. These three circles intersect at $W+$. Similarly, for the internal Fermat triangles new points A'', B'', and C'', we form the circles $AB''C''$, $A''BC''$, and $A''B''C$. These three circles intersect at $W-$.

The next point we will discuss is the Napoleon point, named after the French general Napoleon Bonaparte (1769 – 1821), who became the Emperor of France. The experts are in disagreement on whether or not Napoleon knew enough math to have discovered this point. Apparently, Napoleon excelled at mathematics at military school, so maybe it is true. However, the earliest writings found about this theorem originate after Napoleon's death, so that is certainly a problem. Nevertheless, we will not enter into that dispute in this book. Napoleon's theorem states that if equilateral triangles are erected outwardly on the sides of a given triangle, then their centers form another equilateral triangle we call the outer Napoleon triangle. The vertices of the outer Napoleon triangle are labeled A', B', and C', which are opposite the original triangle's vertices A, B, and C, respectively. We then get the three lines AA', BB', and CC'. These three lines intersect at Napoleon's point, also known as the first Napoleon point. If we find the circle that circumscribes the outer Napoleon triangle, we get the outer Napoleon circle with radius

$$r = \frac{\sqrt{a^2 + b^2 + c^2 + 4\sqrt{3} \cdot Area(\triangle ABC)}}{3\sqrt{2}}.$$ Its center is the centroid. A property of the three

circles that circumscribe the three outwardly erected equilateral triangles is that they meet at the first Fermat point, which Napoleon is said to have proved.

As with the Fermat point, we can construct the equilateral triangles inwardly. The centers of the three inward equilateral triangles make what is known as the inner Napoleon triangle, which is equilateral. The vertices of the inner Napoleon triangle are labeled A', B', and C', which are opposite the original triangle's vertices A, B, and C, respectively. We then get the three lines AA', BB', and CC'. These three lines intersect at the second Napoleon point. Constructing the circle around the inner Napoleon triangle

yields the inner Napoleon circle with radius $r = \dfrac{\sqrt{a^2 + b^2 + c^2 - 4\sqrt{3} \cdot Area(\triangle ABC)}}{3\sqrt{2}}$. The

center of the inner Napoleon circle is the centroid. Thus, the inner and outer Napoleon circles are concentric. A property of the three circles that circumscribe the three inward equilateral triangles is that they meet at the second Fermat point. Regarding the two Napoleon triangles, we have Area(outer Napoleon triangle) – Area(inner Napoleon triangle) = Area($\triangle ABC$). Formulas for the areas of the Napoleon triangles are Area =

$\dfrac{1}{2}\triangle ABC \pm \dfrac{\sqrt{3}}{24}(a^2 + b^2 + c^2)$, where the plus version is the area of the outer Napoleon

triangle, and the minus version is the area of the inner Napoleon triangle.

223

We turn our attention now to the Vecten points. The outer Vecten point V_1 arises when we erect squares outwardly on the three sides of the original triangle, and connect the vertices A, B, and C to the centers of the squares erected on the opposite side to it in turn. The centers of these squares are labeled A', B', and C', respectively. The three lines AA', BB', and CC' intersect at the Vecten point, also called the outer Vecten point. The centers of the three outwardly erected squares form a triangle known as the outer Vecten triangle. The area of the outer Vecten triangle is $Area = Area(\triangle ABC) + \dfrac{1}{8}(a^2 + b^2 + c^2)$.

The side lengths of the outer Vecten triangle are the following:

$$a' = \sqrt{\dfrac{Area(\triangle ABC) + b^2 + c^2}{2}}, b' = \sqrt{\dfrac{a^2 + Area(\triangle ABC) + c^2}{2}}, c' = \sqrt{\dfrac{a^2 + b^2 + Area(\triangle ABC)}{2}},$$

where a', b', and c' are opposite A', B', and C' on the outer Vecten triangle. The line segment AA' is perpendicular to $B'C'$. In symbols, we write $AA' \perp B'C'$. Also, $BB' \perp A'C'$ and $CC' \perp A'B'$. In addition, $AA' = B'C'$. Also, $BB' = A'C'$ and $CC' = A'B'$.

As with the Fermat and Napoleon points, we can erect the Vecten squares inwardly. Doing so results in the three points A'', B'', and C'' at the center of the squares, which are the vertices of a triangle called the inner Vecten triangle. The three lines AA'', BB'', and CC'' intersect at the inner Vecten point V_2. The area of the inner Vecten triangle is $Area = Area(\triangle ABC) - \dfrac{1}{8}(a^2 + b^2 + c^2)$. The side lengths of the inner Vecten triangle are the following:

$$a'' = \sqrt{\dfrac{-Area(\triangle ABC) + b^2 + c^2}{2}}, b'' = \sqrt{\dfrac{a^2 - Area(\triangle ABC) + c^2}{2}}, \text{ and}$$

$$c'' = \sqrt{\dfrac{a^2 + b^2 - Area(\triangle ABC)}{2}},$$

where a'', b'', and c'' are opposite A'', B'', and C'' on the inner Vecten triangle. Note that the perpendicular relationships and distance relationships that were established with the outer Vecten square center points are not true with the inner Vecten square center points.

To find the center points A' and A'' of the inward and outward Vecten squares, a single system of equations can be set up involving the perpendicular bisector of side BC and a circle centered at the midpoint of side BC of radius equal to half the length of side

BC. We discussed this setup a little differently in the *Some Square Erecting Theorems* chapter, but in that chapter we were only interested in outward erected squares. Doing it the way suggested in this paragraph leaves no extraneous solutions behind because both solutions have meaning.

A line that we will call the Vecten line or the Vecten axis is the line between the two Vecten points V_1 and V_2. Also on this line is the nine-point center, which can be used as a check point. The Euler line also contains the nine-point center, as we learned earlier. Thus, the Euler and Vecten lines intersect at the nine-point center.

The Soddy line of a triangle is the line joining the incenter I and the Gergonne point *Ge*. The Euler-Gergonne-Soddy triangle is the triangle composed of the Euler, Gergonne, and Soddy lines. This triangle is always a right triangle because the Soddy line and Gergonne line are always perpendicular. One of the three vertices of this triangle is the Evans point, *Ev*, which is the intersection of the Euler line and Gergonne line (we have already mentioned this). Another vertex is called the Fletcher point, *Fl*, which is the intersection point of the Gergonne line and the Soddy line. The third vertex of the Euler-Gergonne-Soddy triangle is the de Longchamps point, *L*, which is the intersection point of the Euler and Soddy lines (and we have already mentioned this point as well, but we did not say it was the intersection point of two important lines).

I would like to conclude this chapter with some things that we will not discuss further beyond their mention here. Morley's triangle is the triangle arrived at by trisecting the angles of the original triangle and connecting the outer lines. Finding Morley's triangle involves a general cubic equation. Kiepert's and Feuerbach's hyperbolas are rectangular hyperbolas going through four points each. However, they were not included because they would involve rotation of axes—a topic in the second term of calculus requiring knowledge of trigonometry. Kiepert's parabola also involves rotation of axes. An interesting point is the Kenmotu point (the center of the Kenmotu circle), named after Shoto Kenmotu (1790 – 1871), a Japanese mathematician. We get the Kenmotu point by inscribing three congruent squares internally to the triangle, a very difficult mathematical task. The interested student can search for any of these "not discussed" topics on the Internet.

It is my hope that this excursion into *Triangle Geometry* shows the students that mathematicians over time have seen a lot of things when they stare at three points in the plane. The next time the student sees a triangle, maybe some of these things will come to mind. I am impressed that these mathematicians have seen so many things from three simple dots on a page.

Exercises

Note that this exercise set seems small, but the chapter's concepts are enormous. A real exercise set would be almost overwhelming, so at least for this first edition of the book, I will keep this exercise set as it is. I will endeavor to expand it later in the next edition with high quality exercises. A great deal of thought went into the layout of this chapter. But, perhaps this chapter should be split into two or three chapters. Then, real justice could be done with these concepts.

1. Find the equation of the Euler line for the triangle with vertices at $A(0, 4)$, $B(-3, -8)$, and $C(6, -5)$. Confirm the points that are on it as shown in the text.

2. Combine Carnot's and Viviani's theorems. Explain specifically what happens when we do this. Draw any relevant conclusions by deducing the appropriate equations. Solve the equations for each variable present.

3. Writing in mathematics: Plot an 8-15-17 Pythagorean triangle (the author's favorite Pythagorean triangle), and find out as much as you can about it utilizing the ideas in this chapter. If the student desires a triangle with all sides tilted, consult *The Circle* chapter. Otherwise, use the method for plotting a triangle presented in this chapter. Your instructor may narrow down the topics to include in your paper because this chapter contains such a wealth of information. Answers will vary.

4. Writing in mathematics: Plot a 9-10-17 Heronian triangle (the author's favorite Heronian triangle), and find out as much as you can about it utilizing the ideas in this chapter. If the student desires a triangle with all sides tilted, consult *The Circle* chapter. Otherwise, use the method for plotting a triangle presented in this chapter. Your instructor may narrow down the topics to include in your paper because this chapter contains such a wealth of information. Answers will vary.

The Symmedian Point

The symmedian point is also called the Lemoine point, named after Emile Michel Hyacinthe Lemoine (1840 – 1912), a French geometer and engineer, who discussed it extensively in 1873. However, the point was also studied by L Huillier in 1809. Grebe studied it in 1847, and others had studied it as well before Lemoine did in 1873. However, it was the paper presented to the Association Francaise pour l'Avancement des Sciences in Lyon by Lemoine in 1873 that helped to lay the foundations of the modern geometry of the triangle. Because the point had been studied before by others, and some naming problems resulted in various countries wanting to name it after their mathematician, Robert Tucker solved the naming problem by introducing the name symmedian point after the already established term symmedian line. The name "symmedian point" has largely been accepted by the mathematical community.

What does symmedian mean? A symmedian line is a reflection of a median line across its corresponding angle bisector. Thus, it is a blending of the words symmetric and median. We usually denote the symmedian point with the letter K.

Theorem: The three symmedians of a triangle concur at the symmedian point K.

We can find the symmedian point K by solving the system of equations involving two of the three symmedian lines. We can check our work by plugging the coordinates of the symmedian point into the third symmedian line equation. If we get an identity, then we know we have found the symmedian point. If we get a contradiction, then something is wrong somewhere in our work. We will work an example to find the symmedian point of a given triangle.

Example: Find the symmedian point of the triangle with vertices at $A(9, 1)$, $B(5, 4)$, and $C(1, 1)$.

The first thing we need to do is to find the pairs of median lines and angle bisector lines. So we will set to work on that. We will start with vertex A. The median line through point A also goes through the midpoint of the opposite side BC. We will compute that: $M_{BC} = \left(\frac{1}{2} \cdot (5+1), \frac{1}{2} \cdot (4+1) \right) = \left(3, \frac{5}{2} \right)$. Thus, the slope of the median line through $A(9, 1)$ is $m = \frac{y_2 - y_1}{x_2 - x_1} = \frac{\frac{5}{2} - 1}{3 - 9} = \frac{3}{-12} = \frac{-1}{4}$. We can find the equation of the median line through A with the point slope form of the equation of a line. We can denote the median line through A as median$_A$ or even med$_A$. We compute as follows:

$$y - y_1 = m(x - x_1)$$

$$y - 1 = \frac{-1}{4}(x - 9)$$

$$y = \frac{-1}{4} \cdot x + \frac{9}{4} + 1 = \frac{-1}{4}x + \frac{13}{4}$$

Next, we need to find the equation of the angle bisector through the point A. We can denote the angle bisector through point A as angle bisector$_A$ or even bis$_A$. Finding this line can be accomplished with the equations to find the incenter because all three of the angle bisectors go through the incenter. Recall from the *Triangle Geometry* chapter that the incenter is located at (h, k) where $h = \dfrac{ax_a + bx_b + cx_c}{a + b + c}$, $k = \dfrac{ay_a + by_b + cy_c}{a + b + c}$. So to find (h, k), we need to know the lengths of three sides of the triangle, and use them in the appropriate places in the equations above. We compute the lengths of the three sides:

$$a = BC = \sqrt{(5-1)^2 + (4-1)^2} = 5$$

$$b = AC = \sqrt{(9-1)^2 + (1-1)^2} = 8$$

$$c = AB = \sqrt{(9-5)^2 + (1-4)^2} = 5$$

We observe that we have a Heronian triangle on our hands. We can now compute (h, k) to find that it is the point $(5, 7/3)$.

$$h = \frac{5 \cdot 9 + 8 \cdot 5 + 5 \cdot 1}{5 + 8 + 5} = \frac{90}{18} = 5 \qquad k = \frac{5 \cdot 1 + 8 \cdot 4 + 5 \cdot 1}{5 + 8 + 5} = \frac{42}{18} = \frac{7}{3}$$

At this point, we are in a position to find the angle bisector through point A. We compute the slope of the angle bisector as $m = \dfrac{\frac{7}{3} - 1}{5 - 9} = \dfrac{4}{-12} = \dfrac{-1}{3}$. Next, we send a line with this slope through point A to get $y = 1 + \dfrac{-1}{3}(x - 9) = \dfrac{-1}{3}x + 4$.

We may now state our next problem as reflect med_A: $y = \dfrac{-1}{4}x + \dfrac{13}{4}$ across bis_A: $y = \dfrac{-1}{3}x + 4$. We already know that these lines meet at the point $A(9,1)$. We follow the advice in *The Deluxe Toolkit* chapter to accomplish the reflection feat. We choose a point P on the line that is not the mirror (the mirror in this problem is the angle bisector) that is not the intersection point A. If we let $x = 1$ on the median line, we get $y = 3$. Hence, we choose $P(1, 3)$ as a nice integer point. Next, we find the perpendicular line to the mirror (the angle bisector) through the chosen point $P(1, 3)$. The slope of this line is the negative reciprocal of -1/3, which is $m = 3$. Hence, we get the line $y = 3 + 3(x - 1) = 3x$.

We now want to find the point Q, which is the intersection point of the mirror and this perpendicular line. Thus, we need to solve the following system of equations:

$$\left\{ \begin{array}{l} y = \dfrac{-1}{3}x + 4 \\ y = 3x \end{array} \right\}$$

We thus find that the point Q is (1.2, 3.6). Next, we need to reflect $P(1, 3)$ across $Q(1.2, 3.6)$ to get the point P'. Again, perusing the advice in *The Deluxe Toolkit* chapter, we find the equation of the circle having Q as its center and P as a point on it. Well, the only concern with this is how far is it from P to Q? This is the radius r of the circle. We compute the distance from P to Q as $d = \sqrt{(1.2 - 1)^2 + (3.6 - 3)^2} = \sqrt{0.4} = r$. We can construct the equation of the desired circle as follows: $(x - 1.2)^2 + (y - 3.6)^2 = 0.4$. Note that we square r in the circle equation. We can reuse the equation of that perpendicular line. Sometimes, we get to save a step in math because we don't have to redo something that we already have at our disposal. Thus, the equation of the perpendicular line was $y = 3x$. We need to find the two solutions to the system of equations involving the circle and the line, but we already know one of the solutions, namely $P(1, 3)$. The system is the following:

$$\left\{ \begin{array}{l} (x - 1.2)^2 + (y - 3.6)^2 = 0.4 \\ y = 3x \end{array} \right\}$$

A remark is in order about this system here. Observe that placing the $3x$ in for y in the circle equation yields $(3x - 3.6)^2$ for that term. We factor a 3 out as a 9 because of the squaring. This turns this term into $9(x - 1.2)^2$, which can be added to the first term. It may turn out that you get something like $9(1.2 - x)^2$, which is still OK because opposites squared are equal. Hence, these quadratic terms will combine enabling the solution to occur with the square root method without having to resort to the quadratic formula. Also, observe that 0.4 is not a perfect square, but when we divide it by the new coefficient of 10 on the quadratic term, it is a perfect square. This is made possible because P is a rational point on the circle, and we are finding its diametric opposite P'. As a result, I expected this system to produce a factorization of some sort, but I did not expect it to yield a matching quadratic term.

It is beautiful gems like this that make mathematicians smile when they are figuring out stuff. Our material has its tedium, but a surprising twist producing a shortcut is nice. It's like an archeologist at a dig site finding a fossil specimen completely intact without having to piece it together like a jigsaw puzzle from scattered fragments all over the place. The archeologist would smile at such a find. It is so with mathematics, too.

We proceed to finish solving this system to find that the point P' is (1.4, 4.2). Finally, we find the equation of the line through $A(9, 1)$ and $P'(1.4, 4.2)$. This is the symmedian line through A. We can denote that by sym_A. We compute it to be sym_A: $y = \dfrac{-8}{19}x + \dfrac{91}{19}$.

Looking at vertex B, we see that the median is a vertical line through B. Hence, we conclude med_B: $x = 5$. The angle bisector is also accomplished by inspection because of the symmetry. Hence, we immediately write bis_B: $x = 5$. We may casually observe that the x-coordinate of the incenter is 5 because of the symmetry, so we wipe our sweaty brows and take a deep breath being thankful that we didn't have to work too hard to get those. To get the symmedian through B, we reflect the line $x = 5$ across itself. Well, that just yields the original line. Hence, we breezed right through the B vertex to get its symmedian line sym_B: $x = 5$.

The C vertex is not as nice as the B vertex. It requires as much work as the A vertex to get the symmedian line. However, we already have the incenter, so that part of the work does not have to be repeated. The student will be asked to show that the symmedian line through point C is sym_C: $y = \dfrac{8}{19}x + \dfrac{11}{19}$ in the exercises.

We now have the three symmedian lines. According to the theorem, the three symmedian lines concur at the symmedian point. So we set up and solve the system of equations involving two of the symmedian lines. Then, we plug our candidate symmedian point into the third symmedian line equation to see if we get an identity. If we do, then that shows that we have found the symmedian point. If not, then we pore over our work to find the error(s). Thus, we choose to solve the system with sym_A and sym_B.

$$\begin{cases} y = \dfrac{-8}{19}x + \dfrac{91}{19} \\ x = 5 \end{cases}$$

This yields the point (5, 51/19). We plug this point into sym_C: $y = \dfrac{8}{19}x + \dfrac{11}{19}$ to see that 51/19 = 51/19, an identity. So our work is correct. We conclude that the symmedian point is $K(5, 51/19)$. The situation is captured in the below diagram.

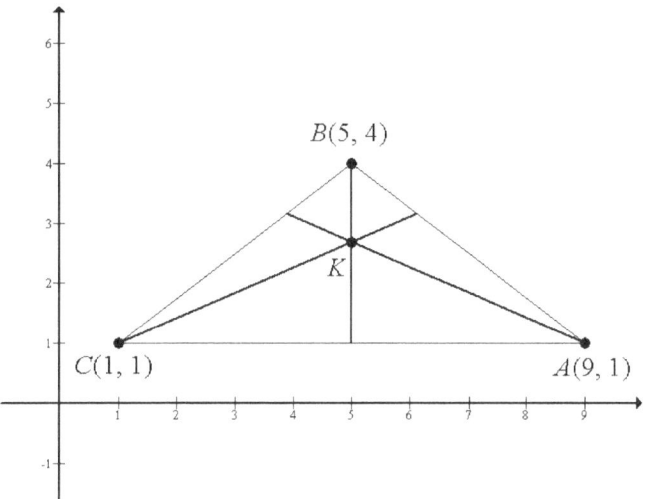

Figure 1: The Symmedian Point K for the Given Triangle

Another way exists to find the symmedian point K. It is through the concurrence of the Schwatt lines. A Schwatt line goes through the midpoint of a side and the midpoint of the altitude having its foot on that side, making appropriate extensions if necessary.

Now that we can find the symmedian point, we are in a position to find some other things. The Brocard line goes through the symmedian point and the circumcenter. The Brocard circle goes through the symmedian point and the circumcenter and has the Brocard line as a diameter. Two special points on this circle are the Brocard points, which we will not discuss beyond their mention here. The interested student can look these up if desired. Pierre René Jean-Baptiste Henri Brocard (1845–1922) was a French mathematician.

The points where the symmedian lines intersect the sides of the triangle form a triangle that we will call the symmedial triangle. The symmedial triangle and the original triangle ABC are in perspective. Hence, they have an axis of perspectivity. The perspector is, of course, the symmedian point. The perspectrix is called the Lemoine line.

231

Another special point is called the retrocenter R. We can find this point through two methods. One way is to find the isotomic conjugate of the orthocenter. The other way is to find the antimedial triangle of ABC (the triangle for which ABC is the medial triangle), and then find the symmedian point of this new triangle. Finding the antimedial triangle involves constructing three systems of equations with a parallel line to a side through the opposite vertex and a circle centered at that vertex with radius twice that side. Once this triangle is found, just get its symmedian point. The symmedian point of the antimedial triangle is the same point as the retrocenter of the original triangle.

We will work an acute triangle example for the isotomic conjugate of the orthocenter; however, strictly speaking, we need an acute triangle to find an isotomic conjugate because the orthocenter of an obtuse triangle is outside the triangle. Thus, confronted with an obtuse triangle, the antimedial triangle method is the only way to get the retrocenter. Example: Find the retrocenter of the triangle with vertices at $A(1, 2)$, $B(8, 12)$, and $C(9, 1)$.

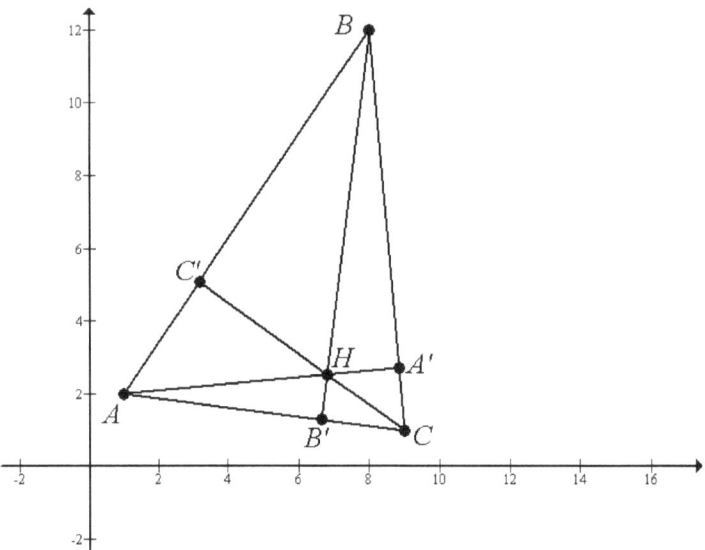

Figure 2: The Orthocenter H of Our Triangle with the Feet of the Altitudes

We will first need to find the orthocenter of the triangle. Finding the necessary perpendicular lines through the appropriate vertices, we construct the system of equations to find the orthocenter H.

$$\begin{cases} y = \dfrac{-7}{10}x + \dfrac{73}{10} \\ y = 8x - 52 \\ y = \dfrac{1}{11}x + \dfrac{21}{11} \end{cases}$$

Solving this system yields the orthocenter $H\left(\dfrac{593}{87}, \dfrac{220}{87}\right)$. We also would like to compute the feet of the altitudes A', B', and C'. Figure 2 captures our situation thus far.

The feet of the altitudes are found by finding the intersection point of an altitude with the side of the triangle that the altitude is perpendicular. Thus, we find the equations of the sides of the triangle first. Then, we set up and solve the following systems of equations. We note that the three points A', B', and C' form the orthic triangle.

$$A': \begin{cases} y = -11x + 100 \\ y = \dfrac{1}{11}x + \dfrac{21}{11} \end{cases} \qquad B': \begin{cases} y = 8x - 52 \\ y = \dfrac{-1}{8}x + \dfrac{17}{8} \end{cases} \qquad C': \begin{cases} y = \dfrac{10}{7}x + \dfrac{4}{7} \\ y = \dfrac{-7}{10}x + \dfrac{73}{10} \end{cases}$$

Solving these systems, we obtain $A'\left(\dfrac{1079}{122}, \dfrac{331}{122}\right)$, $B'\left(\dfrac{433}{65}, \dfrac{84}{65}\right)$, $C'\left(\dfrac{471}{149}, \dfrac{758}{149}\right)$.

Next, we need to find the isotomic conjugate of the orthocenter. This point will be the retrocenter. This will be accomplished by first finding the points A'', B'', and C'', which are the points A', B', and C' reflected across their respective triangular side midpoints. Then, the retrocenter will be the intersection of the lines AA'', BB'', and CC''. Points on a side of a triangle that are equidistant from the midpoint on that side are called isotomic points. The concurrence of the cevians through the three isotomic points produce the point known as the isotomic conjugate. Thus, from these definitions, we have that A' and A'' are isotomic points. Similarly, B' and B'' are isotomic points, and C' and C'' are isotomic points. The cevians through A'', B'', and C'' yield the retrocenter, which is the isotomic conjugate to the orthocenter. Hence, the orthocenter and retrocenter are isotomic conjugates.

We will find the isotomic point A'' to A' and leave B'' and C'' for the student to find. In order to find A'', we need to know the midpoint of side BC. Using the midpoint formula, we find that $M_{BC} = \left(\frac{1}{2}(8+9), \frac{1}{2}(12+1) \right) = \left(\frac{17}{2}, \frac{13}{2} \right)$. Next, we need to compute

the distance from $A'\left(\frac{1079}{122}, \frac{331}{122} \right)$ to $M_{BC}\left(\frac{17}{2}, \frac{13}{2} \right)$. We get the following computation:

$$d = \sqrt{\left(\frac{17}{2} - \frac{1079}{122} \right)^2 + \left(\frac{13}{2} - \frac{331}{122} \right)^2} = \frac{21\sqrt{122}}{61}$$

We now construct a system of equations involving a circle centered at M_{BC} with radius d and the line of side BC. The solution to this system is the pair of isotomic points A' and A''. However, we already know one solution to this system, namely A'. Thus, the resulting quadratic equation in x will factor. Our system is the following:

$$\left\{ \begin{array}{l} \left(x - \frac{17}{2} \right)^2 + \left(y - \frac{13}{2} \right)^2 = \left(\frac{21\sqrt{122}}{61} \right)^2 \\ y = -11x + 100 \end{array} \right\}$$

Inserting the $-11x + 100$ into the top equation for y yields the following polynomial equation upon simplification.

$$14884x^2 - 253028x + 1073605 = 0$$

Let's talk about this equation for a bit. Yes, we could use the quadratic formula. However, the discriminant would involve some numerical acrobatics. A better plan is to use the fact that we know one of the roots already, which is the x-coordinate of $A'\left(\frac{1079}{122}, \frac{331}{122} \right)$. Hence, since this equation is in x, we know that $x = \frac{1079}{122}$ is a root of this quadratic equation. We will construct the factor from this root by getting zero on one side. Thus, $122x - 1079$ is one of the two factors of this quadratic equation. The other is found by dividing the leading coefficient by 122 and the constant term by -1079. As a precaution, it is always wise to check the middle term. Performing this, we find that the other root is $122x - 995$, and the middle term checks out because $122x(-1079 - 995) = -253028x$. Hence, the x-coordinate of A'' is 995/122. The y-coordinate is computed more easily with the line than with the circle. Hence, we find that its y-coordinate is

$y = -11(995/122) + 100 = 1255/122$. We now have the point $A''\left(\frac{995}{122}, \frac{1255}{122} \right)$. Of course,

we could have obtained A'' through the unit circle process in *The Circle* chapter.

Repeating this process for B' and C' to yield B'' and C'', respectively, yields the following: $B''\left(\dfrac{217}{65}, \dfrac{111}{65}\right)$ and $C''\left(\dfrac{870}{149}, \dfrac{1328}{149}\right)$.

We form the equations of the lines AA'', BB'', and CC'', and put them into a system of equations next. Our system is the following:

$$\begin{cases} y = \dfrac{337}{291}x + \dfrac{245}{291} \\[2mm] y = \dfrac{223}{101}x - \dfrac{572}{101} \\[2mm] y = \dfrac{-393}{157}x + \dfrac{3694}{157} \end{cases}$$

To solve this over determined system, we use two of the equations to get a candidate solution. Then, we plug the candidate solution into the third equation to see if we get an identity. If we do, then we have shown that that is the solution to the system. Doing this, we obtain the solution $R\left(\dfrac{347}{56}, \dfrac{449}{56}\right)$. Thus, the orthocenter $H\left(\dfrac{593}{87}, \dfrac{220}{87}\right)$ and the retrocenter $R\left(\dfrac{347}{56}, \dfrac{449}{56}\right)$ are isotomic conjugates. Our situation is captured in the following diagram.

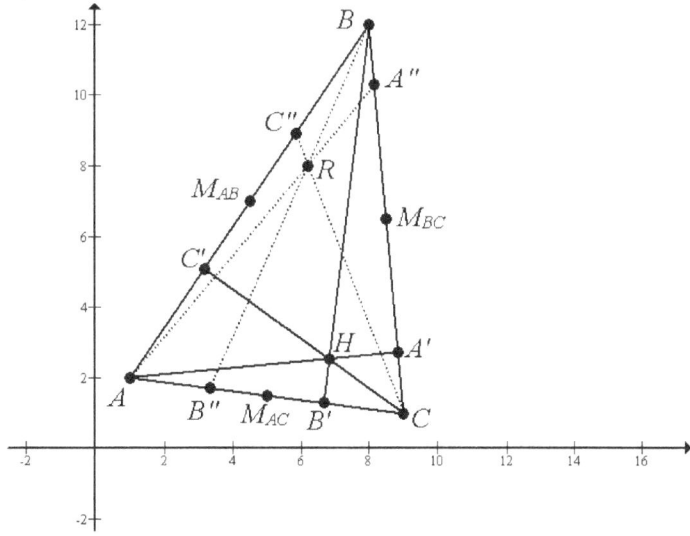

Figure 3: The Isotomic Conjugates H and R Depicted

The Gergonne point *Ge* and the Nagel point *Na* are isotomic conjugates. The centroid *G* is its own isotomic conjugate. A line that goes through the symmedian point *K* is the van Aubel line. Also on this line are the orthocenter *H* and the symmedian point of the orthic triangle.

Exercises

1. Show that for the first example problem in this chapter that sym$_C$: $y = \dfrac{8}{19}x + \dfrac{11}{19}$.

2. Find the coordinates where the three symmedian lines intersect the sides of the triangle in the first example problem in this chapter. This yields the symmedial triangle. Use the symmedial triangle and the symmedian point *K* to find the equation of the Lemoine line.

3. Find the symmedian point *K* for the triangle with vertices at $A(0, 0)$, $B(12, 0)$, and $C(12, 5)$, and find the coordinates where the three symmedian lines intersect the sides of the triangle. Show that the three Schwatt lines also intersect at *K*.

4. Find the symmedian point *K* for the triangle with vertices at $A(2, 6)$, $B(0.2, 8.4)$, and $C(3.4, 10.8)$, and find the coordinates where the three symmedian lines intersect the sides of the triangle. Show that the three Schwatt lines also intersect at *K*.

5. Find the symmedian point *K* for the triangle with vertices at $A(2, 1)$, $B(5, 1)$, and $C(5, 5)$. Also, find the coordinates where the three symmedian lines intersect the sides of the triangle. Show that the three Schwatt lines also intersect at *K*.

6. In the retrocenter example, complete the problem by computing $B''\left(\dfrac{217}{65}, \dfrac{111}{65}\right)$ and $C'''\left(\dfrac{870}{149}, \dfrac{1328}{149}\right)$.

7. Find the equations of the Lemoine and Brocard lines for the triangle with vertices at $A(4, 5)$, $B(8, 8)$, and $C(4, 1/2)$. Show your work.

8. Find the equations of the Brocard line and circle for the triangle in question 5.

9. Writing in mathematics: Use the advice in *The Circle* chapter to construct a triangle with three different integer lengths such that none of its sides are vertical or horizontal on the Cartesian plane. Find this triangle's three vertices, and proceed to find its symmedian point. Be professional in your documentation and in the labeling of your algebra.

10. For the triangle in question 7, find the van Aubel line. Show all work.

Concyclic Points

Concyclic points are points that lie on the same circle. Two points at the endpoints of a diameter determine a unique circle, and three points determine a unique circle. So we are not interested in three concyclic points in this chapter. The stakes are higher. We insist on a minimum of four concyclic points for our consideration here.

We have a number of situations to discuss in this chapter. They all pertain to finding the equation of a special circle meeting the criteria specified. We first present two concyclic points theorems relating to the parabola. The second one is by Harvey. We saw one of his theorems earlier in the *Two Special Quadrilaterals* chapter.

Theorem: The three intersection points A, B, and C of any three tangent lines to a parabola are concyclic with the focus F of the parabola.

Harvey's Theorem: Given a parabola and a point P, if three normal lines from P can be drawn to the parabola creating the intersection points A, B, and C, then the points A, B, and C are concyclic with the vertex of the parabola.

We will work an example of Harvey's theorem. Given the parabola $y = x^2$ and the point $P\left(-12, \dfrac{15}{2}\right)$, show that the three intersection points of the three normal lines to the parabola from P with the parabola are concyclic with the vertex of the parabola.

A normal line through P will intersect the parabola at some (x, y) point. Really, we know that we will have three such intersection points, but using the variables x and y in general will yield three such solutions. We set up the first equation with the understanding that it looks like the slope formula, but we are building a normal slope.

$$m_{\text{norm}} = \frac{y_2 - y_1}{x_2 - x_1} = \frac{\dfrac{15}{2} - y}{-12 - x} = \frac{\dfrac{15}{2} - x^2}{-12 - x} = \frac{15 - 2x^2}{-24 - 2x} = \frac{2x^2 - 15}{2x + 24}$$

Next, we will implement the perpendicular slope function. The slope of a tangent line to the parabola at any given value of x is given by the slope function $m(x) = 2x$. We discussed the slope function in *The Deluxe Toolkit* chapter. Therefore the slope of normal line to the parabola is given by the perpendicular slope function $m_\perp(x) = \dfrac{-1}{m(x)} = \dfrac{-1}{2x}$.

These two slopes are the same. Thus, we can equate them to generate an algebraic equation in the variable x. Then, we can solve this equation to get three solutions for x. These would correspond to the three x-coordinates of the three intersection points. We would just have to compute the y-coordinates to obtain the three points A, B, and C.

$$\frac{2x^2 - 15}{2x + 24} = \frac{-1}{2x} \quad \text{Solve for x.}$$

$$2x(2x^2 - 15) = -1(2x + 24)$$

$$4x^3 - 30x + 2x + 24 = 0$$

$$4x^3 - 28x + 24 = 0$$

$$x^3 - 7x + 6 = 0 \quad \text{Use the Rational Roots Theorem } x \in \{\pm 1, \pm 2, \pm 3, \pm 6\}.$$

We graph the equation in the y-editor of a graphing calculator, and the roots appear to be 1, 2, and –3. We perform the synthetic division to find that this is so. We find that the three points A, B, and C on the parabola then are $A(1, 1)$, $B(2, 4)$, and $C(-3, 9)$. These three points are concyclic with the vertex $(0, 0)$ of the parabola. The equation of the circle through these four points is $(x + 3)^2 + (y - 4)^2 = 25$. We will show a graph of our situation, Figure 1.

We will next revisit a problem that we discussed in the *Some Tangency Problems* chapter. Apollonius, the ancient Greek geometer, discovered some distance relationships pertaining to the tangent line of a hyperbola and its intersection points P and Q with the asymptote lines of it. We discussed this situation in the chapter and studied it with problem number 17 in the exercise set. This situation has a unique circle. We present it with our first theorem in this chapter.

Theorem: Given a tangent line to a hyperbola, it has two intersection points with the asymptote lines. Call these two intersection points P and Q. Then, the two foci of the hyperbola and P and Q are concyclic.

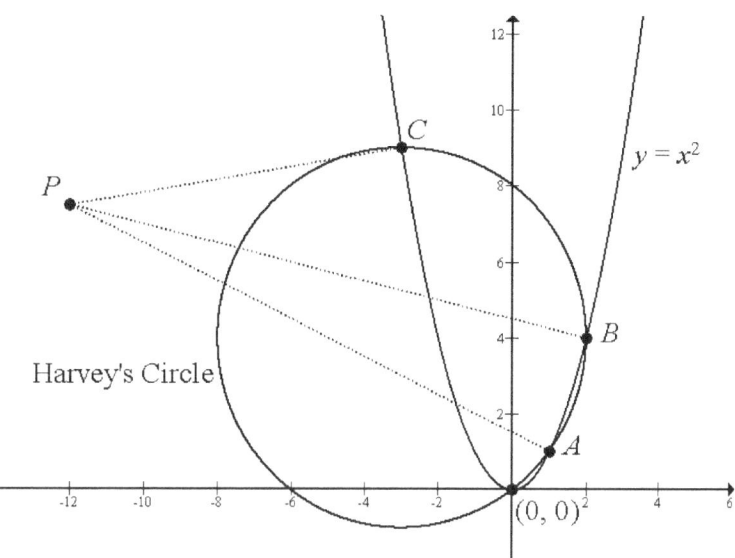

Figure 1: Harvey's Circle for Point P wrt Parabola $y = x^2$

We have another pair of concyclic points theorems involving hyperbolas similar to the previous theorem. We present them next. Note as a reminder that the transverse axis is the line segment between the two vertices of a hyperbola, and the conjugate axis is the line segment of length $2b$ perpendicular to the transverse axis and bisected by the center of the hyperbola. The representative rectangle for a hyperbola is cut in half by the transverse axis and by the conjugate axis.

Theorem: Given a tangent line to a hyperbola, it has two intersection points with the asymptote lines. Call these two intersection points P and Q. Let T be the point of tangency. The normal line through T will intersect the axes of the hyperbola (i.e. the transverse and conjugate axes) at points M and N. Then, the points P, Q, M, and N are concyclic.

Theorem: Given an equilateral hyperbola and three points on it P, Q, and R, that are concyclic with the center C of the hyperbola, the vertices of the triangle formed by the tangent lines at P, Q, and R are also concyclic with the center of the hyperbola.

Our next concept in this chapter will be on antiparallel lines. We will provide two definitions; these two definitions will suit our present purposes.

Antiparallel (1st definition): Let line ℓ_1 intersect the sides of an angle A at points K and L. Let line ℓ_2 intersect the sides of angle A at points M and N. Let *bis* be the bisector of angle A. If ℓ_1 and ℓ_2 make the same angle, in the opposite sense of parallel lines, with the angle bisector, then ℓ_1 and ℓ_2 are said to be antiparallel lines with respect to the sides of angle A.

We can remark on this definition. Recall in *The Deluxe Toolkit* chapter the arguments for two parallel lines cut by a transversal. We got eight angles. If the transversal cut the two parallel lines at right angles, then all eight of the angles formed were right angles. If the transversal cut the two parallel lines by any other angle, then the eight angles were split into two groups: four equal acute angles and four equal obtuse angles. These angles are placed in the same relative positions for the two parallel lines. With antiparallel lines, we still get eight angles, but the placement of the four angles for one of the antiparallel lines is in the opposite orientation or sense than the four angles for the other antiparallel line. Figure 2 illustrates the concept.

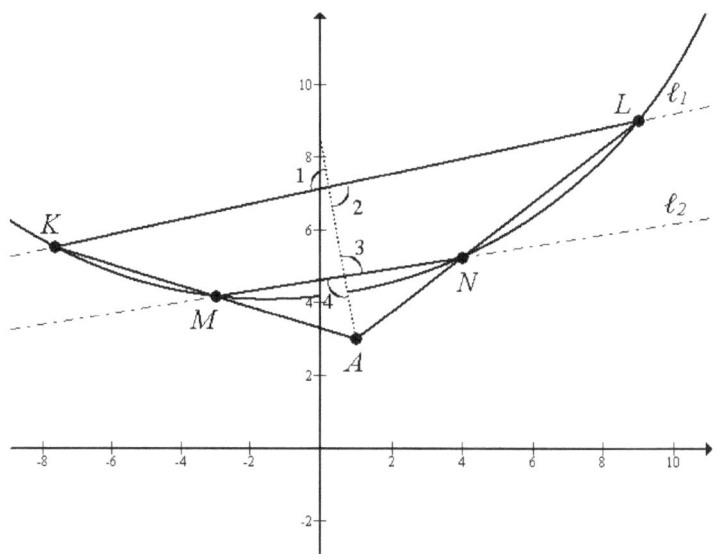

Figure 2: Antiparallel Lines ℓ_1 and ℓ_2

This diagram shows the four obtuse angles. They are labeled 1, 2, 3, and 4. With parallel lines, the second pair of obtuse angles are in the same orientation as the first pair. Observe that angles 3 and 4 are in the opposite orientation as angles 1 and 2 are in the diagram. Hence, the name antiparallel means they are opposite to how they are with parallel lines, which suggests that something has been switched. Also, observe in the figure that the points K, L, M, and N are cyclic, which means that they lie on a circle. It's a big circle that does not fit onto the diagram, but it is a circle nonetheless. This observation on the cyclic nature of the four points suggests the second definition for antiparallel lines. The unlabeled dotted line angle bisector for angle A goes from point A

240

to the y-axis just above eight. This line serves as the transversal line to the antiparallel lines ℓ_1 and ℓ_2.

Note that the antiparallel lines ℓ_1 and ℓ_2 will intersect at some point to the left of the diagram. However, antiparallel lines could actually be parallel if the transversal cut at right angles. It is always true that parallel lines will never intersect regardless of the angle cut by a transversal. The next definition shows us the difference in what we mean here. We know a square has two pairs of opposite sides that are parallel, but a square is a cyclic quadrilateral. So according to the following definition, the pairs of opposite sides of a square are antiparallel to one another. When one pair of lines is perpendicular to another pair of lines, each pair is both parallel and antiparallel to one another in that pair. Thus, we are not wrong when we say that the opposite sides of a square (rectangle or cyclic trapezoid also) are antiparallel to one another.

Antiparallel (2nd definition): Given a cyclic quadrilateral, one pair of opposite sides is antiparallel to the other pair of opposite sides.

Thus, given four points, we can use Ptolemy's theorem stated in the *Two Special Quadrilaterals* chapter to test whether or not a quadrilateral is cyclic. If it is, then we know that the pairs of opposite sides are antiparallel to one another. Also, if we can establish that two lines are antiparallel with respect to a given angle, then the intersection points of the two lines with that angle are cyclic. Having one means having the other.

Our first application of the antiparallel definition is the following: The line joining the feet of two altitudes of a triangle is antiparallel to the third side of the triangle. Hence, the two vertices on the third side of the triangle and the feet of the two altitude lines not on that side are cyclic. It seems that no name exists for this circle.

Theorem: Every antiparallel line to a given side of a triangle is parallel to the tangent line to the circumcircle that goes through the opposite vertex of that triangle.

We can easily find the slope of an antiparallel line to a given side of a triangle with the above theorem. The negative reciprocal of the slope through the circumcenter O and the vertex of the opposite side yields the desired slope of the antiparallel lines to the given side. This is because a tangent line at a point on a circle is perpendicular to the radius at that point.

Theorem: A symmedian line through a given vertex of a triangle bisects all antiparallel lines with respect to the angle of that vertex.

If we find that a symmedian line bisects a line segment that has its two endpoints on the sides of the triangle, then we have shown that that line segment is antiparallel to the side that the foot of the symmedian line hits. Admittedly, this is not the first thing we would think of when determining if a given line is antiparallel to a side of the triangle. Nevertheless, this can be done. We now present three specific concyclic points theorems, and we will work an example of the third. The first two theorems depict circles that do not seem to have names, but the third theorem depicts a circle that does have a name.

Theorem: Orienting from angle A of triangle ABC, if M is the midpoint of the side opposite A, A' is the foot of the altitude from vertex A, and P and Q are the feet of the perpendiculars from the vertices B and C, respectively, to the angle bisector of angle A, then M, A', P, and Q are concyclic. This is true of a similar argument orienting from the vertices B and C as well.

Theorem: Orienting from angle A of triangle ABC, if A' is the foot of the altitude from vertex A, and P and Q are the feet of the perpendiculars from A' to the sides AB and AC respectively, then B, C, P, and Q are concyclic points. This is true of a similar argument orienting from the vertices B and C as well.

Lester's Circle: In a scalene triangle ABC, the two Fermat points F_1 and F_2, the nine-point center N, and the circumcenter O all lie on a circle called Lester's circle. In addition, the points always occur in the order O-N-F_1-F_2 either clockwise or counter-clockwise. It is orthogonal to the orthocentroidal circle—meaning the two circles intersect each other at right angles. This can be demonstrated by finding the tangent lines to the two circles at their intersection points and observing that their slopes are negative reciprocals.

We will work an example for Lester's circle. Find Lester's circle for the triangle with vertices at $A(3.2, 4.8)$, $B(1.8, 0)$, and $C(0, 2.4)$. We will assume that the student can find the equation of the circumcircle, the circumcenter O, and the nine-point center N. These are $\left(x - \dfrac{5}{2}\right)^2 + \left(y - \dfrac{12}{5}\right)^2 = \dfrac{25}{4}$, $\left(\dfrac{5}{2}, \dfrac{12}{5}\right)$, and $\left(\dfrac{5}{4}, \dfrac{12}{5}\right)$, respectively. It remains to find the two Fermat points F_1 and F_2, and the equation of Lester's circle.

Recall from *The Deluxe Toolkit* chapter that certain relationships existed for the lengths of the three sides of the 30-60-90 special triangle. This special triangle is half of an equilateral triangle. To find the Fermat points, we need to construct equilateral triangles on the sides of the given triangle ABC. The length of the altitude of an equilateral triangle is $\dfrac{\sqrt{3}}{2}$ times the length of a side. An altitude line is also a perpendicular bisector of a side for an equilateral triangle. We can use these two facts to construct both the perpendicular bisector for the equilateral triangle and the equation of the circle with radius $\dfrac{\sqrt{3}}{2}$ times the length of that side of the given triangle ABC. The bonus of this system of equations involving this line and circle is that the extraneous solution is not extraneous—it is the coordinates of the other apex of the equilateral triangle erected inwardly. Thus, we get to save a step when finding F_2.

We need the equations of the perpendicular bisectors of the three sides. To do this we need the midpoints of the sides as well as the three perpendicular slopes relative to each of the three sides. We put these in $y = mx + b$ form and get the following:

$$bis_{\perp}(AB): y = \frac{-7}{24}x + \frac{751}{240}$$

$$bis_{\perp}(AC): y = \frac{-4}{3}x + \frac{86}{15}$$

$$bis_{\perp}(BC): y = \frac{3}{4}x + \frac{21}{40}$$

Next, we need to construct the equations of the three circles centered at the midpoints of the sides with radius equal to $\dfrac{\sqrt{3}}{2}$ times that particular side. We will just show the three systems of equations obtained utilizing the perpendicular bisectors and appropriate circle equations.

$$\begin{cases} y = \dfrac{-7}{24}x + \dfrac{751}{240} \\ \left(x - \dfrac{5}{2}\right)^2 + \left(y - \dfrac{12}{5}\right)^2 = \dfrac{75}{4} \end{cases} \qquad \begin{cases} y = \dfrac{-4}{3}x + \dfrac{86}{15} \\ (x - 1.6)^2 + (y - 3.6)^2 = 12 \end{cases}$$

$$\left\{ \begin{array}{l} y = \dfrac{3}{4}x + \dfrac{21}{40} \\ \\ (x-0.9)^2 + (y-1.2)^2 = \dfrac{27}{4} \end{array} \right\}$$

The solutions to these three systems are the six vertices comprising the apices of the six equilateral triangles needed to find the Fermat points. Three of the apices are for the outwardly erected equilateral triangles, which will help us find F_1. The other three apices are for the three inwardly erected equilateral triangles, which will help us find F_2. Figure 3 illustrates how we find F_1 pictorially.

When we solve the above three systems, we need to determine which of the two solutions to a given system is for the outwardly erected apex of an equilateral triangle. Of course, the other solution would be for the inwardly erected apex of the equilateral triangle. We designate the outward ones with a single prime, and the inward ones will be designated with a double prime. Thus, for example to solve the third system shown above, we will get A' and A''.

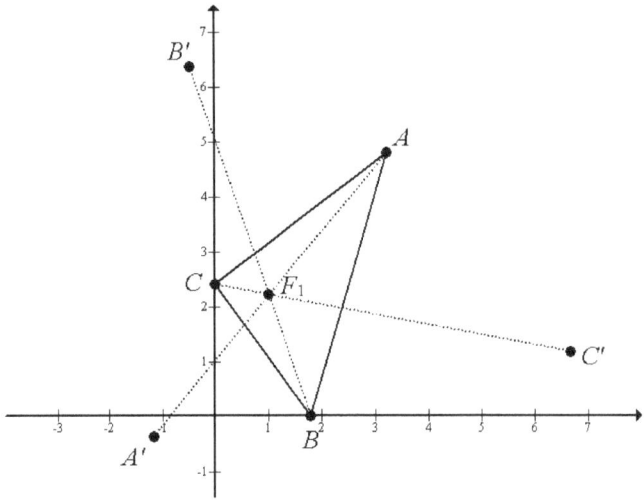

Figure 3: The First (or Outer) Fermat Point F_1 of Our Triangle

244

We find that the solutions to this system are $\left(\dfrac{9+12\sqrt{3}}{10},\dfrac{12+9\sqrt{3}}{10}\right)$ and

$\left(\dfrac{9-12\sqrt{3}}{10},\dfrac{12-9\sqrt{3}}{10}\right)$. Looking at Figure 2, these two points are on either side of the line BC. We look at the x-coordinates of each. The smaller x-coordinate must be for the outer equilateral triangle and the larger x-coordinate must be for the inner equilateral triangle.

Thus, the point $\left(\dfrac{9-12\sqrt{3}}{10},\dfrac{12-9\sqrt{3}}{10}\right)$ must be A', and the other one is A''.

Repeating this analysis for the solutions of the other two systems, we find that the remaining four points are $B'\left(\dfrac{8-6\sqrt{3}}{5},\dfrac{18+8\sqrt{3}}{5}\right)$, $B''\left(\dfrac{8+6\sqrt{3}}{5},\dfrac{18-8\sqrt{3}}{5}\right)$,

$C'\left(\dfrac{25+24\sqrt{3}}{10},\dfrac{24-7\sqrt{3}}{10}\right)$, and $C''\left(\dfrac{25-24\sqrt{3}}{10},\dfrac{24+7\sqrt{3}}{10}\right)$.

From here, we construct the lines AA', BB', and CC'. We form a system of equations with these three lines, and the solution to this system is the point F_1. Next, we construct the lines AA'', BB'', and CC''. We form a system of equations with these three lines, and the solution to this system is the point F_2. We form the systems and solutions below.

$$\begin{cases} AA': y = \dfrac{504-225\sqrt{3}}{107}x - \dfrac{1968-2025\sqrt{3}}{535} \\[2mm] BB': y = \dfrac{-(126+100\sqrt{3})}{107}x + \dfrac{1134+900\sqrt{3}}{535} \\[2mm] CC': y = \dfrac{-504+175\sqrt{3}}{1103}x + \dfrac{12}{5} \end{cases}$$

$$\therefore F_1\left(\dfrac{2(900-239\sqrt{3})}{965},\dfrac{6(211+84\sqrt{3})}{965}\right)$$

$$\left\{ \begin{array}{l} AA'': y = \dfrac{9(56+25\sqrt{3})}{97}x - \dfrac{5736+3600\sqrt{3}}{485} \\[3mm] BB'': y = \dfrac{-2(63-50\sqrt{3})}{107}x + \dfrac{18}{535}(63-50\sqrt{3}) \\[3mm] CC'': y = \dfrac{-7(72+25\sqrt{3})}{1103}x + \dfrac{12}{5} \end{array} \right\}$$

$$\therefore F_2\left(\dfrac{2(900+239\sqrt{3})}{965}, \dfrac{6(211-84\sqrt{3})}{965} \right)$$

We observe in passing that F_1 and F_2 look like solutions to a pair of intersecting circles like Plücker's circles. Now that we have the two Fermat points F_1 and F_2 to go along with the circumcenter O and the nine-point center N, we can construct the equation of Lester's circle. We will use O, N, and F_1 to find the equation of Lester's circle. Then, we will plug F_2 into Lester's circle equation to see if we get an identity. If we do, then we will know that we do indeed have the equation of Lester's circle. We will use the advice in *The Circle* chapter to find the equation of Lester's circle. We will construct the matrix using the points O, N, and F_1 in the equation $x^2 + y^2 + Dx + Ey + F = 0$. Doing this, and obtaining the reduced row echelon form of the matrix yields the following:

$$\left[\begin{array}{ccc|c} \dfrac{5}{2} & \dfrac{12}{5} & 1 & \dfrac{-1201}{100} \\[3mm] \dfrac{5}{4} & \dfrac{12}{5} & 1 & \dfrac{-2929}{400} \\[3mm] \dfrac{2(900-239\sqrt{3})}{965} & \dfrac{6(211+84\sqrt{3})}{965} & 1 & \dfrac{-48(679-48\sqrt{3})}{4825} \end{array} \right] \xrightarrow{\ rref\ }$$

$$\left[\begin{array}{ccc|c} 1 & 0 & 0 & \dfrac{-15}{4} \\[3mm] 0 & 1 & 0 & \dfrac{-4439}{1680} \\[3mm] 0 & 0 & 1 & \dfrac{5189}{1400} \end{array} \right]$$

246

Therefore $D = -15/4$, $E = \dfrac{-4439}{1680}$, and $F = \dfrac{5189}{1400}$. Observe that the ugly square roots of three have vanished. So we plug these values back into the template equation $x^2 + y^2 + Dx + Ey + F = 0$ to get Lester's circle. Thus, the equation of Lester's circle is $x^2 + y^2 - \dfrac{15}{4}x - \dfrac{4439}{1680}y + \dfrac{5189}{1400} = 0$. We plug the coordinates of F_2 into this equation to get $0 = 0$, which shows that we have indeed found Lester's circle.

The ambitious student can proceed to find the orthocentroidal circle and show that it is orthogonal to Lester's circle. For the record, the equation of the orthocentroidal circle turns out to be $\left(x - \dfrac{5}{6}\right)^2 + \left(y - \dfrac{12}{5}\right)^2 = \dfrac{25}{36}$. We show a graph of Lester's circle with O-N-F_1-F_2, and the dotted circle is the orthocentroidal circle, Figure 4. Observe that the orthocentroidal circle appears to be orthogonal to Lester's circle—this is a graphical demonstration that the two circles are orthogonal.

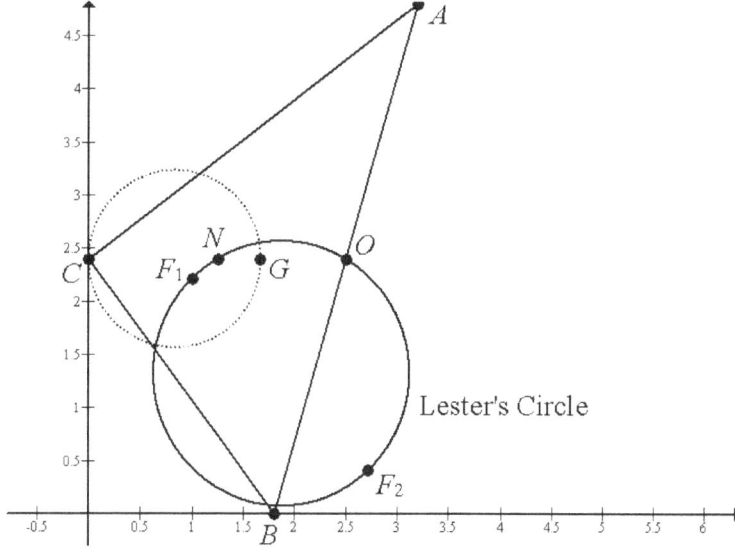

Figure 4: Lester's Circle and the Orthocentroidal Circle are Orthogonal

Before we move to the six concyclic points part of this chapter, we have a theorem to determine if six points are concyclic—Furhmann's theorem. This theorem does for six points what Ptolemy's theorem does four four points.

Fuhrmann's Theorem: Label a hexagon's sides in order, such that it is convex, producing the sides a, b, c, d, e, and f. A long diagonal of the hexagon connects vertices with two free vertices on either side of the long diagonal. Thus, label the hexagon's long diagonals as p, q, and r, such that p has two free vertices on side a. The long diagonal q has two free vertices on side b, and the long diagonal r has two free vertices on side c. Then, the following distance relationship holds for a cyclic hexagon (or if the relationship is true, then the hexagon is cyclic):

$$pqr = adp + beq + cfr + ace + bdf$$

Our next theorem is a general theorem that provides infinitely many circles for a given triangle ABC. To compare, a triangle has many Desargues' lines because it depends on the point chosen to be the perspector, and Gergonne's line is a specific example of a Desargues' line because it comes from a specific perspector. In the same vein, we have a general theorem yielding a circle for a given triangle that depends on the boundary point chosen as a starting point. The theorem following this general theorem gives a specific starting point that yields a specific circle for a given triangle.

Tucker Circle Theorem: From a given triangle ABC, choose a point P on the boundary of the triangle that is not a vertex. Draw a parallel line from P to one of the remaining sides. Call this point Q. Draw an antiparallel line from Q to the remaining side. Call this point R. Continue around the triangle in the same direction drawing alternating parallel and antiparallel lines creating three more points S, T, and U. The next step from U always produces the original chosen point P. The six points P, Q, R, S, T, and U are concyclic. Another Tucker circle exists by starting the process with a parallel line from P in the other direction. Thus, a point P provides two Tucker circles for a given triangle ABC.

Taylor Circle Theorem: We are given the triangle ABC and the feet of the altitudes A', B', and C', respectively. From A' the feet of the perpendiculars to the other two sides produce two new points. From B' the feet of the perpendiculars to the other two sides produce an additional two new points. Finally, from C' the feet of the perpendiculars to the other two sides produce the third pair of new points. Then, the six new points produced are concyclic in what is known as Taylor's circle. The Taylor circle is a specific example of a Tucker circle.

The next three theorems have named circles. These three theorems depict six points of intersection with the original triangle that are concyclic. These three theorems also tell us the center of the circle.

First Lemoine Circle: Three lines through the symmedian point K that are parallel to the sides of triangle ABC cut the triangle in six points all lying on a circle called the First Lemoine circle. Its center is the midpoint of the symmedian K and the circumcenter O.

Second Lemoine Circle: Three lines through the symmedian point K that are antiparallel to the sides of triangle ABC cut the triangle in six points all lying on a circle called the Second Lemoine circle. Its center is the symmedian point K.

Adams Circle Theorem (1843): When three lines parallel to the Gergonne triangle's sides go through the Gergonne point, they intersect the original triangle in six places. These six points of intersection all lie on a circle known as Adams Circle. The center of Adams Circle is the same as the incenter I of the original triangle. Thus, Adams circle is concentric with the incircle, but it has a larger radius.

We have another named circle presented next whose center is the incenter. This circle is called the Conway circle. Thus, it is concentric with Adams circle and the incircle, but it has a larger radius than either.

Conway Circle Theorem: We are given triangle ABC. Extend the sides of the triangle from vertex A outwardly a distance of a units. This produces two new points. Extend the sides of the triangle from vertex B outwardly a distance of b units. This also produces two new points. Finally, extend the sides of the triangle from vertex C outwardly a distance of c units. This produces the last pair of new points. The six points produced in this fashion all lie on a circle called Conway's circle. The center of this circle is the incenter I of the original triangle.

We will make a few observations on Conway's circle. One circle theorem is known as the equally distant chords theorem, which states that chords equally distant from the center are congruent. The extension of the sides of triangle ABC in Conway's circle yields three congruent chords. The mutual chord length is $a + b + c$. Hence, their distance from the center I must be the same. What is this same distance? Since the incenter I is the center of the circle, and the three congruent chords are the sides (extended), then this mutual distance from I is r, the inradius. The distance from a point to a line equation for all three of these lines with the point I would yield the same value r.

We next discuss the concept of isogonal lines. We have already met them, but we didn't have a name for them until now. An isogonal line with respect to another line is the line formed that creates the same angle as the original two lines. This feat is accomplished by reflecting the line we want to find its isogonal across the other line. We called this other line the mirror in *The Deluxe Toolkit* chapter.

In a triangle *ABC*, a line through a vertex has an isogonal line by reflection across the angle bisector through that vertex. The angle bisector serves as the mirror. Given a point *P* and a triangle *ABC*, we can find the three lines *AP*, *BP*, and *CP*. These three lines can be reflected across the angle bisectors through the vertices *A*, *B*, and *C*, respectively, to obtain the three isogonal lines to *AP*, *BP*, and *CP*. Of course, the incenter, *I*, is a point on all three angle bisectors for the triangle, so its importance cannot be understated here.

Theorem: Given a triangle *ABC*, a point *P*, and the three lines *AP*, *BP*, and *CP*, the three isogonal lines to *AP*, *BP*, and *CP* concur in a point *P′*—the isogonal conjugate of *P*.

As an example, the isogonal conjugate of the nine-point center *N* is the Kosnita point *Ko*. The Kosnita point is found by forming the three circumcircles through two vertices and the circumcenter of the original triangle and connecting the three vertex to opposite new circumcenter points. These three vertex to opposite new circumcenter lines are concurrent in the Kosnita point *Ko*. Also, since the isogonal conjugate of the nine-point center *N* is the Kosnita point *Ko*, we can find the isogonal conjugate of the nine-point center *N* to confirm that it is the Kosnita point *Ko*. Two more examples of isogonal conjugates are the following: the orthocenter *H* and the circumcenter *O*, and the centroid *G* and the symmedian point *K*. The incenter *I* is the only point that is its own isogonal conjugate.

Theorem: Given a triangle *ABC*, a point *P*, and its isogonal conjugate *P′*, then the six points comprising the two pedal triangles of *P* and *P′* are concyclic. Note that we discussed pedal triangles in the *Triangle Geometry* chapter.

Our final topic in this chapter is another type of conjugate. We have seen isotomic conjugates in *The Symmedian Point* chapter, and we have just learned about isogonal conjugates. We have another type of conjugate to present next—the cyclocevian conjugate.

We have seen many times now in a triangle *ABC* that three cevian lines meeting a certain criterion concur at a point *P* in the interior of the triangle. The three intersection points of the cevians with the sides of the triangle are usually labeled *A′*, *B′*, and *C′*, and the lines *AA′*, *BB′*, and *CC′* are the cevian lines associated with the point *P*. The points *A′*, *B′*, and *C′* have a circumcircle associated with them since three noncollinear points determine a unique circle. This circle is not called the circumcircle of those points. Rather, we call this circle the cevian circle of point *P*. For example, the cevian circle of the centroid of a triangle is the nine-point circle. A cevian circle is always smaller than the circumcircle, so it will intersect the triangle in three other places that we will denote as A'', B'', and C'', which are on the sides opposite the vertices A, B, and C, respectively. An amazing fact is that the cevians AA'', BB'', and CC'' will also concur at a point we will denote as P'. The points P and P' are called cyclocevian conjugates. A pair of cyclocevian conjugates share the same cevian circle. This fact suggests the following theorem.

Theorem: The six feet of the cevian lines of a pair of cyclocevian conjugates P and P' are concyclic. This is to say that P and P' share the same cevian circle.

One well known cyclocevian conjugate pair is the centroid G and orthocenter H. The only point that is its own cyclocevian conjugate is the Gergonne point Ge.

Exercises

1. For the Harvey's Theorem problem worked out in the text, show that the three intersection points of the tangent lines through A, B, and C are concyclic with the focus F of the parabola by finding the equation of the circle through A, B, C, and F.

2. Find Harvey's Circle for the parabola $y = x^2$ with the point $P(-40, 43/2)$.

3. Revisit problem 17 in the *Some Tangency Problems* chapter. This time, use points P, Q, and the two foci of the hyperbola to find the equation of the circle through these four points.

4. Using problem 3, find the normal line through T. Find the points M and N, which are the intersection points of the normal line with the transverse axis and conjugate axis, respectively. Next, find the unique circle equation through the points M, N, P, and Q.

5. Plot the 9-10-17 Heronian triangle with vertices $A(-9, 0)$, $B(8, 0)$, and $C(1/17, 72/17)$. Show that the incircle is $x^2 + (y-2)^2 = 4$. From this show that the angle bisector line for angle C is $y = 38x + 2$. Find the feet of the perpendiculars P and Q from the vertices A and B, respectively. Find the equation of the line PQ. By definition, the line PQ is anti-parallel to the line $y = 0$ with respect to angle C. Therefore, the points A, B, P, and Q are concyclic. Show that the circle going through A, B, P, and Q is $\left(x + \dfrac{1}{2}\right)^2 + y^2 = \dfrac{289}{4}$.

6. Find Lester's Circle and the orthocentroidal circle for the triangle with vertices $A(4, 0)$, $B(0, 3)$, and $C(0, 0)$. Draw and label a complete diagram.

7. Find the two Tucker circles for the triangle in problem 6 using $P(0, 1)$.

8. Find the Taylor circle for the triangle in problem 5.

9. Find the first and second Lemoine circles for the triangle in problem 6.

10. Find Adams circle for the triangle with vertices $A(-4, 0)$, $B(4, 0)$, and $C(0, 3)$.

11. Find Conway's circle for the triangle in problem 10.

12. For the triangle in problem 10, show that the orthocenter H and the circumcenter O are isogonal conjugates. In addition, show that the centroid G and symmedian point K are isogonal conjugates.

13. For the triangle in problem 10, show that the centroid G and orthocenter H are cyclocevian conjugates by showing that they share the same cevian circle.

Lines and Conic Centers

We will now ask the basic question: What is the slope of the line traveling through each of the centers of the ellipses and hyperbolas $\frac{x^2}{a^2} \pm \frac{y^2}{b^2} = 1$ and $\frac{y^2}{a^2} \pm \frac{x^2}{b^2} = 1$ that has a distance d between intersection points? See Figures 1 through 4.

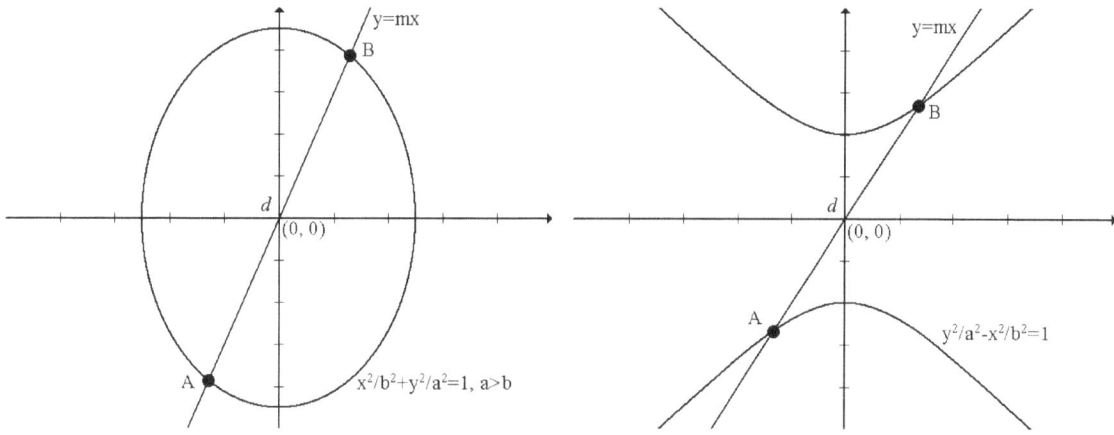

Figure 1: A Tall Ellipse   Figure 2: An Up-Down Hyperbola

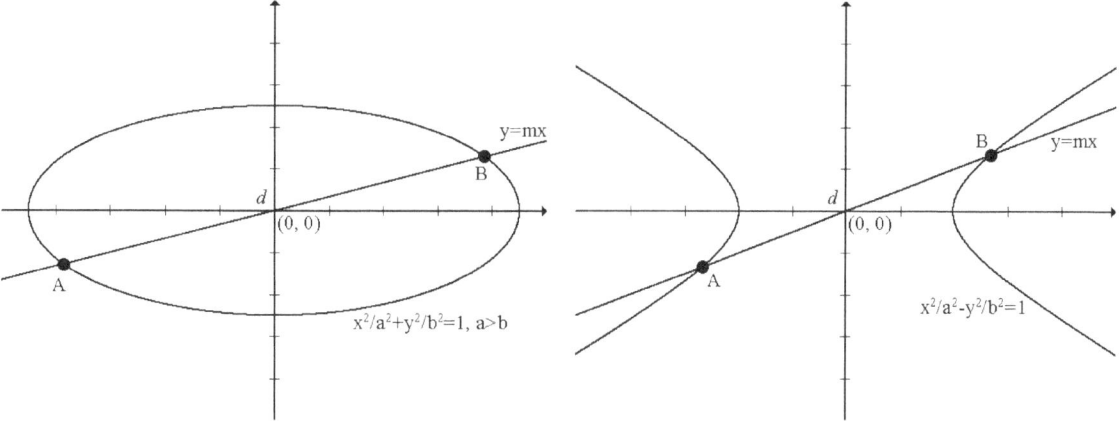

Figure 3. A Wide Ellipse   Figure 4. A Left-Right Hyperbola.

We can do all four of these derivations in the same chapter because they all follow the same format. In addition, there are no surprises algebraically. Thus, one just generates the preliminary slope and distance with an x_0; however, the distance from point A to point B is twice the distance from either point to the origin. Then, one solves the distance

253

equation for x_0. Finally, one inserts the expression generated for x_0 into the preliminary slope equation and simplifies. This generates the desired slope formula. With the proper slope formula in hand, one may then solve for the other variables a^2, b^2, and d.

All of the derivations in this section will be omitted because they all follow the exact same procedure as outlined in the previous paragraph. However, one note is in order, and that is that a^2 and b^2 are positive but do not have the absolute value bars attached when a subtraction is present to force the subtracted quantity to be positive. Why is this? It is because it can be proven from the surrounding equations to that equation that the algebraic quantity is always positive. There are no other surprises in the algebra. Thus, only the results for each situation will be given. At the end, an exercise set will be included. Hopefully, the problems will be scintillating.

The first set of results will apply to the tall ellipse in Figure 1. Observe the discriminant in the slope equation. The peculiar placement of the d's in the numerator and denominator means that the logically connected inequalities $d^2 - 4b^2 \geq 0$ and $4a^2 - d^2 > 0$ must be satisfied. In other words, the distance must be at least $2b$ for the line segment to reach across to the left and right vertices, and the distance must be less than $2a$ so that the line segment won't protrude beyond the boundary of the top and bottom vertices. If d is $2a$, then the slope becomes undefined. Observe that there is no discriminant in the distance equation because everything is positive. Also observe that each of the binomials in the slope equation have a positive term and a negative term, and so a cosmetic difference in this equation can be achieved by multiplying the top and bottom binomials by negative one, which switches the places of the signs in each binomial. However, from a purist point of view, the rendering as depicted is preferred because each binomial must be positive. For example, is $d^2 - 4b^2 > 0$. This factors as $(d + 2b)(d - 2b) > 0$. We can divide both sides of this inequality by $d + 2b$ (a positive quantity). This yields $d - 2b > 0$. We add $2b$ to both sides to yield $d > 2b$, which must be true as per previous comments. A similar result happens with the other binomial. With the a^2 and b^2 equations, set the denominator greater than zero. Replace d^2 with the square of the d equation contents. Simplify, and it is seen that these denominators are always greater than zero in the form they are depicted.

$$m = \pm\sqrt{\frac{a^2(d^2 - 4b^2)}{b^2(4a^2 - d^2)}} \qquad d = \sqrt{\frac{4a^2b^2(m^2 + 1)}{a^2 + b^2m^2}}$$

$$a^2 = \frac{b^2d^2m^2}{4b^2(m^2 + 1) - d^2} \qquad b^2 = \frac{a^2d^2}{4a^2(m^2 + 1) - d^2m^2}$$

254

The second set of results will apply to the up-down hyperbola in Figure 2. Notice that there is a discriminant in the slope equation, which means that $d > 2a$. This must be true for the line segment to stretch across to both branches of the hyperbola. The discriminant in the distance equation means that $|m| > a / b$, which is another way of saying that the magnitude of the slope must be greater than the slope of the asymptote line so that it will eventually hit the hyperbola. The denominator of the b^2 equation can be set greater than zero with a^2 replaced to show that it is always positive.

$$m = \pm\sqrt{\frac{a^2(4b^2 + d^2)}{b^2(d^2 - 4a^2)}} \qquad\qquad d = \sqrt{\frac{4a^2b^2(m^2 + 1)}{b^2m^2 - a^2}}$$

$$a^2 = \frac{b^2d^2m^2}{d^2 + 4b^2(m^2 + 1)} \qquad\qquad b^2 = \frac{a^2d^2}{d^2m^2 - 4a^2(m^2 + 1)}$$

The third set of results will apply to the wide ellipse in Figure 3. Notice that there is a discriminant in the slope formula. Again, because of the peculiar placement of the terms in the numerator and denominator, we deduce that the compound inequality $2b < d \le 2a$ holds, which is the same sort of geometrical consideration as in the tall ellipse. Also, everything is positive under the radical in the distance formula, so there is no discriminant here. The denominators of a^2 and b^2 can be shown to be positive by the methods that served in the earlier groups.

$$m = \pm\sqrt{\frac{b^2(4a^2 - d^2)}{a^2(d^2 - 4b^2)}} \qquad\qquad d = \sqrt{\frac{4a^2b^2(m^2 + 1)}{a^2m^2 + b^2}}$$

$$a^2 = \frac{b^2d^2}{4b^2(m^2 + 1) - d^2m^2} \qquad\qquad b^2 = \frac{a^2d^2m^2}{4a^2(m^2 + 1) - d^2}$$

The fourth and final set of results apply to the left-right hyperbola in Figure 4. The discriminant in the slope formula here shows us that $d \ge 2a$. Also, the discriminant in the distance formula tells us that a lesser-sloped line than the asymptote lines will eventually hit the left-right hyperbola. The denominator of the b^2 equation is always positive as the methodology of demonstration in the earlier groups holds here as well.

$$m = \pm\sqrt{\frac{b^2(d^2 - 4a^2)}{a^2(d^2 + 4b^2)}} \qquad\qquad d = \sqrt{\frac{4a^2b^2(m^2 + 1)}{b^2 - a^2m^2}}$$

$$a^2 = \frac{b^2 d^2}{d^2 m^2 + 4b^2(m^2 + 1)} \qquad\qquad b^2 = \frac{a^2 d^2 m^2}{d^2 - 4a^2(m^2 + 1)}$$

Exercises

As in the other chapters, here are some exercises to try. Note that $|x_0|$ can be written as $\sqrt{x_0^2}$. This will enable it to be pulled under the radical. Bear this in mind when deriving the slope equations.

1. Derive the slope equation for the situation in Figure 1. Solve this slope equation for the other variables d, a^2, and b^2.

2. Derive the slope equation for the situation in Figure 2. Solve this slope equation for the other variables d, a^2, and b^2.

3. Derive the slope equation for the situation in Figure 3. Solve this slope equation for the other variables d, a^2, and b^2.

4. Derive the slope equation for the situation in Figure 4. Solve this slope equation for the other variables d, a^2, and b^2.

5. Compare the contents of the radical in the slope formula for the tall ellipse to the contents of the radical in the slope formula for the wide ellipse. Are these expressions reciprocals? If they are not, provide an explanation of why they are not. [Hint: They are not. A similar phenomenon exists in the slope formulas for the hyperbolas.] What would need to happen for them to be reciprocals?

6. The center of a left-right hyperbola is $(-1, 7\tfrac{1}{8})$. A line with a slope of 2/15 travels through the center of it and intersects it twice with a distance between the intersection points of fourteen units. The distance between the vertices is ten units. Find the equation of this hyperbola, and locate its foci.

7. A wide ellipse has an eccentricity of 0.82. Its center is at the point $(-4, 3)$. A line of slope $m = 5/7$ travels through the center of the ellipse with a distance between intersection points of twelve units. Find the equation of the ellipse, and locate its foci.

8. An up-down hyperbola has an eccentricity of $\sqrt{3}$. Its center is at the point $(6, 9)$. A line of slope fifty travels through the center of the hyperbola with a distance between the intersection points of eleven units. Find the equation of the hyperbola and locate its foci.

9. The length of each of the two focal chords of an ellipse is $\ell = \dfrac{2b^2}{a}$. Of course, the minimum distance possible for a line segment, that passes through the center of the ellipse, to reach across to the closer vertices is $2b$ units. What percent increase must the length of the focal chord undergo to achieve the minimum distance necessary of $2b$? Use this formula to ascertain the minimum percent increase necessary if an ellipse has the values of $a = 5.3$ and $b = 3.5$.

10. A clever designer of problems wants it so that the focal chord of an ellipse would need to be increased by exactly one percent to reach the minimum distance of $2b$ units that occurs in these types of problems. He also wants the eccentricity of the ellipse to be 0.81 because his grandfather is 81 years old. Is this possible? If it is possible, find the equations of the ellipses (one tall ellipse equation and one wide ellipse equation) centered at the origin. If it is not possible, explain why it is not possible.

11. Find the equation of the ellipse $\dfrac{x^2}{9} + \dfrac{y^2}{16} = 1$ if we want a line of slope $m = 2$ to travel through the center of this ellipse with a distance $d = 6$ units between intersection points. Does this problem inform us of anything regarding a formula that does not have a discriminant? We might say that a formula solved for a variable squared has an implied discriminant. What is the range of possible values for d such that this problem can work because six is out of the required interval? Find the resulting equations of the two wide ellipses when we change the given distance of six to the minimum and maximum possible distances. [Hint: Use a sign analysis for when a rational function is greater than zero.]

12. Show that the denominator of $a^2 = \dfrac{b^2 d^2 m^2}{4b^2(m^2+1) - d^2}$ is positive by replacing d^2 with the square of $d = \sqrt{\dfrac{4a^2 b^2 (m^2+1)}{a^2 + b^2 m^2}}$ and simplifying.

13. Show that the denominator of $b^2 = \dfrac{a^2 d^2}{4a^2(m^2+1) - d^2 m^2}$ is positive by replacing d^2 with the square of $d = \sqrt{\dfrac{4a^2 b^2 (m^2+1)}{a^2 + b^2 m^2}}$ and simplifying.

14. Show that the denominator of $b^2 = \dfrac{a^2 d^2}{d^2 m^2 - 4a^2(m^2 + 1)}$ is positive by replacing a^2 with $a^2 = \dfrac{b^2 d^2 m^2}{d^2 + 4b^2(m^2 + 1)}$ and simplifying.

15. Show that the denominator of $a^2 = \dfrac{b^2 d^2}{4b^2(m^2 + 1) - d^2 m^2}$ is positive by replacing d^2 with the square of $d = \sqrt{\dfrac{4a^2 b^2(m^2 + 1)}{a^2 m^2 + b^2}}$ and simplifying.

16. Show that the denominator of $b^2 = \dfrac{a^2 d^2 m^2}{4a^2(m^2 + 1) - d^2}$ is positive by replacing d^2 with the square of $d = \sqrt{\dfrac{4a^2 b^2(m^2 + 1)}{a^2 m^2 + b^2}}$ and simplifying.

17. Show that the denominator of $b^2 = \dfrac{a^2 d^2 m^2}{d^2 - 4a^2(m^2 + 1)}$ is positive by replacing a^2 with $a^2 = \dfrac{b^2 d^2}{d^2 m^2 + 4b^2(m^2 + 1)}$ and simplifying.

18. Find the minimum value for d in the hyperbola $\dfrac{x^2}{a^2} - \dfrac{y^2}{16} = 1$ if we want a line of slope $m = 6$ to travel through the center of the hyperbola. With that value of d, find the equation of the hyperbola.

19. Find the range of possible slopes for a line traveling through the center of the ellipse $\dfrac{x^2}{9} + \dfrac{y^2}{16} = 1$ if we want a distance d between intersection points between seven and seven and a half units.

20. Find the minimum absolute value of the slope of the line traveling through the center of the hyperbola $\dfrac{x^2}{9} - \dfrac{y^2}{b^2} = 1$ with a distance d between intersection points of nine units.

21. The hyperbola $\dfrac{x^2}{9} - \dfrac{y^2}{b^2} = 1$ has an eccentricity of 1.1. A line of slope four travels through the center of it. What is the distance between the intersection points?

22. For the ellipse $\dfrac{x^2}{a^2} + \dfrac{y^2}{10} = 1$, a line of slope two travels through the center. Find a^2 such that the distance between the intersection points of the line and the ellipse is an integer. How many such solutions are possible?

23. We saw in *The Polar Line and Conic Section Diameters* chapter that the slopes of two conjugate diameters is $m_1 m_2 = \dfrac{-b^2}{a^2}$ for a wide ellipse and $m_1 m_2 = \dfrac{-a^2}{b^2}$ for a tall ellipse. Use this and the results of the two ellipses in this chapter to find equations relating the lengths of conjugate diameters of an ellipse.

The Eyeball Theorem

The name of this theorem is due to the way the image looks; it looks like a pair of eyes looking at one another. The very name "eyeball" gives pause for thought. We will eyeball what we have covered thus far in this course. The first thing we discussed was the Pappus line—what a marvel that was!

We have increased our numerical skills, and we now know about things like Midy's theorem and Farey sequences. We found out how to get rational points on first and second degree equations as long as they were nice enough to let us. We also learned to denest radicals—a skill taught once upon a time. We explored new methods for finding points and lines, such as reflecting a point across a line or a triangle.

We then learned about Desargues' theorem. That was a treat in that it opened the doors to later things. We learned what a perspector was and a perspectrix. We even studied a few, such as the centroid, the Prasolov point, and the orthic axis.

We then learned about Pythagorean triples and how to compute them using a variety of approaches. One approach dates back to Euclid, and another happened just a few short years ago. Sometimes, a topic in math can stretch through the centuries before someone takes it up again to find something new about it.

We then learned about the circle and how to resize and move them around. We also employed our friends the Pythagorean triples to find rational points on circles. We also learned to use circles to plot triangles having specified side lengths.

Next, we learned about the Wallace-Simson line. Later, we found out that a pedal triangle from a point on the circumcircle results in a degenerate triangle—i.e. a triangle with collinear vertices. Nevertheless, we did discover some interesting properties about the Wallace-Simson line, such as that the intersection point of two Wallace-Simson lines from diametrically opposed points on the circumcircle lies on the nine-point circle of the original triangle.

Then, we saw the first two chapters of brand new math material that the author discovered. We now understand what he meant when he told the math department head, "There's really something here!" when he was getting his senior project approved. In the second chapter of the author's new material we found a secondary result from his equation: *The Rational Slope and Focal Chord Pieces* theorem, which stated that if we wanted to have two rational lengths for the two pieces of the focal chord as well as a rational slope for the line making up the focal chord, then the product of the two focal chord pieces had to be a perfect square. That's something you can wow your friends with at your next cocktail party.

We learned a few chapters preliminary to the big *Triangle Geometry* chapter after this. Then, we got to see some interesting tangency problems. These were all able to be solved with algebra. Finding tangencies without calculus is a big deal. We could deduce that because we had several chapters dealing with that concept. The present chapter is not excluded.

Finally, we arrived at the *Triangle Geometry* chapter. It is amazing that mathematicians have seen so much from looking at three dots on a grid. The math just does not quit. If the student has gone to Clark Kimberling's website to peruse the *Encyclopedia of Triangle Centers*, then the student *really* understands how deep this subject is. Also, the subject is live, which means that mathematicians are making discoveries about new triangle centers right now.

Most recently, we saw the author's third installment of his new math material. The lines through the centers of the conic sections. For the ellipses, these distances are lengths of the diameters. For the hyperbolas, we can just say that we have found the relationship of the slope, distance, and hyperbolic constants.

Well, that's an eyeball view of what the student has learned thus far in this course. We have this highlight to go along with two more highlights. We have the remaining four chapters of the author's own work to plumb as well. We see that we have come far, but we still have a little way to go to get to the end of the book. Let's press on.

The eyeball theorem is a counter-intuitive theorem like the midpoint quadrilateral theorem. It has a conclusion that does not seem reasonable. Yet, we can compute it to see for ourselves that the counter-intuitive conclusion is true. We now present the eyeball theorem for you to eyeball. With that, the comedian is asked to leave the building.

The Eyeball Theorem: Given that we have two circles such that the center of either is not inside the other, then the four tangent lines from the centers to the other circle cuts the circles exactly the same distance apart. Note that we are not saying the distance between the two points of tangency; rather, we are saying the distance between the cut points— where the tangent lines for one circle *cut* the other circle—is the same.

We illustrate this with a worked out example. Test the conclusion of the eyeball theorem on the circles C_1: $x^2 + y^2 = 1$ and C_2: $(x-3)^2 + y^2 = 4$.

The circles are tangent at the point $(1, 0)$, but that is not of any concern to the theorem. It only states that the center of one circle cannot be inside the other circle. We now try to figure out how to find the tangent lines from the center of the circles to the other circle. Our old friend the polar line will work. We may consider the center of a circle as the pole for the other circle.

We transform the equation of C_1 to $xx_0 + yy_0 = 1$ and put in the pole $P_1(3, 0)$ for x_0 and y_0 to get $3x = 1$ or $x = 1/3$. Thus, the polar line for C_1 is $x = 1/3$. Similarly, we transform the equation of C_2 to $x^2 - 3(2x) + 9 + y^2 = 4$ and make the substitutions into it to get $xx_0 - 3(x + x_0) + yy_0 = -5$. Then, using the pole $P_2(0, 0)$, we get that the polar line for C_2 is the vertical line $x = 5/3$.

We now proceed to solve the two systems of equations involving a circle and its polar line. This will yield the four points of tangency on the two circles. We construct the two systems below.

$$\begin{cases} x^2 + y^2 = 1 \\ x = \dfrac{1}{3} \end{cases} \qquad \begin{cases} (x - 3)^2 + y^2 = 4 \\ x = \dfrac{5}{3} \end{cases}$$

The solution to the first system is $\left(\dfrac{1}{3}, \dfrac{\pm 2\sqrt{2}}{3} \right)$, and the solution to the second system is $\left(\dfrac{5}{3}, \dfrac{\pm 2\sqrt{5}}{3} \right)$. Next, we will use these points of tangency to find the equations of the two tangent lines for each circle. Thus, for C_1, we find that the equation of the tangent lines through $\left(\dfrac{1}{3}, \dfrac{\pm 2\sqrt{2}}{3} \right)$ and $(3, 0)$ are $y = \dfrac{\pm \sqrt{2}}{4} x \mp \dfrac{3\sqrt{2}}{4}$. Note: This is two equations. We get the top version of the signs and the bottom version of the signs.

Similarly, for C_2 we find that the equations of the tangent lines through $\left(\dfrac{5}{3}, \dfrac{\pm 2\sqrt{5}}{3} \right)$ and $(0, 0)$ are $y = \dfrac{\pm 2\sqrt{5}}{5}$. This is even easier to see the two tangent line equations than the ones for C_1.

Now that we have the equations of the tangent lines, we next need to find the coordinates of the four cut points that we will label U_1, U_2, U_3, and U_4. We have already used the letter C for circle, so we should pick another letter to avoid confusion. We will put U_1 and U_2 on C_1, and we will put U_3 and U_4 on C_2. Then, armed with the coordinates of the two cut points on each circle, we can use the distance formula to find the distance between each pair of cut points. Of course, the eyeball theorem tells us that this is supposed to be the same distance.

In this particular example, we can make use of some symmetry. Specifically, the y-coordinates of U_1 and U_2 will be opposite in sign. Also, the y-coordinates of U_3 and U_4 will be opposite in sign. So to find U_1, we solve the system of equations involving C_1 and the tangent line containing the point T_3. And to find U_3, we solve the system of equations involving C_2 and the tangent line containing the point T_1. We get the following systems.

$$\begin{cases} x^2 + y^2 = 1 \\ y = \dfrac{2\sqrt{5}}{5} x \end{cases} \qquad \begin{cases} (x-3)^2 + y^2 = 4 \\ y = \dfrac{\sqrt{2}}{4} x - \dfrac{3\sqrt{2}}{4} \end{cases}$$

The solution to the first system is $\left(\dfrac{\sqrt{5}}{3}, \dfrac{2}{3} \right)$. Thus, U_1 and U_2 are $\left(\dfrac{\sqrt{5}}{3}, \dfrac{\pm 2}{3} \right)$. The solution to the second system is $\left(\dfrac{9 - 4\sqrt{2}}{3}, \dfrac{2}{3} \right)$. Thus, U_3 and U_4 are $\left(\dfrac{9 - 4\sqrt{2}}{3}, \dfrac{\pm 2}{3} \right)$.

Because the y-coordinates match, we may safely conclude that the distances between the cut points is exactly the same. However, in less clear situations, we could use the distance formula to compute $d(U_1, U_2)$ and $d(U_3, U_4)$ to see if they were in fact the same. We present all of our efforts with a diagram. See Figure 1. It looks like two eyes touching. We can say for sure now that the two principals see eye to eye. (I couldn't resist one more joke).

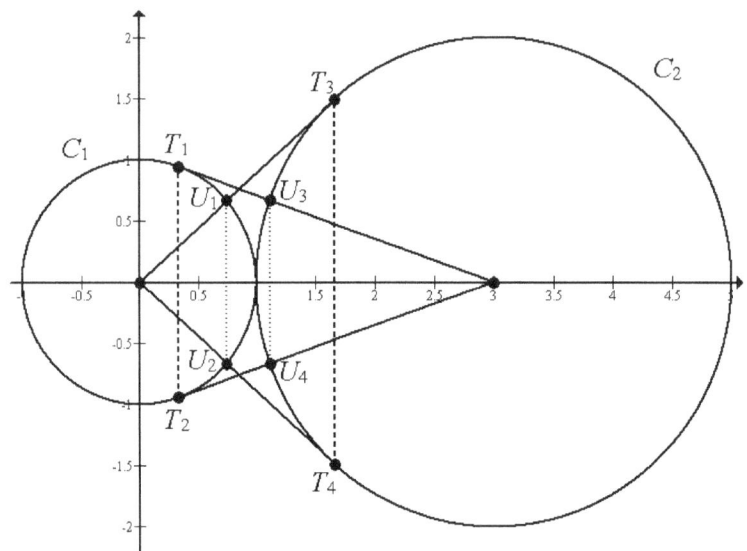

Figure 1: The Eyeball Theorem

Exercises

1. Show that the eyeball theorem is true for the circles C_1: $x^2 + y^2 = 1$ and C_2: $x^2 + (y-6)^2 = 9$.

2. Show that the eyeball theorem is true for the circles C_1: $(x-5)^2 + (y+7)^2 = 16$ and C_2: $(x+9)^2 + (y-6)^2 = 169$.

Lines and Ellipses: Part 1

We shift our focus (no pun intended) to ellipses. For our first derivation with ellipses, we ask the basic question, "What is the slope of the line going through a left or right vertex of the tall ellipse $\dfrac{x^2}{b^2} + \dfrac{y^2}{a^2} = 1$ that touches the ellipse again a distance of d units away?" The situation is captured in the below diagram. Incidentally, the author is aware that an ellipse technically only has only two vertices that are on the major axis, but a need exists to call all four of the four outermost points of the ellipse by some name. Until a satisfactory remedy is at hand, the most obvious name of vertex will be used.

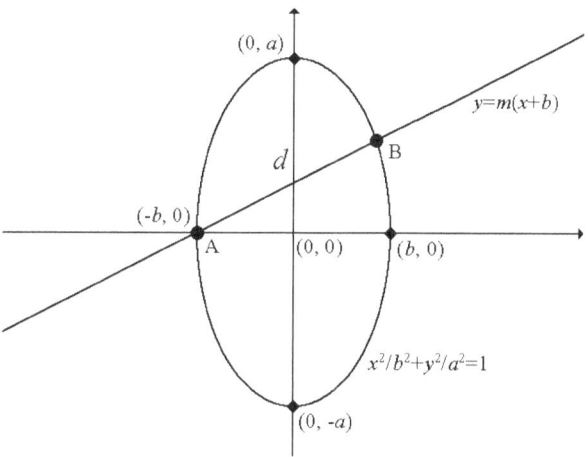

Figure 1: Through Left Vertex of Tall Ellipse

To begin the derivation, we center the ellipse at the origin because having nonzero h and k values complicate the derivation too much. What are the coordinates of the points A and B? Well, the point A is easy because it is just the point $(-b, 0)$. The point B must be in terms of some value x_0. So, we plug the x_0 into the equation for the ellipse as x and solve for y. Then, the coordinates of point B are $\left(x_0, \ \pm\sqrt{\dfrac{a^2(b^2 - x_0^2)}{b^2}} \right)$. The binomial $(b^2 - x_0^2)$ is guaranteed to be positive since in the picture $x_0 < b$. Thus, the contents of the radical will never be negative. Next, we need to know the preliminary slope of the line through the points A and B. It is the following:

$$m = \frac{y_2 - y_1}{x_2 - x_1} = \frac{0 - \pm\sqrt{\dfrac{a^2(b^2 - x_0^2)}{b^2}}}{-b - x_0} = \frac{\pm\sqrt{\dfrac{a^2(b^2 - x_0^2)}{b^2}}}{b + x_0} = \pm\sqrt{\dfrac{a^2(b - x_0)}{b^2(b + x_0)}}$$

We also need a preliminary distance from point A to point B.

$$d = \sqrt{(y_2 - y_1)^2 + (x_2 - x_1)^2} = \sqrt{(x_0 + b)^2 + \left(\pm\sqrt{\dfrac{a^2(b^2 - x_0^2)}{b^2}} - 0\right)^2}$$

$$d = \sqrt{(x_0 + b)^2 + \dfrac{a^2(b^2 - x_0^2)}{b^2}}$$

We need to solve this distance for x_0 to plug into the m formula. This will change the preliminary slope to a derived slope because it will be in terms of a, b, and d. However, there is a lot of algebra required to do this; or as Robert Frost might say, "And I have miles to go before I sleep."

$$d^2 = (x_0 + b)^2 + \frac{a^2(b^2 - x_0^2)}{b^2}$$
$$b^2 d^2 = b^2(x_0 + b)^2 + a^2(b^2 - x_0^2)$$
$$b^2 d^2 = b^2(x_0^2 + 2bx_0 + b^2) + a^2 b^2 - a^2 x_0^2$$
$$b^2 d^2 = b^2 x_0^2 + 2b^3 x_0 + b^4 + a^2 b^2 - a^2 x_0^2$$
$$(a^2 - b^2)x_0^2 + (-2b^3)x_0 + (b^2 d^2 - a^2 b^2 - b^4) = 0$$

We observe that $c^2 = a^2 - b^2$ from the equation to find the coordinates of the focus. Also, this equation is quadratic in form, so we can use the quadratic formula.

$$x_0 = \frac{2b^3 \pm \sqrt{4b^6 - 4c^2(b^2 d^2 - a^2 b^2 - b^4)}}{2c^2} = \frac{b^3 \pm b\sqrt{b^4 + c^2(a^2 + b^2 - d^2)}}{c^2}$$

Next, we plug this into the preliminary slope equation to eliminate the x_0. This will, of course, leave the equation with just the variables a, b, c, d, and m. We will get c out of the equation at some point to leave just the variables a, b, d, and m.

$$m = \pm\sqrt{\frac{a^2(b - x_0)}{b^2(b + x_0)}} = \pm\sqrt{\frac{a^2\left(b - \dfrac{b^3 \pm b\sqrt{b^4 + c^2(a^2 + b^2 - d^2)}}{c^2}\right)}{b^2\left(b + \dfrac{b^3 \pm b\sqrt{b^4 + c^2(a^2 + b^2 - d^2)}}{c^2}\right)}}$$

$$m = \pm\sqrt{\frac{a^2[bc^2 - b^3 \mp b\sqrt{b^4 + c^2(a^2 + b^2 - d^2)}]}{b^2[bc^2 + b^3 \pm b\sqrt{b^4 + c^2(a^2 + b^2 - d^2)}]}}$$

$$m = \pm\sqrt{\frac{a^2[(a^2 - b^2) - b^2 \mp \sqrt{b^4 + c^2(a^2 + b^2 - d^2)}]}{b^2[(a^2 - b^2) + b^2 \pm \sqrt{b^4 + c^2(a^2 + b^2 - d^2)}]}}$$

$$m = \pm\sqrt{\frac{a^2[a^2 - 2b^2 \mp \sqrt{b^4 + c^2(a^2 + b^2 - d^2)}]}{b^2[a^2 \pm \sqrt{b^4 + c^2(a^2 + b^2 - d^2)}]}}$$

At this point, we need to rationalize the denominator under the big radical. To this end, we let $w = b^4 + c^2(a^2 + b^2 - d^2)$. This will simplify the appearance of the algebra in the next few steps.

$$m = \pm\sqrt{\frac{a^2 - 2b^2 \mp \sqrt{w}}{b^2(a^2 \pm w)} \cdot \frac{(a^2 \mp \sqrt{w})}{(a^2 \mp \sqrt{w})}} = \pm\sqrt{\frac{a^2(a^4 \mp a^2\sqrt{w} - 2a^2b^2 \pm 2b^2\sqrt{w} \mp a^2\sqrt{w} + w)}{b^2(a^4 - w)}}$$

$$m = \pm\sqrt{\frac{a^2(a^4 \mp 2a^2\sqrt{w} \pm 2b^2\sqrt{w} - 2a^2b^2 + w)}{b^2(a^4 - w)}}$$

$$m = \pm\sqrt{\frac{a^2(a^4 \mp 2(a^2 - b^2)\sqrt{w} - 2a^2b^2 + w)}{b^2(a^4 - w)}}$$

$$m = \pm\sqrt{\frac{a^2(a^4 \mp 2c^2\sqrt{w} - 2a^2b^2 + w)}{b^2(a^4 - w)}}$$

Now, we can let w return back to the mess we let it equal.

$$m = \pm\sqrt{\frac{a^2(a^4 \mp 2c^2\sqrt{b^4 + c^2(a^2 + b^2 - d^2)} - 2a^2b^2 + b^4 + c^2(a^2 + b^2 - d^2))}{b^2(a^4 - [b^4 + c^2(a^2 + b^2 - d^2)])}}$$

267

$$m = \pm \sqrt{\frac{a^2(a^4 - 2a^2b^2 + b^4 \mp 2c^2\sqrt{b^4 + c^2(a^2 + b^2 - d^2)} + c^2(a^2 + b^2 - d^2))}{b^2(a^4 - [b^4 + (a^2 - b^2)(a^2 + b^2 - d^2)])}}$$

$$m = \pm \sqrt{\frac{a^2((a^2 - b^2)^2 \mp 2c^2\sqrt{b^4 + c^2(a^2 + b^2 - d^2)} + c^2(a^2 + b^2 - d^2))}{b^2(a^4 - [b^4 + (a^2 - b^2)(a^2 + b^2 - d^2)])}}$$

$$m = \pm \sqrt{\frac{a^2(c^4 \mp 2c^2\sqrt{b^4 + (a^2 - b^2)(a^2 + b^2 - d^2)} + c^2(a^2 + b^2 - d^2))}{b^2(a^4 - [b^4 + a^4 + a^2b^2 - a^2d^2 - a^2b^2 - b^4 + b^2d^2])}}$$

$$m = \pm \sqrt{\frac{a^2(c^4 \mp 2c^2\sqrt{b^4 + a^4 + a^2b^2 - a^2d^2 - a^2b^2 - b^4 + b^2d^2} + c^2(a^2 + b^2 - d^2))}{b^2d^2(a^2 - b^2)}}$$

$$m = \pm \sqrt{\frac{a^2(c^4 \mp 2c^2\sqrt{a^4 - a^2d^2 + b^2d^2} + c^2(a^2 + b^2 - d^2))}{b^2c^2d^2}}$$

Note that each of the terms under the big radical has a c^2. This enables us to perform a major cancellation.

$$m = \pm \sqrt{\frac{a^2(a^2 - b^2 \mp 2\sqrt{a^4 - a^2d^2 + b^2d^2} + a^2 + b^2 - d^2)}{b^2d^2}}$$

$$m = \pm \sqrt{\frac{a^2(2a^2 - d^2 \mp 2\sqrt{a^4 - a^2d^2 + b^2d^2})}{b^2d^2}}$$

$$\therefore m = \pm \frac{a}{bd}\sqrt{2a^2 - d^2 \pm 2\sqrt{a^4 - d^2(a^2 - b^2)}} \quad \text{Next, we will solve this for } d.$$

$$d = \frac{2a^2b\sqrt{m^2 + 1}}{a^2 + b^2m^2}, b > 0$$

This is the derived slope of the line that we have been seeking, along with having this equation solved for d, one of its other variables. Since d is a positive quantity, we force it to be so by restricting b to be positive; however, we have been assuming a and b to be positive from the outset. So explicitly labeling $b > 0$ is really more of a reminder. We can also observe that the $\pm$ and $\mp$ in the slope formula are independent, so that's why they can both be $\pm$ without any subscript symbols like we saw earlier in one of the problems we were working through in the *Some Tangency Problems* chapter.

We come now to the next phase of our work in this chapter. We will first observe that the distance from the left vertex to the opposite vertex is $2b$. We will ask a natural question: Where else is the distance from the vertex $2b$? We can answer that by setting the new distance equation equal to $2b$ and solving for m. This would yield the slope (or slopes) of any and all lines meeting this criterion. The notation $m|_{d=2b}$ denotes "<u>when</u> d is $2b$."

$$d = \frac{2a^2 b\sqrt{m^2+1}}{a^2+b^2 m^2} = 2b$$

$$2a^2 b\sqrt{m^2+1} = 2b(a^2+b^2 m^2)$$

$a^2\sqrt{m^2+1} = (a^2+b^2 m^2)$ Next, square both sides.

$a^4(m^2+1) = (a^2+b^2 m^2)^2 = a^4+2a^2 b^2 m^2+b^4 m^4$ Note: m is to the 4th power.

$a^4 m^2 + a^4 = a^4 + 2a^2 b^2 m^2 + b^4 m^4$

$a^4 m^2 = 2a^2 b^2 m^2 + b^4 m^4$

$0 = 2a^2 b^2 m^2 + b^4 m^4 - a^4 m^2$

$b^4 m^4 + 2a^2 b^2 m^2 - a^4 m^2 = m^2(b^4 m^2 + 2a^2 b^2 - a^4) = 0$

$\therefore m^2 = 0, b^4 m^2 + 2a^2 b^2 - a^4 = 0$ Note: $m = 0$ is a root of multiplicity 2.

$$m = \pm\sqrt{\frac{a^4 - 2a^2 b^2}{b^4}} = \frac{\pm a}{b^2}\sqrt{a^2 - 2b^2} = m|_{d=2b}$$ Note: These are the other 2 roots.

Well, our investigation has revealed that when $m = 0$, our distance from point A to point B will be $2b$. However, we already knew that. But, it has also shown us a single positively sloped line in terms of a and b that also has a distance of $2b$. Because of the presence of the radical, this positively sloped line only exists when $a^2 \geq 2b^2$. When we plot this line on the ellipse that has it, it will sever the top half of the ellipse into two regions. In the region containing the vertex (i.e. point A), all of the distances will be less than $2b$. In the other region (i.e. the one containing the opposite vertex), all of the distances will be greater than $2b$. The only places where the distance is exactly equal to $2b$ are at the intersection point of this positively sloped line with the ellipse (i.e. point B) and at the opposite vertex. Let us call the region of lesser distances R_1 and the other region R_2. Let us call the line that separates R_1 and R_2 the *demarcation* line.

The distance from point A to the top (or bottom) vertex of the ellipse is the length of the hypotenuse of the right triangle formed with a short leg of b and a long leg of a. This hypotenuse length is computed to be $\sqrt{a^2 + b^2}$ using the Pythagorean theorem. Thus, we let $d = \sqrt{a^2 + b^2} = 2b$. This shows us that when $a^2 = 3b^2$, the demarcation line will run right through the top vertex. So R_1 and R_2 will be, respectively, the top left and top right quarter arcs of the ellipse. Consequently, when $2b^2 < a^2 < 3b^2$, the demarcation line will fall on the opposite quarter arc of the ellipse containing the originating vertex. When $a^2 > 3b^2$, then the demarcation line will fall on the same quarter arc of the ellipse.

We would next like to evaluate the interior discriminant $a^4 - d^2(a^2 - b^2)$ in the slope formula.

$$a^4 - d^2(a^2 - b^2) \geq 0$$
$$-d^2(a^2 - b^2) \geq -a^4$$
$$d^2 \leq \frac{a^4}{a^2 + b^2} = \left(\frac{a^2}{c}\right)^2$$
$$\therefore d \leq \frac{a^2}{c}$$

Well, suppose we like testing the limits of things and let $d = a^2/c$. What would happen to the slope then? Note that we call $d = a^2/c$ the maximum distance or d_{max}.

$$m = \pm \frac{a}{bd}\sqrt{2a^2 - d^2 \pm 2\sqrt{a^4 - d^2(a^2 - b^2)}} \quad \text{Let } d = \frac{a^2}{c}.$$

$$m = \frac{\pm a}{b \cdot \dfrac{a^2}{c}} \cdot \sqrt{2a^2 - \left(\frac{a^2}{c}\right)^2 \pm 2\sqrt{a^4 - \left(\frac{a^2}{c}\right)^2(a^2 - b^2)}}$$

$$m = \frac{\pm c}{ab}\sqrt{2a^2 - a^4/c^2 \pm 0} = \frac{\pm\sqrt{a^2 - b^2}}{ab}\sqrt{\frac{2a^2c^2 - a^4}{a^2 - b^2}} = \frac{\pm\sqrt{2a^4 - 2a^2b^2 - a^4}}{ab}$$

$$\therefore m = \frac{\pm\sqrt{a^2 - 2b^2}}{b}$$

We were able to find out a couple of nice things here. The first is that we found an upper bound for d in terms of a and b. This is the maximum possible d, called d_{max} or max d. The second is that when we let d attain this maximum value of a^2/c, we were able to discover the corresponding slope equation. This m we will call m_{dmax} or $m_{max\,d}$. Regarding m_{dmax}, we observe the same contents under the radical as in the $m|_{d=2b}$

equation. Thus, the values for $d_{\max}$ and $m_{d\max}$ will also only exist when $a^2 \geq 2b^2$. Sure, $d_{\max}$ can always be computed, but it won't have any physical meaning or interpretation for the ellipse when $a^2 < 2b^2$.

The last thing we will say about the tall ellipse in this section is that when $a^2 \leq 2b^2$, we will always get just one positively sloped line of a requested distance within the bounds of the ellipse, of course. When $a^2 > 2b^2$, we will get one or two positively sloped lines of a requested distance—it depends on the requested distance's size in relation to $2b$. If $d < 2b$, then only one positively sloped line will occur in R_1. We can say that the max d line splits R_2 into two regions, and we can call the region of greater ordinates "Upper R_2" and the region of lesser ordinates "Lower R_2." The word ordinates is a fancy term for y-coordinates. If $2b < d < d_{\max}$, then the two positively sloped lines in R_2 will be placed around the max d line and between the $m = 0$ and demarcation lines. In other words, one line will fall in Upper R_2, and the other line will fall in Lower R_2. If $d = a^2/c = d_{\max}$, then only one positively sloped line will occur in R_2 on the max d line. Lastly, if $d = 2b$, then one line is $m = 0$, and the other line is the demarcation line. We will delineate an ellipse with the relevant parts we've discussed. See Figure 2.

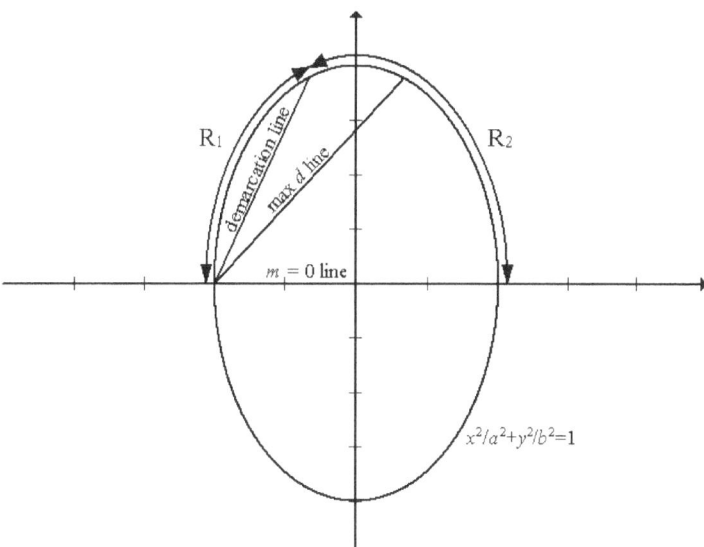

Figure 2: Tall Ellipse Showing the Demarcation, $m = 0$, and Max d Lines

Starting with the distance equation $d = \dfrac{2a^2 b\sqrt{m^2 + 1}}{a^2 + b^2 m^2}, b > 0$, we observe that all quantities are positive since b is assumed to be positive. No absolute value bars are

271

necessary in this equation because of the nature of its layout. When we solve this for a^2, we get the following:

$$a^2 = \frac{b^2 dm^2}{2b\sqrt{m^2+1} - d}$$

From the methodology in the *Lines and Conic Centers* chapter, we saw how to use the related equations from a group of equations to show that a denominator was always positive. We can employ that same methodology here. Set the denominator greater than zero. Substitute d from the equation for d we just got a moment ago and simplify. We may now reach a conclusion about d.

$$2b\sqrt{m^2+1} - d > 0 \quad \therefore d < 2b\sqrt{m^2+1}.$$

Regarding these two terms in the denominator, we must observe that $0 \le d \le a^2/c$, and so d is at a disadvantage with $2b\sqrt{m^2+1}$ because this term is always greater than $2b$. If we let d attain its maximum value of a^2/c, we must also use the $m_{d\max}$ expression for m. What happens then to our inequality?

$$d \le 2b\sqrt{m^2+1}$$

$$\frac{a^2}{c} \le 2b\sqrt{\frac{a^2 - 2b^2}{b^2} + 1}$$

$$\frac{a^2}{c} \le 2b\sqrt{\frac{a^2 - 2b^2 + b^2}{b^2}}$$

$$\frac{a^2}{c} \le 2\sqrt{a^2 - b^2}$$

$$\frac{a^2}{c} \le 2c$$

$$a^2 \le 2c^2 = 2a^2 - 2b^2$$

$$\therefore a^2 \ge 2b^2 \qquad\qquad \text{Hence, this is only possible with an ellipse having } a^2 \ge 2b^2.$$

We tried to make d as big as possible here. Our effort to make the denominator negative only shows us something that we already knew. Namely, the only time we can have that d max value is when $a^2 \ge 2b^2$. Our inequality has not been falsified, which means that it is always true. The results of this investigation show us that we do not need absolute value bars around the denominator of the a^2 formula. Next, we will solve the distance equation for the b^2 variable.

$$b^2 = \frac{a^2}{d^2 m^4}\left(a\sqrt{m^2+1} \pm \sqrt{a^2+(a^2-d^2)m^2}\right)^2$$

We do see some subtraction signs floating around here. Do we need absolute value bars anywhere? From the presence of a subtraction sign in the radical, we see a discriminant. Hence, $a^2+(a^2-d^2)m^2 \geq 0$ implies that $d \leq \frac{a}{|m|}\sqrt{m^2+1}$. Is this max value for d larger than our theoretical maximum of a^2/c? The student will be asked to show that it is. Since it is, then we know that the second radical will never have its contents go negative by replacing $d = \frac{a}{|m|}\sqrt{m^2+1}$ and simplifying. Hence, absolute value bars are not necessary within the second radical of the b^2 equation.

What about having two distinct values for b^2? The answer to this question must be left aside for now because upon substituting d in the b^2 equation, we do wind up clearing both radicals. The plus version then boils down to b^2, and the minus version boils down to $a^4/(b^2 m^4)$. The b^2 result for the plus version is fine, obviously. The only requirement we have that pertains to the minus version result is that we know that $a^2 > b^2$. Thus, we set $a^4/(b^2 m^4) < a^2$ and proceed to evaluate this. We conclude that $m^2 > a/b$, which might be saying something relevant. For now, we will just have to leave in the minus version.

We now turn our attention to the wide ellipse with a line going through the left or right vertex. Can we find the slope of the line given the distance? It turns out that, because the only difference in the initial setup for the tall and wide ellipses is that the a's and the b's are switched, all we have to do is switch the a's and the b's in the final formulas. The equation for a wide ellipse is the following:

$$\frac{x^2}{a^2} + \frac{y^2}{b^2} = 1$$

The picture of a wide ellipse with a line going through its left vertex intersecting the ellipse again d units away is depicted in Figure 3. The coordinates of point A are $(-a, 0)$. Thus, everywhere there was an a, there is now a b and vice versa, so we are justified in making the switch at the final step. So, the formula for the slope changes from this:

$$m = \pm\frac{a}{bd}\sqrt{2a^2 - d^2 \pm 2\sqrt{a^4 - d^2(a^2-b^2)}}$$ This is m for a tall ellipse.

To this: $m = \pm \dfrac{b}{ad} \sqrt{2b^2 - d^2 \pm 2\sqrt{b^4 - d^2(b^2 - a^2)}}$

$\therefore m = \pm \dfrac{b}{ad} \sqrt{2b^2 - d^2 \pm 2\sqrt{b^4 + d^2(a^2 - b^2)}}$ This is m for a wide ellipse, but hold on!

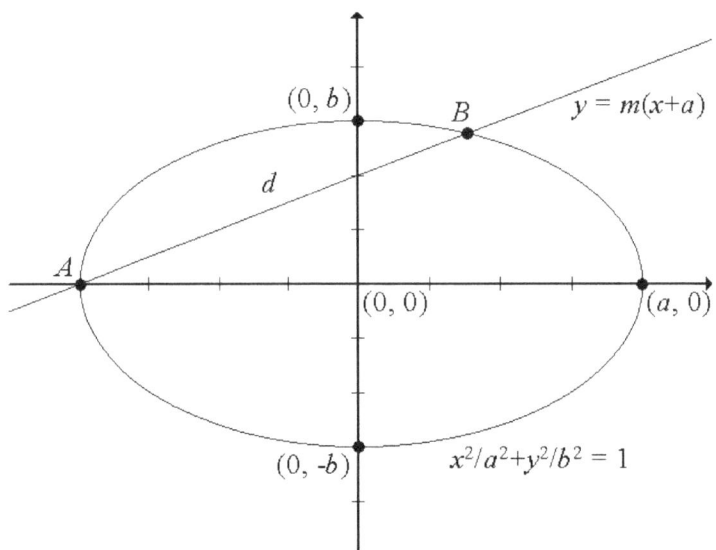

Figure 3: A Line Through the Left Vertex of a Wide Ellipse

If we play the old switcheroo game for the other three equations and simplify, it leads us to the following:

$$d = \frac{2ab^2\sqrt{m^2 + 1}}{a^2 m^2 + b^2} \qquad\qquad b^2 = \frac{a^2 dm^2}{2a\sqrt{m^2 + 1} - d}$$

$$a^2 = \frac{b^2}{d^2 m^4}\left(b\sqrt{m^2 + 1} \pm \sqrt{b^2 + (b^2 - d^2)m^2}\right)^2$$

From the "hold on" comment above, note that we are able to eliminate the interior minus version of the slope equation. The computation is horrendous. But it can be shown that setting the minus version of the contents of the outer radical (which includes the inner radical as well as its contents) greater than zero, and replacing d with the equation above after squaring, we get the following: $\dfrac{-4b^2(a^4 m^2 + b^2(2a^2 - b^2))}{(a^2 m^2 + b^2)^2} > 0$. This is false

274

because $2a^2 - b^2$ is positive, so only the plus version of the contents of the big radical is true. Thus, we can declare that our slope formula is now the following (cue the trumpet fanfare):

$$\therefore m = \pm \frac{b}{ad} \sqrt{2b^2 - d^2 + 2\sqrt{b^4 + d^2(a^2 - b^2)}}$$ This is m for a wide ellipse.

In a way, we get some relief because we see only one positively sloped line is possible for a wide ellipse for a line going through its left or right vertex intersecting the ellipse again a specified distance d away. Could it really have been true for a second such line? The proof of this says no. No such second line is possible. Our intuition tells us there would not have been such a second positively sloped line, but demonstrating it was quite a computational feat. This exercise won't be assigned because of its difficulty, but the ambitious student may attempt it, of course.

There is no maximum distance formula because it is a wide ellipse. The maximum distance possible, $2|a|$, is all the way across to the opposing vertex. So, in this type of ellipse, only one positively sloped line is possible. The equation for b^2 has a subtraction sign suggesting the possibility of its value going negative, but the student will be asked in the exercises to demonstrate that that is not a concern. The a^2 equation has a minus version that resists attempts to show it is always less than b^2. Hence, both versions of the a^2 equation must be reported for now.

We have two more situations with the ellipse to cover in this chapter. We can have a line going through the top or bottom vertices of the tall and wide ellipses. Let us do the line going through the top or bottom vertex of the tall ellipse next. Because the vertex position has changed from the x-axis to the y-axis, we need to implement the idea of the inverse function. A slope, m, between two points on one function will have a slope $1/m$ between those points with the coordinates switched on the inverse function. Hence, all we need to do is to recognize that the top vertex on a tall ellipse is the inverse point of the right vertex on a wide ellipse. Therefore, if we take the reciprocal of the slope equation that we most recently obtained, we will have the slope equation for the top and bottom vertices of the tall ellipse. Doing this, of course, saves us a lot of derivation work.

$$m = \pm \frac{b}{ad} \sqrt{2b^2 - d^2 + 2\sqrt{b^4 + d^2(a^2 - b^2)}}$$ This is right/left vertex on a wide ellipse.

$$\therefore m = \frac{\pm ad}{b\sqrt{2b^2 - d^2 + 2\sqrt{b^4 + d^2(a^2 - b^2)}}}$$ This is top/bottom vertex on a tall ellipse.

Replacing m with $1/m$ in the other three equations and simplifying produces the following three equations for the tall ellipse with line through the top/bottom vertex:

$$a^2 = \frac{b^2 m^2}{d^2}\left(b\sqrt{m^2+1} \pm \sqrt{b^2(m^2+1)-d^2}\right)^2$$

$$b^2 = \frac{a^2 d}{2a\,|m|\,\sqrt{m^2+1}-dm^2} \qquad\qquad d = \frac{ab^2\,|m|\,\sqrt{m^2+1}}{a^2+b^2 m^2}$$

We get more inequality demonstrations to inspect because we see subtraction signs aplenty. These will appear in the exercises. We carried the plus version only from the slope equation from the last set. However, we reintroduce the minus version to get two possible positive slopes for the last set of equations.

From here, we switch a and b in all four of these equations (with the $\pm$ reinserted in the slope equation as just discussed) to produce the four equations pertaining to a wide ellipse with a line going through the top/bottom vertex.

$$d = \frac{2a^2 b\,|m|\,\sqrt{m^2+1}}{a^2 m^2 + b^2} \qquad\qquad m = \frac{\pm bd}{a\sqrt{2a^2-d^2 \pm 2\sqrt{a^4-d^2(a^2-b^2)}}}$$

$$a^2 = \frac{b^2 d}{2b\,|m|\,\sqrt{m^2+1}-dm^2} \qquad\qquad b^2 = \frac{a^2 m^2}{d^2}\left(a\sqrt{m^2+1} \pm \sqrt{a^2(m^2+1)-d^2}\right)^2$$

All of the concepts presented in the tall ellipse with a line through the left/right vertex are also here in this last set of four equations that pertain to the wide ellipse with a line going through the top/bottom vertex. Thus, we have a demarcation line, a max d line, the max d distance, the regions R_1 and R_2, and the rest. The reason is that the picture is just a cookie cutter image of the other just rotated ninety degrees. The difference is only in the manifestation of things like Upper R_2 and Lower R_2 becoming Right R_2 and Left R_2.

The last item we will mention is that these four equations are sort of dancing in a square dance. The singer for the square dance calls out the dance moves like "inverse slope" and "switch a and b." The equations do the dance moves and change appearance during the dance. If we do one more step from where we are at now by doing the inverse slope, we will wind up at the tall ellipse with a line going through the left/right vertex, which completes the square dance by winding back up where the dancers started.

Exercises

1. Find the equation of the negatively sloped line going through the right vertex of the ellipse $\dfrac{(x-1)^2}{11} + \dfrac{(y+2)^2}{7} = 1$ that intersects the ellipse again 21 units away.

2. Determine that the denominator of $a^2 = \dfrac{b^2 dm^2}{2b\sqrt{m^2+1} - d}$ is always positive using

$d = \dfrac{2a^2 b\sqrt{m^2+1}}{a^2 + b^2 m^2}$ and simplifying the inequality.

3. Use the maximum value of d from $d \le \dfrac{a}{|m|}\sqrt{m^2+1}$, and show that this value is larger

than $\dfrac{a^2}{c}$.

4. Regarding the ellipse in problem #1, a line perpendicular to the line $5x - 6y = 4$ goes through the left vertex. What is the distance between the intersection points of the line and ellipse?

5. The ellipse $3x^2 + ax + 5y^2 - 2y = 12$ has a line of slope 2 traveling through its right vertex. The distance between the intersection points is 8 units. Find the value(s) of a.

6. Show that the denominator of $b^2 = \dfrac{a^2 dm^2}{2a\sqrt{m^2+1} - d}$ is positive by using

$d = \dfrac{2ab^2\sqrt{m^2+1}}{a^2 m^2 + b^2}$ and simplifying.

7. Find the equations of the two positively sloped lines going through the right vertex of the ellipse $\dfrac{(x+3)^2}{2} + \dfrac{(y-4)^2}{5} = 1$ that intersect the ellipse 2.85 units away. Before you find these lines, predict the quarter-arcs where they will have their intersection points.

8. Show that the denominator of $b^2 = \dfrac{a^2 d}{2a\,|m|\,\sqrt{m^2+1} - dm^2}$ is always positive by using

$d = \dfrac{ab^2\,|m|\,\sqrt{m^2+1}}{a^2 + b^2 m^2}$ and simplifying.

9. Regarding the ellipse in problem #7, find the equation of the line going through the left vertex that generates the maximum distance between intersection points. What is this maximum distance?

10. Solve the slope formula for the tall ellipse for d. Show all work.

11. Solve the distance formula for a wide ellipse for a^2. Show all work.

12. Find the equation of the two positively sloped lines going through the right vertex of the ellipse $\dfrac{(x-3)^2}{3} + \dfrac{(y-2)^2}{10} = 1$ that intersect the ellipse 3.5 units away. Before you find these lines, predict the quarter-arcs where they will have their intersection points.

13. Regarding the ellipse in problem #12, find the equation of the line going through the left vertex that generates the maximum distance between intersection points. What is this maximum distance?

14. Show that the denominator of $a^2 = \dfrac{b^2 d}{2b\,|m|\,\sqrt{m^2+1} - dm^2}$ is always positive by

using $d = \dfrac{2a^2 b\,|m|\,\sqrt{m^2+1}}{a^2 m^2 + b^2}$ and simplifying.

15. Show that the inner discriminant of $m = \dfrac{\pm bd}{a\sqrt{2a^2 - d^2 \pm 2\sqrt{a^4 - d^2(a^2 - b^2)}}}$ is

positive by using $d = \dfrac{2a^2 b\,|m|\,\sqrt{m^2+1}}{a^2 m^2 + b^2}$ and simplifying.

278

Lines and Hyperbolas: Part 1

For our next derivation, we ask the basic question, "What is the equation of the line that goes through a vertex of the hyperbola $\dfrac{x^2}{a^2} - \dfrac{y^2}{b^2} = 1$ and touches the hyperbola again d units away?" However, we do not need to derive this equation formally because it is almost identical to the ellipse derivation in the previous chapter. The similarities in the structure of these two slope formulas are telling. The situation is captured in the below diagram. See Figure 1.

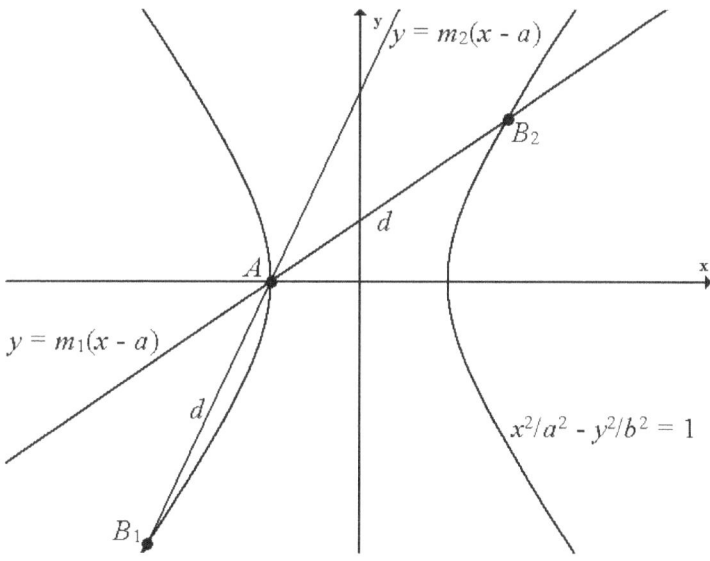

Figure 1: Line Through Hyperbola's Left Vertex Showing Two Possible Lines with $m > 0$

The first thing to notice is that there are two distinct positively sloped lines. This will always happen when the requested distance is large enough to get you over to the other branch on the hyperbola. Of course, the minimum requested distance to produce two distinct non-negatively sloped lines is the distance $2|a|$ between the two vertices. As long as d is larger than that, we'll see two distinct positively sloped lines. So, writing this in symbols, $d > 2|a|$ will do this. If $d < 2|a|$, there will only be one positively sloped line that has both intersection points on the same branch of the hyperbola. Of course, a is assumed to be greater than zero, so the absolute value bars are really more of a reminder. We produce the relevant slope equation.

$$m = \frac{\pm b}{ad}\sqrt{2b^2 + d^2 \pm 2\sqrt{b^4 + d^2(a^2 + b^2)}}$$

As with the slope formula going through a left/right vertex through a tall ellipse, we see nested radicals. The inner radical does not have a discriminant because of only additions and squarings. However, because of the subtraction portion of the $\pm$ symbol, there is a discriminant within the outer radical. Thus, the student can show that the minus version of this discriminant set greater than or equal to zero evaluates to $d \geq 2|a|$. Next, let's put in d as $2|a|$ into the slope formula. Then, we have the following:

$$m = \frac{\pm b}{ad}\sqrt{2b^2 + d^2 \pm 2\sqrt{b^4 + d^2(a^2 + b^2)}}$$

$$m = \frac{\pm b}{a(2|a|)}\sqrt{2b^2 + (2|a|)^2 \pm 2\sqrt{b^4 + (2|a|)^2 \cdot (a^2 + b^2)}} = \ldots = \frac{\pm b}{a^2}\sqrt{2a^2 + b^2} , 0$$

This shows us that at a distance d of $2|a|$ units, we have three lines because the expression under the radical is always positive. Of course, if $d > 2|a|$ units, we get two pairs of lines, which are the two lines of opposite slope on the branch of the hyperbola containing the vertex as well as the two lines of opposite slope that reach to the other branch of the hyperbola.

We can solve the slope equation for d, and this equation for d can be solved for the other variables a^2 and b^2. We will not show the work, however, because these questions will be asked in the exercises at the end. Note that there are two values for both a^2 and b^2 when given the rest of the necessary values. This means that there are two different hyperbolas meeting those criteria. It is clear that the denominator of the d equation can go negative because any magnitude slope is allowed. Thus, we need absolute value bars around this denominator. However, it is unclear if the denominator of b^2 can go negative, so we place the absolute value bars around the denominator to ensure that it will not.

$$d = \frac{2ab^2\sqrt{m^2 + 1}}{\left|a^2m^2 - b^2\right|} \qquad b^2 = \frac{a^2dm^2}{\left|d \pm 2a\sqrt{m^2 + 1}\right|}$$

$$a^2 = \frac{b^2}{d^2m^4}\left[b\sqrt{m^2 + 1} \pm \sqrt{(b^2 + d^2)m^2 + b^2}\right]^2$$

We turn our attention to the up-down hyperbola. To find these four equations, we will employ our old friend, the inverse slope. Thus, everywhere there is an m in the above four equations, we will replace it with $1/m$ and simplify. When we perform this task, we will get the four equations pertaining to an up-down hyperbola with a line going through one of the vertices. We arrive at the following:

$$m = \frac{\pm ad}{b\sqrt{2b^2 + d^2 \pm 2\sqrt{b^4 + d^2(a^2 + b^2)}}} \qquad d = \frac{2ab^2|m|\sqrt{m^2 + 1}}{|a^2 - b^2m^2|}$$

$$a^2 = \frac{b^2m^2}{d^2}\left[b\sqrt{m^2 + 1} \pm \sqrt{(m^2 + 1)b^2 + d^2}\right]^2 \qquad b^2 = \frac{a^2d}{|dm^2 \pm 2a|m|\sqrt{m^2 + 1}|}$$

Exercises

As always, here are some exercises to try with these equations based on the hyperbola. We have omitted the derivations of these formulas in the text so that the student may obtain the results asked for in the exercises. Note that they follow closely the ellipse derivation studied in the *Lines and Ellipses Part 1* chapter.

1. Derive the equation $m = \frac{\pm b}{ad}\sqrt{2b^2 + d^2 \pm 2\sqrt{b^4 + d^2(a^2 + b^2)}}$ from scratch by following the method in the *Lines and Ellipses Part 1* chapter.

2. Show that for the equation $m = \frac{\pm b}{ad}\sqrt{2b^2 + d^2 \pm 2\sqrt{b^4 + d^2(a^2 + b^2)}}$, its discriminant set greater than or equal to zero for the minus version yields $d \geq 2|a|$, which is the same thing we concluded by looking at the picture.

3. When we replace d with $2a$ in the slope equation, we obtained $m = \frac{\pm b}{a^2}\sqrt{2a^2 + b^2}$, 0. Fill in the missing steps to this conclusion.

4. Find the equation of the *positively* sloped line traveling through the left vertex of the hyperbola $2x^2 + 7y - 11 = 5y^2$ and intersects the hyperbola again six units away. If there are two such equations, find both.

5. Given the hyperbola $5y^2 - 12 = kx^2 + 7y$, what is the value of k if we request that a line of slope three goes through the bottom vertex of the hyperbola and intersects the hyperbola again nine units away? If there are two such values for k, find both. Write the equations for the obtained hyperbola(s) in standard form. [Hint: There are two such hyperbolas. Use both the plus and minus version of the appropriate equation.]

6. Solve $m = \dfrac{\pm b}{ad}\sqrt{2b^2 + d^2 \pm 2\sqrt{b^4 + d^2(a^2 + b^2)}}$ for d.

7. Solve $d = \dfrac{2ab^2\sqrt{m^2 + 1}}{\left|a^2m^2 - b^2\right|}$ for a^2.

8. Solve $d = \dfrac{2ab^2\sqrt{m^2 + 1}}{\left|a^2m^2 - b^2\right|}$ for b^2.

9. Show that $d = \dfrac{2ab^2\sqrt{m^2 + 1}}{\left|a^2m^2 - b^2\right|}$ becomes $d = \dfrac{2ab^2|m|\sqrt{m^2 + 1}}{\left|a^2 - b^2m^2\right|}$ by replacing m in the first equation with $1/m$ and simplifying.

10. Show that $a^2 = \dfrac{b^2}{d^2m^4}\left[b\sqrt{m^2 + 1} \pm \sqrt{(b^2 + d^2)m^2 + b^2}\right]^2$ becomes the equation $a^2 = \dfrac{b^2m^2}{d^2}\left[b\sqrt{m^2 + 1} \pm \sqrt{(m^2 + 1)b^2 + d^2}\right]^2$ by replacing m in the first equation with $1/m$ and simplifying.

11. Show that $b^2 = \dfrac{a^2dm^2}{\left|d \pm 2a\sqrt{m^2 + 1}\right|}$ becomes the equation $b^2 = \dfrac{a^2d}{\left|dm^2 \pm 2a|m|\sqrt{m^2 + 1}\right|}$ by replacing m in the first equation with $1/m$ and simplifying.

12. Given the hyperbola $\dfrac{(y-3)^2}{11} - \dfrac{(x+2)^2}{13} = 1$, what is the equation of the negatively sloped line traveling through the top vertex of the hyperbola that intersects it again five units away. If there are two such equations, find both.

13. The hyperbola $7x^2 - 5 = 2x + ky^2$ has a line of slope $\dfrac{\sqrt{3}}{5}$ traveling through its right vertex that intersects it again one unit away. Find the value or values of k that make this happen. Put the equation(s) for the hyperbola(s) in standard form. Use a graphical software package to depict the situation(s), and verify that the distance is one unit.

14. The hyperbola $3y^2 + 2 = 5y + kx^2$ has a line of slope 0.2 traveling through its left vertex that intersects the hyperbola again eight units away. Find the value or values of k that make this happen. Put the equation(s) of the hyperbola in standard form. Use a graphical software package to depict the situation(s), and verify that the distance is eight units.

15. The hyperbola $2 - ky^2 = 3x - 4x^2$ has a line perpendicular to the line $3x - 9y = 2$ traveling through its bottom vertex and intersects the hyperbola again 12.34 units away. Find the value or values of k that make this happen. Put the equation(s) of the hyperbola or hyperbolas in standard form. Use a graphical software package to depict the situation(s), and verify that the distance is 12.34 units.

16. The hyperbola $5x^2 - 7x + 4 = 2y^2 + 3y - 8$ has a line of slope 17 traveling through one of its vertices. How far away is the other intersection point?

17. A square hyperbola (also called equilateral hyperbola or rectangular hyperbola) is one where $a^2 = b^2$ in $\dfrac{(x-h)^2}{a^2} - \dfrac{(y-k)^2}{b^2} = 1$ and $\dfrac{(y-k)^2}{a^2} - \dfrac{(x-h)^2}{b^2} = 1$. What is the value or values of k for the hyperbola $ky^2 - kx^2 = 7.55$ if a line of slope two goes through a vertex and intersects the hyperbola again four units away? Put the equation(s) of the hyperbola(s) in standard form, and use a graphical software package to depict the situation(s). Verify that the distance is four units.

18. The hyperbola $\dfrac{x^2}{9} - \dfrac{y^2}{16} = 1$ has a line traveling through its left vertex intersecting the hyperbola again at the point (3.75, 3). Find the equation of the other positively sloped line going through the left vertex that has the same distance between the intersection points. Show by direct computation that when the values of these two slopes are inserted into the

$$d = \frac{2ab^2\sqrt{m^2+1}}{\left|a^2m^2 - b^2\right|}$$ equation that the same distance is obtained.

19. The hyperbola $\dfrac{x^2}{9} - \dfrac{y^2}{16} = 1$ has a line of slope 1/3 traveling through the right vertex. How far away is the other intersection point?

20. The hyperbola $\dfrac{x^2}{a^2} - \dfrac{y^2}{16} = 1$ has a line of slope 2 going through its left vertex and intersects the hyperbola again at the distance asked for in question 19. Of course, one answer for a^2 is 9. What is the other possibility for a^2? Show a complete graph with all relevant details.

21. The hyperbola $5x^2 - 7x + 4 = 2y^2 + 3y - 8$ has a line of slope 17 traveling through one of its vertices. The distance between intersection points is the same as that for problem 16. Put this equation for the hyperbola in standard form. Replace the contents for a^2 with the variable a^2. Find the other value of a^2 for this problem in the style of problem 20. As in problem 20, show a complete graph with all relevant details.

22. The hyperbola $\dfrac{x^2}{9} - \dfrac{y^2}{b^2} = 1$ has a line of slope 2 going through its left vertex and intersects the hyperbola again at the distance asked for in question 19. Of course, one answer for b^2 is 16. What is the other possibility for b^2? Show a complete graph with all relevant details.

23. The hyperbola $5x^2 - 7x + 4 = 2y^2 + 3y - 8$ has a line of slope 17 traveling through one of its vertices. The distance between intersection points is the same as that for problem 16. Put this equation for the hyperbola in standard form. Replace the contents for b^2 with the variable b^2. Find the other value of b^2 for this problem in the style of problem 20. As in problem 20, show a complete graph with all relevant details.

24. Theorem: If a triangle's vertices are on a rectangular hyperbola, then the orthocenter H is also on the hyperbola. Find three rational points on the hyperbola $\dfrac{x^2}{9} - \dfrac{y^2}{9} = 1$ to make a triangle, and show that the orthocenter of your triangle is on the hyperbola.

Johnson's Theorem

Roger A. Johnson (1890 – 1954), an American geometer, discovered this theorem in 1916. I guess the ancient Greek mathematicians missed the bus on this one. Johnson wrote an influential book *Modern Geometry – An Elementary Treatise on the Geometry of the Triangle and the Circle*. This book is being reprinted by Dover, and it can be purchased at book stores that carry Dover books; but it may need to be ordered. Note that this is a graduate level text dealing with proofs on theorems from triangle geometry.

Johnson's Theorem: Given three circles of equal radii r that all pass through a common point H, then the circle through their other three intersections has the same radius r.

We will work an example first. Then, we will discuss this in further detail. Our example with three circles all going through the point $H(2, 3)$ is the following:

$$\left\{\begin{array}{l} C_1 : (x+1)^2 + (y-3)^2 = 9 \\ C_2 : (x-2)^2 + y^2 = 9 \\ C_3 : (5x-19)^2 + (5y-3)^2 = 225 \end{array}\right\}$$

The student should verify that the requirements of Johnson's theorem are met before proceeding to find Johnson's circle. In other words, confirm that these three circles go through the common point (2, 3) and that they all have the same radius of $r = 3$.

To begin, it looks like C_2 is the best place to start. We will solve C_2 for y, and then insert it into C_1 and C_3. This will transform C_1 and C_3 into one-variable equations in x that can be solved. C_2 solved for y is the following: $y = \pm\sqrt{9-(x-2)^2}$. Inserting this into C_1, expanding everything out and simplifying produces the quadratic equation $x^2 - x - 2 = 0$. This factors and has the roots $x = 2$ and $x = -1$. Of course, we knew that $x = 2$ was a root because of the common intersection point. Plugging $x = -1$ into C_1 produces $y = 6$ or $y = 0$. Plugging $x = -1$ into C_2 produces $y = \pm 0$, which is a root of multiplicity 2. Since both the C_1 and C_2 equations produce $y = 0$, we conclude that when $x = -1$, then $y = 0$. We form the ordered pair (-1, 0). This is the "other" intersection point between C_1 and C_2. This point is on Johnson's circle.

Next, we insert $y = \pm\sqrt{9-(x-2)^2}$ into C_3 and repeat. We arrive at the polynomial equation $5x^2 - 29x + 38 = 0$, which factors. We get the roots $x = 2$ and $x = 19/5$. Again, we knew that $x = 2$ already, so we put $x = 19/5$ into both the C_2 and C_3 equations to find that the mutual y-coordinate produced is $y = $ -12/5. Hence, the ordered pair (19/5, -12/5) is the point on Johnson's circle from circles C_2 and C_3.

We have one more point to find on Johnson's circle. It is the other intersection point between C_1 and C_3. It looks better to solve C_1 for y to place into C_3. So we get that C_1 produces $y = 3 \pm \sqrt{9 - (x+1)^2}$. Placing this into C_3, expanding out and simplifying produces the polynomial equation $5x^2 - 14x + 8 = 0$. This factors, and we get the two roots $x = 2$ and $x = 4/5$. We knew the first root, so we plug the second root into C_1 and C_3 to look for the mutual y-coor-dinate. We find that it is $y = 3/5$. We form the ordered pair (4/5, 3/5) and conclude that this is the third point on Johnson's circle that we needed.

Use any method to find the equation of the circle through the set of points {(-1, 0), (19/5, -12/5), (4/5, 3/5)}. Several are presented in *The Circle* chapter. One warning is in order here. Do not assume the radius is the same as the radius of each of the three given circles. Let the three points tell us what that radius is. That way, we confirm Johnson's theorem instead of assuming it. Performing this feat, we find that Johnson's circle is

$$\left(x - \frac{4}{5} \right)^2 + \left(y + \frac{12}{5} \right)^2 = 9.$$ The picture of our situation is captured below in Figure 1.

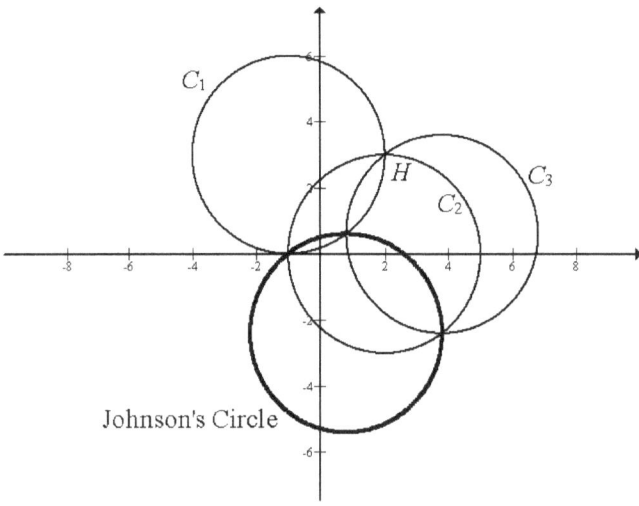

Figure 1: The Three Given Circles, *H*, and Johnson's Circle

A property of the three given circles in a Johnson's circle problem is that their centers form another circle with the same radius. For our example problem, this circle is depicted in Figure 2 with a dashed circle, and H is the center of this circle. The three given circle centers form a triangle called the Johnson triangle; thus, the dashed circle is known as the

286

circumcircle of the Johnson triangle. The Johnson triangle is shown in the figure with solid lines, and it is the smaller of the two triangles with solid lines.

The three other intersection points form what is called the reference triangle. This triangle is depicted in the figure with a dotted line. The reason H is chosen as the letter representing the common intersection point of the three given circles is that H is the orthocenter of the reference triangle. The orthocenter H for our example is clearly outside the dotted reference triangle shown in the figure. A circle centered at H (the common intersection point of the three given circles) with radius $2r$ is tangent to each of the three given circles. This circle is the big circle shown in the figure. Each point of tangency to the big circle for a given circle is the diametric opposite to H for that circle. Reflect H across the center of each circle in turn, and one can see graphically that this is true. The three points of tangency form what is called the doubled Johnson triangle. The doubled Johnson triangle is also called the anticomplementary triangle. This is the bigger triangle shown in the figure with solid lines.

The three other intersection points of the three given circles are midpoints of a side of the doubled Johnson triangle. In other words, the vertices of the reference triangle are the midpoints of the sides of the doubled Johnson triangle. If we form perpendicular lines through the midpoints of the doubled Johnson triangle, these three lines all pass through H.

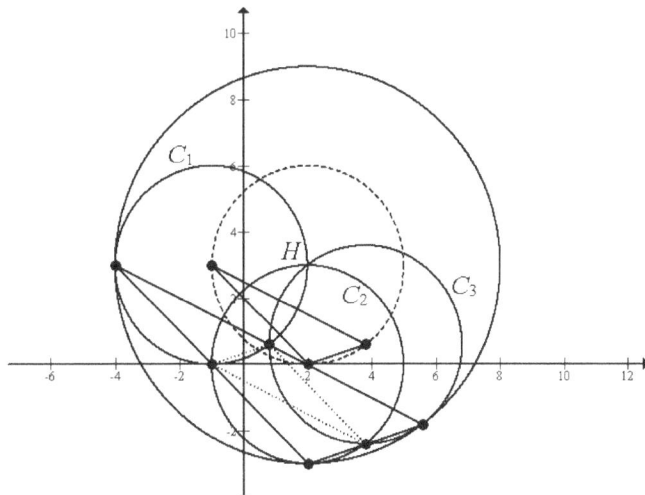

Figure 2: A Depiction of Several Properties of the Three Given Circles

The Johnson triangle and the reference triangle have a lot in common. They are congruent triangles. They have the same nine-point circle and the same Euler line. These two triangles are also in perspective with the nine-point center as the perspector.

We will demonstrate an interesting way to find the points diametrically opposite to H for the three given circles to conclude this chapter. This is the method outlined in *The Circle* chapter. We will illustrate the method just for C_3.

The point $H(2, 3)$ is a rational point on C_3: $(5x - 19)^2 + (5y - 3)^2 = 225$. We can consider this point as (x_{C3}, y_{C3}). The center of C_3 is $(h, k) = (19/5, 3/5)$. In *The Circle* chapter, we focused on translating a rational point on the unit circle to a different circle. We will reverse the process to find out what the point H corresponds to on the unit circle. In order to do that, we need to solve the translation equations $x_C = r(x_u) + h$ and $y_C = r(y_u) + k$ for x_u and y_u. Doing this produces the following:

$$x_u = \frac{x_C - h}{r}, \quad y_u = \frac{y_C - k}{r}$$

We identify that $r = 3$, $x_C = x_{C3} = 2$, $y_C = y_{C3} = 3$. We already identified (h, k) above. With these identifications, we compute $(x_u, y_u) = (-3/5, 4/5)$. It is a snap to find the diametric opposite to (x_u, y_u), which we will notate as (x'_u, y'_u). All we have to do is change the signs of the coordinates of (x_u, y_u). Thus, $x'_u = -x_u$, and $y'_u = -y_u$. We do this to get $(x'_u, y'_u) = (3/5, -4/5)$.

Finally, we translate the rational point (x'_u, y'_u) to C_3 giving us (x'_{H3}, y'_{H3}). Thus, we compute the following:

$$x'_{H3} = r(x'_u) + h = 3\left(\frac{3}{5}\right) + \frac{19}{5} = \frac{28}{5}$$

$$y'_{H3} = r(y'_u) + k = 3\left(-\frac{4}{5}\right) + \frac{3}{5} = -\frac{9}{5}$$

Hence, we conclude that the diametric opposite point to H on C_3 is the following:

$$H'_3 = (28/5, -9/5)$$

We remark that this point could have been obtained with a system of equations involving the line through the center of C_3 and H with the circle C_3, which is the method outlined in *The Deluxe Toolkit* chapter. However, the method just presented involves only first-degree equations without the messiness of quadratics in spite of the fact that we would already know one of the roots.

Exercises

1. Find Johnson's circle for the following three circles intersecting at $H(3, 4)$.

$$\begin{cases} C_1 : (x-7)^2 + (y-1)^2 = 25 \\ C_2 : (x-6)^2 + y^2 = 25 \\ C_3 : (x-3)^2 + (y-9)^2 = 25 \end{cases}$$

2. The parts below pertain to the system of circles in question 1.

   a) Delineate the vertices of the Johnson triangle.

   b) Delineate the vertices of the reference triangle.

   c) Find the equation of the circumcircle of Johnson's triangle. Note that the radius of this circle is the same as the radius of the three given circles.

   d) Show that H is the orthocenter of the reference triangle by finding the orthocenter of the reference triangle and comparing it to H.

   e) Find the equation of the circumcircle of three given circles (which is the same circle as the circumcircle of the doubled Johnson triangle). Delineate the vertices of the doubled Johnson triangle.

   f) Find the midpoints of the doubled Johnson triangle and compare them to the other intersection points of the three given circles.

   g) Find the equations of the perpendicular lines to the sides of the doubled Johnson triangle through their midpoints. Confirm in each case that plugging in H produces an identity. This shows that these perpendicular lines all pass through H.

   h) Find the circumcenter of the Johnson triangle, and compare it to H.

   i) Show that the Johnson triangle and the reference triangle have the same nine-point circle and the same Euler line.

   j) Show that the Johnson triangle and the reference triangle are in perspective by finding the perspector. Show that this is the same as the nine-point center. By Desargues theorem, the perspectrix exists. Explain why there is a problem in finding the perspectrix.

3. Writing in mathematics: Set up a situation of three circles of equal radius intersecting at a common intersection point H. Perform the steps in question 2 on it. Note: Answers will vary.

Lines and Ellipses: Part 2

In this chapter and the next, we will find the equations when our line goes through the focus of the conic. In this chapter, we will study the situation pertaining to the ellipse, and the hyperbola will occupy our attention in the next chapter of the book.

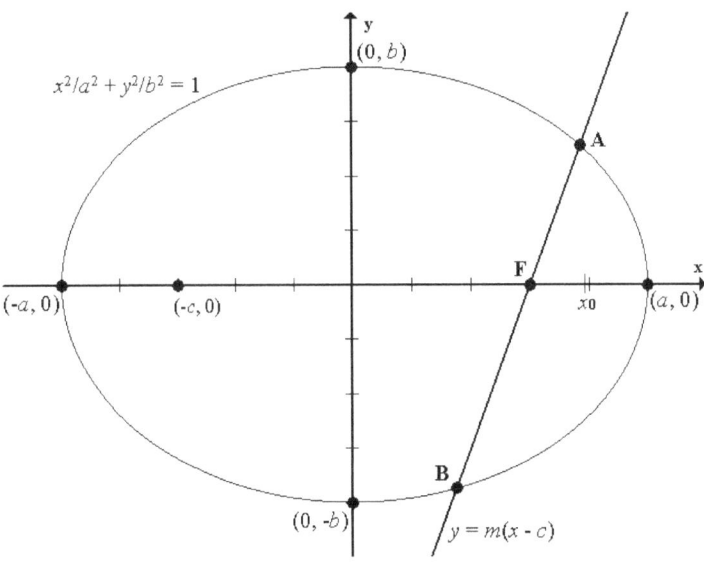

Figure 1: A General Positively Sloped Line Through Right Focus

The subsequent math will be derived from this picture. So let's talk about it in some detail before we proceed. Observe that we have a wide ellipse on our hands because a is associated with the x^2. The coordinates of the focus are $(c, 0)$. We have chosen the positively sloped line (i.e. $m > 0$) to go through the right focus. We have limited x_0 to be between c and a inclusive (i.e. $c \leq x_0 \leq a$). The x-coordinate of point A is x_0, so point A will be on the upper half of the ellipse with the same restrictions on the x-coordinate that x_0 has. Point B will then be on the lower half of the ellipse with the restriction of the x-coordinate to $[-a, c]$. If we let $x_0 = c$, then points A and B will be on the right latus rectum of the ellipse. If we let $x_0 = a$, then point A will be $(a, 0)$ and point B will be $(-a, 0)$.

The first thing we need to do is to solve the ellipse equation for y to get two functions. When we do this, we get $y = \pm \dfrac{a}{b} \sqrt{a^2 - x^2}$. When we plug in x_0 into the top function, we get a preliminary point A of $(x_0, \frac{a}{b}\sqrt{a^2 - x_0^2}$). From this point and the focus $(c, 0)$, we derive the preliminary slope of the line.

$$m = \frac{\frac{b}{a}\sqrt{a^2 - x_0^2} - 0}{x_0 - c} = \frac{b\sqrt{a^2 - x_0^2}}{a(x_0 - c)}$$

Next, we need to solve the system of equations of the ellipse and line for x.

Ellipse: $y = \frac{\pm b}{a}\sqrt{a^2 - x^2}$

Line: $y = \frac{b\sqrt{a^2 - x_0^2}}{a(x_0 - c)}(x - c)$ Note: Equate, since both are equal to y.

$\frac{\pm b}{a}\sqrt{a^2 - x^2} = \frac{b\sqrt{a^2 - x_0^2}}{a(x_0 - c)}(x - c)$ Note: Multiply by a/b first before squaring.

$(a^2 - x^2)(x_0 - c)^2 = (a^2 - x_0^2)(x - c)^2$ After some more work, we get the following:

$(b^2 - 2a^2 + 2cx_0)x^2 + 2c(a^2 - x_0^2)x + (2ax_0 - 2a^2c - b^2x_0)x_0 = 0$

Let's talk about this last line for a bit. Clearly, it looks like a real mess. Do we do the quadratic formula on it and hope for the best? No. Remember, in several chapters now, we discussed that when we know one of the solutions, then the quadratic ought to factor. Well, we know one of the solutions, namely $x = x_0$. So we construct our two factors by dividing the lead term by x and the constant term by $-x_0$. Observe that the constant term has an x_0 conveniently factored out, so all that remains is to change the sign. As a precaution, we should FOIL it out anyway to check that the middle term is OK. The student can check this independently.

$(x - x_0)[(b^2 - 2a^2 + 2cx_0)x - (2a^2x_0 - 2a^2c - b^2x_0)] = 0$

$\therefore x = x_0, x = \frac{2a^2x_0 - 2a^2c - b^2x_0}{2cx_0 - 2a^2 + b^2}$ The second answer is the x-coordinate for point B.

Next, we compute the y-coordinate for point B using the line equation.

$$y = \frac{b\sqrt{a^2 - x_0^2}}{a(x_0 - c)}(x - c) = \frac{b\sqrt{a^2 - x_0^2}}{a(x_0 - c)}\left(\frac{2a^2 x_0 - 2a^2 c - b^2 x_0}{2cx_0 - 2a^2 + b^2} - c\right)$$ After more work, we get:

$$y = \frac{b^3\sqrt{a^2 - x_0^2}}{a(2cx_0 - 2a^2 + b^2)} \qquad \therefore \text{Point } B \text{ is } \left(\frac{2a^2 x_0 - 2a^2 c - b^2 x_0}{2cx_0 - 2a^2 + b^2}, \frac{b^3\sqrt{a^2 - x_0^2}}{a(2cx_0 - 2a^2 + b^2)}\right).$$

We can check points A and B by letting $x_0 = a$, and evaluate all four coordinates of these two points. Point A should evaluate to $(a, 0)$, and point B should evaluate to $(-a, 0)$. We then let $x_0 = c$, and evaluate all four coordinates of these two points. Point A should evaluate to $\left(c, \frac{b^2}{a}\right)$, and point B should evaluate to $\left(c, \frac{-b^2}{a}\right)$. We know that the length of a latus rectum is $\frac{2b^2}{a}$, and the distance between point A and point B when $x_0 = c$ is this length. These little checks on our algebra enable us to proceed with confidence.

In a manner analogous to going through the focus of the parabola, we will split the distance between A and B into two pieces. This works because A, F, and B are collinear.

$$d(A, F) = \sqrt{(x_0 - c)^2 + \left(\frac{b}{a}\sqrt{a^2 - x_0^2}\right)} = \dots = \left|\frac{a^2 - cx_0}{a}\right| = \frac{a^2 - cx_0}{a}$$

$$d(B, F) = \sqrt{\left(\frac{2a^2 x_0 - 2a^2 c - b^2 x_0}{2cx_0 - 2a^2 + b^2} - c\right)^2 + \left(\frac{b^3 \cdot \sqrt{a^2 - x_0^2}}{a(2cx_0 - 2a^2 + b^2)}\right)^2} = \dots = \frac{b^2}{a}\left|\frac{cx_0 - a^2}{2cx_0 - 2a^2 + b^2}\right|$$

$$= \frac{b^2(a^2 - cx_0)}{a(2a^2 - 2cx_0 - b^2)}$$

How did we dispense with the absolute value signs here? We perform an analysis on the expressions based on our original assumption that $c \leq x_0 \leq a$. We wish to determine which of $a^2 - cx_0$ and $cx_0 - a^2$ is positive for our numerator expression in both distances. If we let $x_0 = a$, we conclude that $a^2 - cx_0$ is positive because $a(a - c)$ is positive. For the denominator of the second distance, we select $2a^2 - 2cx_0 - b^2$ as positive because of the following analysis:

292

$$2a^2 - 2cx_0 - b^2$$
$$a^2 + a^2 - b^2 - 2cx_0$$
$$a^2 + c^2 - 2cx_0 \qquad \text{Since an } x_0 \text{ term involves a subtraction we let } x_0 = a\text{—its largest value.}$$
$$a^2 + c^2 - 2ac$$
$$a^2 - 2ac + c^2$$
$$(a - c)^2 \qquad \text{This is clearly positive, which confirms our choice.}$$

Thus, we can now compute $d = d(A, F) + d(B, F)$, and we arrive at the following:

$$d = \frac{a^2 - cx_0}{a} + \frac{b^2(a^2 - cx_0)}{a(2a^2 - 2cx_0 - b^2)} = \ldots = \frac{2(a^2 - cx_0)^2}{a(2a^2 - 2cx_0 - b^2)}$$

We now have a preliminary slope and a preliminary distance both expressed in terms of x_0. It appears that the best route from here is to solve the slope equation for x_0, and insert that expression into the distance equation in the places where there is an x_0. When we do this and simplify (and a major factor winds up canceling), we get the following:

$$d = \frac{2ab^2(m^2 + 1)}{a^2m^2 + b^2}$$

Our efforts have been striving for this equation. Note that a, b, and m are all present in the numerator and denominator. It remains for us to solve this equation for its other variables. When we do this, we get the following:

$$m = \pm\sqrt{\frac{b^2(2a - d)}{a(ad - 2b^2)}} \qquad\qquad b^2 = \frac{a^2dm^2}{2a(m^2 + 1) - d}$$

$$a^2 = \frac{b^2}{d^2m^4}\left(b(m^2 + 1) \pm \sqrt{b^2m^4 + (2b^2 - d^2)m^2 + b^2}\right)^2$$

Some remarks are in order here. First, we prove that the denominator of the b^2 equation is positive because it happens to have a subtraction sign in it. Thus, we get the following:

$$2a(m^2+1)-d>0$$

$$2a(m^2+1)-\frac{2ab^2(m^2+1)}{a^2m^2+b^2}>0$$

$$2a(m^2+1)\left[\frac{a^2m^2+b^2-b^2}{a^2m^2+b^2}\right]>0$$

$$2a(m^2+1)\left[\frac{a^2m^2}{a^2m^2+b^2}\right]>0$$

The last line is clearly true, so the rendering of the b^2 equation is correct. It should be clear that $d \le 2a$ making the numerator of the slope equation always positive. However, the denominator of the slope equation also has a subtraction sign. Thus, it will be left as an exercise for the student to show that the denominator of the slope equation is always positive.

When we set the contents of the radical in the a^2 equation greater than or equal to zero, we begin the proof of determining whether this radical's contents can go negative or not. We replace d^2 with the contents of the d equation and simplify. When we are finished with all of this, we are left with the following:

$$\frac{b^2(a^2m^2-b^2)^2(m^2+1)^2}{(a^2m^2+b^2)^2}\ge 0 \quad \text{True.}$$

Hence, we can conclude that this radical will never go negative.

When attempting to prove that the minus version of the a^2 equation is greater than b^2, no clear conclusion results. Also, when attempting to prove the minus version of the a^2 equation is false by setting the entire factor containing the radical greater than zero (before it is squared) to see if we can get a contradiction, we get $2b^3(m^2+1)/(a^2m^2+b^2)$, which is clearly positive. We cannot get a contradiction this way either. Hence, there may be cases when both versions of the a^2 equation yield a valid ellipse. It may be that more powerful methods are needed to prove the minus version false. Thus, for now both versions of the a^2 equation must be presented.

It turns out that we do not get a rational slope and focal chord pieces theorem with the ellipse (and hyperbola as well) as we did with a parabola. This is because the constant that the sum of the reciprocals equals involves both a and b as the below formula shows.

$$\frac{1}{d_1} + \frac{1}{d_2} = \frac{a}{2b}$$

We now proceed to the tall ellipse. Instead of performing another derivation, which may well be computationally impossible, we will employ a different approach. The approach we will use to find the equations associated with a tall ellipse is to employ the concept of an inverse slope coupled with what happens going from the wide ellipse to the tall ellipse. Given the equation of the line $y = mx$, we can find its inverse line with the inverse slope m^{-1}. Thus, the line $y = m^{-1}x$ is the inverse line to $y = mx$. Of course, the question is begged, what is the specific relationship between m and m^{-1}? It is $m^{-1} = \frac{1}{m}$. We have discussed this relationship a number of times now.

The other idea that we need to implement is that of switching the roles of a and b in equations that pertain to the wide ellipse in order that they work in the tall ellipse. Thus, we start with the slope equation for the wide ellipse. We first invert the slope to make it apply to going through the foci on the other axis, which for the tall ellipse is the y-axis. Then, we switch a and b to obtain our new slope equation for the tall ellipse.

$$m = \pm \sqrt{\frac{b^2(2a-d)}{a(ad-2b^2)}} \Rightarrow m^{-1} = \pm \sqrt{\frac{a(ad-2b^2)}{b^2(2a-d)}} \Rightarrow m = \pm \sqrt{\frac{b(bd-2a^2)}{a^2(2b-d)}}$$

That last equation for m is our slope equation for going through the focus of a tall ellipse. We got away with an easy task on that one. We solve for the other variables.

$$d = \frac{ab^2(m^2+1)}{a^2+b^2m^2} \qquad b^2 = \frac{a^2d}{2a(m^2+1)-dm^2}$$

$$a^2 = \frac{b^2}{d^2}\left[b(m^2+1) \pm \sqrt{b^2m^4+(2b^2-d^2)m^2+b^2}\right]^2$$

As before, we will not attempt to remove the minus version of the a^2 equation. It will be left as an exercise for the student to show that the denominator of the b^2 equation is rendered correctly in the procession of its signs.

Exercises

1. In the wide ellipse equations, several areas were left out with ellipsis (i.e. the three dots) as well as verbiage. Perform a complete analysis, and fill in those details. What is that major factor that winds up canceling in the end to yield the distance equation?

2. The denominator of the slope equation for the wide ellipse is $ad - 2b^2$. Show that this is always positive by replacing either d or b^2 from the wide ellipse equations and simplifying.

3. In the b^2 equation for the tall ellipse, the denominator $2a(m^2 + 1) - dm^2$ must be shown to always be positive. Replace d and simplify to show that this is the case. Note that squaring the slope equation and inserting it will also work.

4. For the wide ellipse solve $d = \dfrac{2ab^2(m^2 + 1)}{a^2m^2 + b^2}$ for a^2, b^2, and m.

5. For the tall ellipse solve $m = \pm\sqrt{\dfrac{b(bd - 2a^2)}{a^2(2b - d)}}$ for a^2, b^2, and d.

6. Given $\dfrac{x^2}{25} + \dfrac{y^2}{16} = 1$ and the line of slope one going through its left focus, find the distance between the intersection points by the following:

a) By finding the two intersection points and using the distance formula.
b) By plugging in the relevant information into our new distance equation in this chapter.

7. Given $\dfrac{x^2}{25} + \dfrac{y^2}{169} = 1$ and a line going through its upper focus.

a) Find the slope of the line when the distance is 50/13.
b) What is the slope when the distance is 26?
c) Express these two extreme distances in terms of a and b as defined by the ellipse constants.
d) Finally, what is the slope of the line when the distance is the average of these two extreme distances? Compute this in two ways: One using the explicit distances given in the problem, and two using the symbolic distances asked for in part c.

8. For the ellipse in problem 7, three lines through the upper focus are utilized. Each of these lines has two focal chord pieces. Show that the sum of the reciprocals of these focal chord pieces is constant (i.e. the same for all three lines) and is equal to $\frac{a}{2b}$.

9. Given $\frac{x^2}{b^2} + \frac{y^2}{169} = 1$ and a line going through its lower focus.

a) Let $m = 0.01$, and analyze the b^2 equation as a function of its remaining variable. The domain of this function is $2b^2/a \le d \le 2a$. Graph the upper bound $a^2 = 169$. Provide any useful comments.
b) Let $m = 1000$ and repeat part a.

10. Given $\frac{x^2}{25} + \frac{y^2}{a^2} = 1$ and a line going through its upper focus.

a) Let $m = 0.01$, and analyze the a^2 equation as a function of its remaining variable. Note: Provide two separate analyses—one for the plus version and one for the minus version. The domain of these functions is $2b^2/a \le d \le 2a$. Graph the lower bound $b^2 = 25$. Provide any useful comments.
b) Let $m = 1000$ and repeat part a.

11. Given $\frac{x^2}{169} + \frac{y^2}{b^2} = 1$ and a line going through its right focus.

a) Let $m = 0.01$, and analyze the b^2 equation as a function of its remaining variable. The domain of this function is $2b^2/a \le d \le 2a$. Graph the upper bound $a^2 = 169$. Provide any useful comments.
b) Let $m = 1000$ and repeat part a.

12. Given $\frac{x^2}{a^2} + \frac{y^2}{25} = 1$ and a line going through its upper focus.

a) Let $m = 0.01$, and analyze the a^2 equation as a function of its remaining variable. Note: Provide two separate analyses—one for the plus version and one for the minus version. The domain of these functions is $2b^2/a \le d \le 2a$. Graph the lower bound $b^2 = 25$. Provide any useful comments.
b) Let $m = 1000$ and repeat part a.

13. Choose an ordered pair from one of the graphs of the previous four problems. Graph the situation that that ordered pair represents. Show that the distance is correct by an analytical demonstration as well as by a graphical one. Note: Answers vary.

14. For a tall ellipse, we are given $m = 2$, $d = 8$, and $b^2 = 36$. Find the value or values of a^2 for this situation. Draw any relevant labeled diagrams.

Lines and Hyperbolas: Part 2

In this chapter, we will find the distance between the intersection points A and B of a line of given slope m going through a focus of a hyperbola. We will begin with the left-right hyperbola with a line through its right focus. The following figure illustrates our problem.

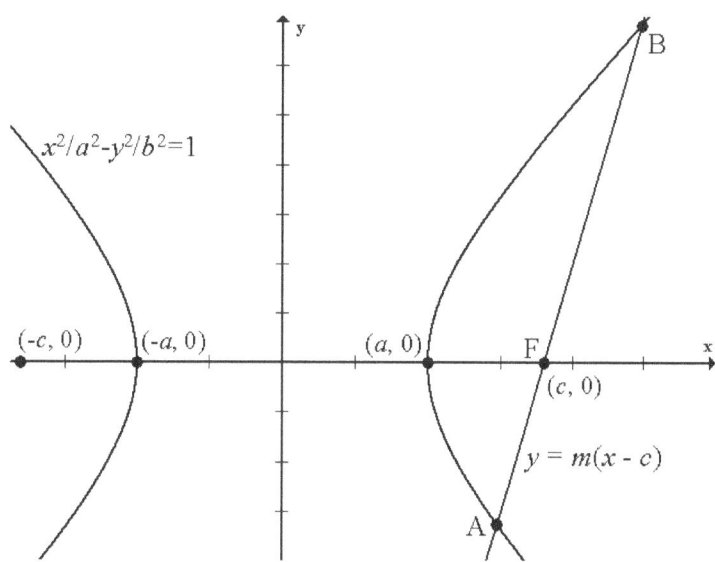

Figure 1: Line Through Right Focus of Left-Right Hyperbola

All of the subsequent mathematics will be derived from this picture, so we will take some pains to describe our situation. We have a left-right hyperbola $\dfrac{x^2}{a^2}-\dfrac{y^2}{b^2}=1$ with its vertices $(-a, 0)$ and $(a, 0)$ pictured. The two foci at $(-c, 0)$ and $(c, 0)$ are also pictured. We have a line of some slope m going through the right focus. This makes the equation of that line $y = m(x-c)$. The line will intersect the hyperbola in two points A and B such that A and B are on the same branch of the hyperbola. In order for the two points to be in their respective positions on the diagram, we need to restrict the values that x can take. We do so with a preliminary x value that we call x_0. We need to have x_0 be between the x-coordinates of the right vertex and the right focus. This translates to $a < x_0 \le c$. We cannot let x_0 be a itself because then point B would hit the other branch of the hyperbola at the other vertex $(-a, 0)$. Using an x_0 in this range yields point A $(x_0, \dfrac{-b}{a}\sqrt{x_0^2-a^2})$. Of course, we solved the hyperbola equation for y and used the lower half with the negative sign to get the y-coordinate.

The next thing that we can do is to compute a preliminary slope of the line.

$$m = \frac{y_2 - y_1}{x_2 - x_1} = \frac{\dfrac{-b}{a}\sqrt{x_0^2 - a^2} - 0}{x_0 - c} = \frac{b\sqrt{x_0^2 - a^2}}{a(c - x_0)}$$

As with the other two derivations going through the focus, we will split the distance into two pieces. Thus, we have enough information to get a preliminary distance from point A to point F.

$$d(A,F) = \sqrt{(x_0 - c)^2 + \left(\tfrac{b}{a}\sqrt{x_0^2 - a^2} - 0\right)^2} = \cdots = \frac{cx_0 - a^2}{a}.$$

The next thing is to solve the system of equations involving the line and the hyperbola for x.

Line: $\quad y = \dfrac{b\sqrt{x_0^2 - a^2}}{a(c - x_0)}(x - c)$

Hyperbola: $y = \dfrac{\pm b}{a}\sqrt{x^2 - a^2}$

We equate since both are equal to y. The fraction $\dfrac{b}{a}$ cancels from both sides. Clear the rest of the fractions by multiplying both sides by $(c - x_0)$. Then, square both sides to get rid of the radical. Expand everything out and collect like terms. Get zero on one side to arrive at the following:

$$(2a^2 - 2cx_0 + b^2)x^2 - (2a^2c - 2cx_0^2)x - (2a^2x_0 + b^2x_0 - 2a^2c)x_0 = 0$$

As in the ellipse derivation, we know what one of the roots is, namely, $x = x_0$. Thus, this quadratic equation ought to factor, so we set up the candidate factoring accordingly:

$$(x - x_0)([2a^2 - 2cx_0 + b^2]x + [2a^2x_0 + b^2x_0 - 2a^2c]) = 0$$

As a precaution, we should check the middle term by FOILing it out. We can now set each factor equal to zero, and solve for x.

$$\therefore x = x_0 = x_A, x = \frac{2a^2c - 2a^2x_0 - b^2x_0}{2a^2 - 2cx_0 + b^2} = x_B$$

Using the line equation, we can compute the y-coordinate of point B.

$$y_B = \frac{b\sqrt{x_0^2 - a^2}}{a(c - x_0)}\left(\frac{2a^2c - 2a^2x_0 - b^2x_0}{2a^2 - 2cx_0 + b^2} - c\right) = \ldots = \frac{b^3\sqrt{x_0^2 - a^2}}{a(2cx_0 - 2a^2 - b^2)}$$

Thus, the coordinates of point B are $\left(\dfrac{2a^2c - 2a^2x_0 - b^2x_0}{2a^2 - 2cx_0 + b^2}, \dfrac{b^3\sqrt{x_0^2 - a^2}}{a(2cx_0 - 2a^2 - b^2)}\right)$.

We now have point A and point B in terms of x_0. We can evaluate points A and B with a couple of specific values of x_0 to see if our coordinates seem correct. The two values we have are $x_0 = a$ and $x_0 = c$. Plugging these into points A and B yield the following:

$$A\big|_{x_0=a} = (a,0), \, A\big|_{x_0=c} = \left(c, \frac{-b^2}{a}\right), \, B\big|_{x_0=a} = (-a,0), \, B\big|_{x_0=c} = \left(c, \frac{b^2}{a}\right)$$

Technically, as we stated before, we cannot let $x_0 = a$ because the intersection points between the line and the hyperbola will not lie on the same branch. Nevertheless, to check to see if our algebraic efforts have any errors, we see that the behavior of the expressions in points A and B is correct. Thus, we move on with confidence to the next phase of the project, which is to find the distance between the focus, F, and point B.

$$d(F, B) = \sqrt{\left(c - \frac{2a^2c - 2a^2x_0 - b^2x_0}{2a^2 - 2cx_0 + b^2}\right)^2 + \left(\frac{b^3\sqrt{x_0^2 - a^2}}{a(2cx_0 - 2a^2 - b^2)}\right)^2} = \ldots = \frac{b^2(cx_0 - a^2)}{a\,|\,2cx_0 - 2a^2 - b^2\,|}$$

No comment was made about the distance between points A and F. However, the numerator has a subtraction sign in it. In the expression for the distance between points F and B, we see the numerator has that same expression as earlier, except that it is a factor here. Furthermore, one of the factors in the denominator has three terms in it that is difficult to tell upon inspection which version is the correct one. Let us perform some analysis to determine which versions of these factors are the ones to use. We will start with the $cx_0 - a^2$ factor.

301

Earlier, we laid out the restrictions on x, and we chose x_0 to represent those restrictions. Thus, $a < x_0 \leq c$. Since both c and x_0 are both greater than a, it should be abundantly clear that $cx_0 - a^2 > 0$ and not $a^2 - cx_0 > 0$. Hence, the correct version to use in this case is $cx_0 - a^2$.

The other expression is $|2cx_0 - 2a^2 - b^2|$. So which version of this absolute value expression is the positive version? This one is more difficult to ascertain. In the ellipse derivation, both of the endpoints were OK to use to check this out, but here we can only use the c endpoint. The other alternative to use if neither endpoint were possible is to use an interior point such as the midpoint. Let us use both. First, we will use the endpoint c.

$2cx_0 - 2a^2 - b^2$ Let $x_0 = c$

$2c \cdot c - 2a^2 - b^2 = 2c^2 - 2a^2 - b^2 = 2a^2 + 2b^2 - 2a^2 - b^2 = b^2 > 0$

$\therefore 2cx_0 - 2a^2 - b^2 > 0$

$2cx_0 - 2a^2 - b^2$ Next, we will use the midpoint by letting $x_0 = \dfrac{a+c}{2}$.

$2c\left(\dfrac{a+c}{2}\right) - 2a^2 - b^2 = ac + c^2 - 2a^2 - b^2$

$= ac + a^2 + b^2 - 2a^2 - b^2 = ac - a^2 = a(c - a) > 0$

$\therefore 2cx_0 - 2a^2 - b^2 > 0$ This is the same conclusion as using the endpoint c.

Now that we know the correct expressions to use, we will find a preliminary distance between the points A and B by adding the distances between points A and F and between points F and B. Incidentally, this works because the points A, F, and B are collinear.

$$d = d(A,F) + d(F,B) = \dfrac{cx_0 - a^2}{a} + \dfrac{b^2(cx_0 - a^2)}{a(2cx_0 - 2a^2 - b^2)} = \ldots = \dfrac{2(cx_0 - a^2)^2}{a(2cx_0 - 2a^2 - b^2)}$$

Recall our preliminary slope that we obtained early on in this derivation. We need to solve that for x_0. This is so that we can plug the expression for x_0 into the preliminary distance equation above. This eliminates the variable x_0 leaving behind our desired formula—after appropriate simplification, of course.

$$m = \dfrac{b\sqrt{x_0^2 - a^2}}{a(c - x_0)}$$ Solving for x_0 yields the following: $x_0 = \dfrac{a(acm^2 \pm b^2\sqrt{m^2 + 1})}{a^2m^2 - b^2}$

302

Admittedly, plugging this nasty expression for x_0 into the distance equation above is a very difficult algebraic task. Carrying along the plus and minus signs along the way is fine because it turns out that a four-term factor common to the numerator and denominator just cancels, and so the plus and minus stuff winds up making just one formula anyway. When all the dust settles (and there's a lot of it to settle), we get the following:

$$d = \frac{2ab^2(m^2+1)(a^2m^2+2a^2+b^2\pm2ac\sqrt{m^2+1})}{(a^2m^2-b^2)(a^2m^2+2a^2+b^2\pm2ac\sqrt{m^2+1})} = \frac{2ab^2(m^2+1)}{a^2m^2-b^2}$$

Do we need absolute value bars around that denominator because of the presence of the subtraction sign? After all, this equation is computing a distance for us, and distances are positive quantities. If b^2 is ever greater than a^2m^2, then the denominator becomes negative, and the whole fraction then becomes negative. This would be bad for a distance computation.

Let us discuss the situation because we just do not want to put absolute value bars on quantities if they are not needed. What we have here is a line going through the focus of the hyperbola. The branches of the hyperbola are asymptotic to the asymptote lines. For the left-right hyperbola, the asymptote lines are $y = \frac{\pm b}{a}x$. If m, the slope of our line through the focus, was equal to either b/a or $-b/a$, then the line through the focus would be parallel to one of the asymptote lines. In either case, the line through the focus would intersect the hyperbola in only one place, so only one of the points A or B would be created. Looking at the hyperbola, we see that if the slope of the line through the focus was less steep than the asymptote lines, points A and B would fall on different branches of the hyperbola. This is clearly not good. Therefore, it must be the case that the slope of the line must be greater than b/a for a positively sloped line through the focus to work or less than $-b/a$ for a negatively sloped line through the focus to work. We can state this more succinctly as the magnitude of the slope has to be greater than b/a. We will put these observations in symbols. The symbol ε is used here to represent a positive quantity.

$$a^2m^2 - b^2 > 0, \; |m| > \frac{b}{a} \Rightarrow \quad a^2m^2 > b^2 \Rightarrow a^2\left(\frac{b+\varepsilon}{a}\right)^2 > b^2 \Rightarrow (b+\varepsilon)^2 > b^2 \quad \text{True.}$$

$$\therefore a^2m^2 - b^2 > 0$$

So it turns out that the denominator will never be negative, and the absolute value bars are unnecessary. In the ellipse version, the denominator had a plus sign, so this issue never came up there. The hyperbola, however, throws us a wrinkle that we must consider.

Solving the distance equation for its other variables yields the following:

$$a^2 = \frac{b^2}{d^2 m^4}\left(b(m^2+1) \pm \sqrt{b^2 m^4 + (2b^2 + d^2)m^2 + b^2}\right)^2$$

$$b^2 = \frac{a^2 d m^2}{2a(m^2+1)+d} \qquad m = \pm\sqrt{\frac{b^2(2a+d)}{a(ad-2b^2)}}$$

We can eliminate the minus version of the a^2 equation here because we can show that the factor $b(m^2+1) - \sqrt{b^2 m^4 + (2b^2 + d^2)m^2 + b^2} < 0$ when we replace b with its contents and simplifying. We were not able to eliminate the minus version in the *Lines and Ellipses: Part 2* chapter, but we can do so here. The final line of the analysis is $\sqrt{\frac{dm^2(a^2(m^2+1))}{2am^2+2a+d}} - \sqrt{\frac{dm^2(a^2(m^2+1)+\varepsilon)}{2am^2+2a+d}} < 0$. This means that $a < 0$ before it became a^2, but we are assuming that $a > 0$. This contradicts our assumption about a. Thus, this contradiction lets us eliminate the minus version of the a^2 equation. Our equation for a^2 then becomes the following:

$$a^2 = \frac{b^2}{d^2 m^4}\left(b(m^2+1) + \sqrt{b^2 m^4 + (2b^2 + d^2)m^2 + b^2}\right)^2$$

We observe that there is no discriminant in the a^2 equation, but the m equation does have one. This will be addressed in the exercises.

We turn our attention to getting the four equations for the up-down hyperbola for a line going through one of the foci. We employ both the inverse slope and switching the roles of a and b to accomplish this feat. Doing this produces the four equations.

$$d = \frac{2ab^2(m^2+1)}{a^2 - b^2 m^2} \qquad a^2 = \frac{b^2}{d^2}\left(b(m^2+1) + \sqrt{b^2(m^2+1)^2 + d^2 m^2}\right)^2$$

$$b^2 = \frac{a^2 d}{2a(m^2+1)+dm^2} \qquad m = \pm\sqrt{\frac{a(ad-2b^2)}{b^2(2a+d)}}$$

Because we eliminated the minus version of a^2 earlier, that benefit carries over to the new a^2 equation. We peruse the equations to see that the denominator of the d equation has a subtraction in it, and the numerator of the m equation also has a subtraction in it. These will be addressed in the exercises.

Exercises

1. A number of details were left out of the derivation for the left-right hyperbola in this chapter. Perform a complete analysis to obtain $d = \dfrac{2ab^2(m^2+1)}{a^2m^2-b^2}$.

2. Solve $d = \dfrac{2ab^2(m^2+1)}{a^2m^2-b^2}$ for a^2, b^2, and m.

3. Show that the denominator of $m = \pm\sqrt{\dfrac{b^2(2a+d)}{a(ad-2b^2)}}$ is strictly positive by using $b^2 = \dfrac{a^2dm^2}{2a(m^2+1)+d}$ and simplifying the inequality.

4. For the four equations present in question 2, replace any m with $1/m$ and also switch the roles of a and b and simplify to produce the four equations depicted for the up-down hyperbola situation in this chapter.

5. Show that the denominator of $d = \dfrac{2ab^2(m^2+1)}{a^2-b^2m^2}$ is always positive by using $b^2 = \dfrac{a^2d}{2a(m^2+1)+dm^2}$ and simplifying.

6. Show that the numerator of $m = \pm\sqrt{\dfrac{a(ad-2b^2)}{b^2(2a+d)}}$ is at least zero by using $b^2 = \dfrac{a^2d}{2a(m^2+1)+dm^2}$ and simplifying.

7. Given the hyperbola $\dfrac{x^2}{25} - \dfrac{y^2}{9} = 1$ and the line of slope 7/5 through its right focus, find the distance between the intersection points of the hyperbola and line by the following:

a) Find the coordinates of the intersection points and plug them into the distance formula.

b) Use the new distance formula developed in this chapter $d = \dfrac{2ab^2(m^2+1)}{a^2m^2 - b^2}$.

8. Given the hyperbola $\dfrac{(y-1)^2}{9} - \dfrac{x^2}{4} = 1$ and the line of slope 0.8 going through its upper focus, find the distance between the intersection points of the hyperbola and line by the following:

a) Find the coordinates of the intersection points and plug them into the distance formula.

b) Use the new distance formula developed in this chapter $d = \dfrac{2ab^2(m^2+1)}{a^2 - b^2m^2}$.

9. Analyze $d = \dfrac{2ab^2(m^2+1)}{a^2m^2 - b^2}$ for when m is undefined, which is to say put a vertical line through the focus. Compare your result with the length of a latus rectum of a hyperbola.

10. Analyze $d = \dfrac{2ab^2(m^2+1)}{a^2 - b^2m^2}$ for when m is zero, which is to say put a horizontal line through the focus. Compare your result with the length of a latus rectum of a hyperbola.

11. Based on the results of 7, 8, 9, and 10, provide a commentary on what these two distance equations are doing.

12. For a left-right hyperbola, we are given that $m = 2$, $d = 9$, and $a^2 = 49$. Find b^2. Draw any relevant labeled diagrams.

13. For a left-right hyperbola, we are given that $m = 3$, $d = 7$, and $b^2 = 16$. Find a^2. Draw any relevant labeled diagrams.

14. For a left-right hyperbola, we are given that $a^2 = 9$, $b^2 = 25$, and $d = 6$. Find the positive value of m for this situation. Draw any relevant labeled diagrams.

15. For an up-down hyperbola, we are given that $m = 4$, $d = 11$, and $b^2 = 16$. Find a^2. Draw any relevant labeled diagrams.

16. For an up-down hyperbola, we are given that $m = 4$, $d = 9$, and $a^2 = 25$. Find b^2. Draw any relevant labeled diagrams.

17. For an up-down hyperbola, we are given that $d = 9$, $a^2 = 25$, and $b^2 = 4$. Find $m < 0$. Draw any relevant labeled diagrams.

Pascal's Theorem

This highlight chapter features a very famous line called Pascal's line. This line is named after Blaise Pascal (1623 – 1662), a French mathematician and philosopher.

Pascal's Theorem (1640): If the six vertices of a hexagon lie on a conic section, then the intersection points of the three pairs of opposite sides lie on a line.

We will note a few details here. The first is that a hexagon does not need to be regular or convex even. In fact, we can even have the sides crossing one another. The only important feature is that the vertices, numbered in any order we please, lie on the conic section. By the way, if a hexagon is regular, its opposite sides are parallel, so we would not be able to find the intersection points. The Pascal line for a regular hexagon does not exist. [Actually, it does exist, and it is called the line at infinity; however, that is for students in projective geometry to know. For us, we will just say that it does not exist.]

To help us cope with the numbering of the vertices situation, we will adopt a convention here. We will settle for some numbering of the six vertices which we will call the *original numbering*. With this numbering scheme, we will declare that the three intersection points A, B, and C of the opposite sides of the hexagon are to be found with the following:

$$A = \overline{P_1 P_2} \cap \overline{P_4 P_5}$$
$$B = \overline{P_2 P_3} \cap \overline{P_5 P_6}$$
$$C = \overline{P_1 P_6} \cap \overline{P_3 P_4}$$

The purpose of the above formulations is to make it easier to find the opposite sides of the hexagon—especially when we allow sides to cross one another. With the vertices numbered from one to six, finding the opposite sides is a snap with the above template. We just find the equation of the line through P_1 and P_2, find the equation of the line through P_4 and P_5, and find the intersection point of these two lines. Then, point A is that intersection point. It's as simple as that. Points B and C are found similarly by following the above template.

It turns out that there are 60 unique ways to permute (i.e. number in a different order) the six vertices of the hexagon. This means that for one hexagon on the conic section we can get up to 60 Pascal lines. The 60 unique numbering schemes are displayed next.

1. 1-2-3-4-5-6-1	2. 1-2-3-4-6-5-1	3. 1-2-3-5-4-6-1	4. 1-2-3-5-6-4-1
5. 1-2-3-6-4-5-1	6. 1-2-3-6-5-4-1	7. 1-2-4-3-5-6-1	8. 1-2-4-3-6-5-1
9. 1-2-4-5-3-6-1	10. 1-2-4-5-6-3-1	11. 1-2-4-6-3-5-1	12. 1-2-4-6-5-3-1
13. 1-2-5-3-4-6-1	14. 1-2-5-3-6-4-1	15. 1-2-5-4-3-6-1	16. 1-2-5-4-6-3-1
17. 1-2-5-6-3-4-1	18. 1-2-5-6-4-3-1	19. 1-2-6-3-4-5-1	20. 1-2-6-3-5-4-1
21. 1-2-6-4-3-5-1	22. 1-2-6-4-5-3-1	23. 1-2-6-5-3-4-1	24. 1-2-6-5-4-3-1
25. 1-3-2-4-5-6-1	26. 1-3-2-4-6-5-1	27. 1-3-2-5-4-6-1	28. 1-3-2-5-6-4-1
29. 1-3-2-6-4-5-1	30. 1-3-2-6-5-4-1	31. 1-3-4-2-5-6-1	32. 1-3-4-2-6-5-1
33. 1-3-4-5-2-6-1	34. 1-3-4-6-2-5-1	35. 1-3-5-2-4-6-1	36. 1-3-5-2-6-4-1
37. 1-3-5-4-2-6-1	38. 1-3-5-6-2-4-1	39. 1-3-6-2-4-5-1	40. 1-3-6-2-5-4-1
41. 1-3-6-4-2-5-1	42. 1-3-6-5-2-4-1	43. 1-4-2-3-5-6-1	44. 1-4-2-3-6-5-1
45. 1-4-2-5-3-6-1	46. 1-4-2-6-3-5-1	47. 1-4-3-2-5-6-1	48. 1-4-3-2-6-5-1
49. 1-4-3-5-2-6-1	50. 1-4-3-6-2-5-1	51. 1-4-5-2-3-6-1	52. 1-4-5-3-2-6-1
53. 1-4-6-2-3-5-1	54. 1-4-6-3-2-5-1	55. 1-5-2-3-4-6-1	56. 1-5-2-4-3-6-1
57. 1-5-3-2-4-6-1	58. 1-5-3-4-2-6-1	59. 1-5-4-2-3-6-1	60. 1-5-4-3-2-6-1

Let us use a regular hexagon to show an example of how to permute the vertices to one of the other numbering schemes. Thus, to change from the original numbering scheme to scheme #43, we change the vertices as in the following diagram below.

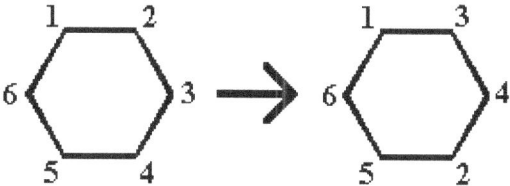

Then, use the template above to find the three intersection points A, B, and C of the lines through the opposite sides of the newly numbered hexagon. We are guaranteed by Pascal's theorem that A, B, and C will be collinear even in this newly numbered hexagon assuming of course that we do not get more than one pair of opposite sides parallel.

A geometry class beyond this course is called *Projective Geometry*. In that class they study not only a particular theorem such as Pascal's line, but they study what is called its dual. A different theorem that is obtained by switching certain words to other words. If a statement is true (such as Pascal's theorem), then its dual is true as well. The dual of Pascal's theorem is Brianchon's theorem, named after Charles Julien Brianchon (1783 – 1864), a French mathematician. The interested student can look up Brianchon's theorem. Any theorem in projective geometry has a dual. It turns out that Desargues' theorem begins with triangles in perspective from a point (the perspector), and then, the theorem concludes with the statement about the perspectrix. The dual of this winds up flipping around the conclusion with the hypothesis. A big discovery of projective geometry is this dualistic principle.

This is the last highlight chapter in this book. It is related to the first highlight chapter because the Pappus line involves two intersecting lines, which is a degenerate hyperbola. A degenerate hyperbola is still a conic section and Pascal's theorem applies. When Pappus formulated his theorem, he did not know that he was making a special case of the later to come Pascal's theorem. Such is the evolution of mathematics. My own equations from the seven chapters with an asterisk will at some point be special cases of some grander theorem uniting the conics. Such is the evolution of mathematics. The math department head at Lawrence Technological University wanted me to generalize my equations into a single theorem. I had to tell him that I did not have the ability to do that. Maybe a mathematician who reads these sentences will be inspired to take up the case and do it. If I am still alive, please show it to me. Maybe I will understand. Gödel did get to see Cohen's proof didn't he? However, maybe like Pappus and Pascal, too much time will have elapsed for me to see.

For our final time together in this book, we will work an example of Pascal's line. We wish to find Pascal's line for the hyperbola $\frac{x^2}{81} - \frac{y^2}{121} = 1$ with the points of the hexagon on it as $P_1(-9, 0)$, $P_2(15, 44/3)$, $P_3(-11.25, 8.25)$, $P_4(23.4, -26.4)$, $P_5(9.75, -55/12)$, and $P_6(9, 0)$.

We proceed to use the template above to find point A. Thus, we find the intersection point of the equations of the lines $\overline{P_1P_2}$ and $\overline{P_4P_5}$. The system of equations for finding point A is the following:

$$\begin{cases} y = \dfrac{11}{18}x + \dfrac{11}{2} \\ y = \dfrac{-187}{117}x + 11 \end{cases}$$

Thus, we solve this system to find that point A is $\left(\dfrac{117}{47}, \dfrac{330}{47}\right)$. We proceed to use the next two lines of the template. That is, we find point B by finding the intersection point of the equations of the lines $\overline{P_2P_3}$ and $\overline{P_5P_6}$. Finally, we find point C by finding the intersection point of the equations of the lines $\overline{P_1P_6}$ and $\overline{P_3P_4}$. The system for finding point B is the left one, and the other system below is for finding point C.

$$\left\{ \begin{array}{l} y = \dfrac{11}{45}x + 11 \\[2mm] y = \dfrac{-55}{9} + 55 \end{array} \right\} \qquad\qquad \left\{ \begin{array}{l} y = 0 \\[2mm] y = -x - 3 \end{array} \right\}$$

Solving the above two systems yields points $B\left(\dfrac{90}{13}, \dfrac{165}{13}\right)$ and $C(-3, 0)$. We will use two of these points, such as A and B, to find the equation of the Pascal line. The third point can be used as a check to verify our work. The equation of the line through points A and B is $y = \dfrac{55}{43}x + \dfrac{165}{43}$. Clearly, point C checks in this equation. Hence, this confirms that we have found the Pascal line. See Figure 1.

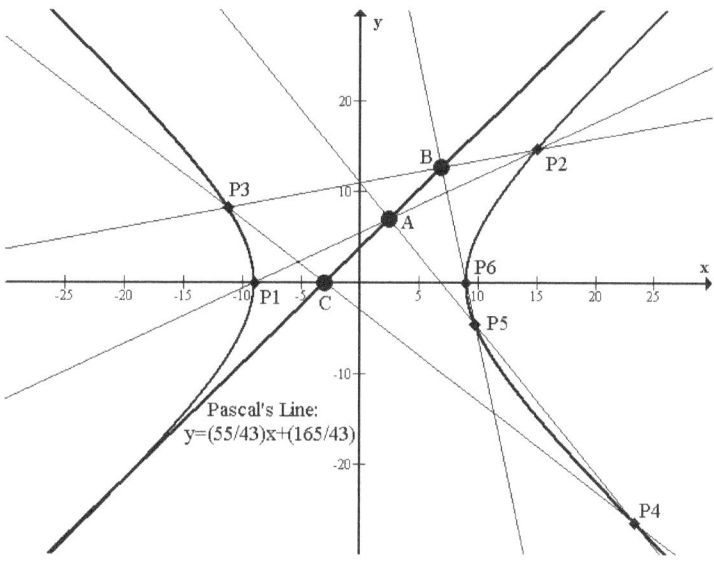

Figure 1: The Pascal Line

Suppose we want to find the equation of the Pascal line for one of the other permutations of the vertices? Let us say we want to find Pascal's line for the 43rd permutation. Then, we renumber the vertices as shown above. The original vertices are $P_1(-9, 0)$, $P_2(15, 44/3)$, $P_3(-11.25, 8.25)$, $P_4(23.4, -26.4)$, $P_5(9.75, -55/12)$, and $P_6(9, 0)$. The first vertex always stays as the first vertex in all 60 permutations. In the 43rd permutation, vertices 5 and 6 do not change. However, vertices 2, 3, and 4 move around. Vertex 2 becomes vertex 3. Vertex 3 becomes vertex 4, and vertex 4 becomes vertex 2.

This produces the new numbering of the vertices $P_1(-9, 0)$, $P_2(23.4, -26.4)$, $P_3(15, 44/3)$, $P_4(-11.25, 8.25)$, $P_5(9.75, -55/12)$, and $P_6(9, 0)$.

Going through the procedure produces the points $A\left(\dfrac{-171}{4}, \dfrac{55}{2}\right)$, B(-27, 220), and

C(-45, 0). The Pascal line for this permutation is the following: $y = \dfrac{110}{9} x + 550$.

For the interested student, a number of mathematicians have studied this system of 60 Pascal lines. A number of important points and lines, named in honor of the mathematicians who discovered them, are the Steiner, Kirkman, and Salmon points along with the Plücker and Cayley lines. We studied the Steiner point, line, and circles in the *Triangle Geometry* chapter; and we studied the Plucker line and points in the Two Special Quadrilaterals chapter. These two men also studied the 60 Pascal lines configuration discussed in this chapter and received more honor by getting their names mentioned in this chapter as well. We have not discussed Kirkman and Salmon before now. I do not know much about Kirkman, but I am aware that Salmon published some math books that are quite advanced in the 1800s.

I would like to thank you for spending some time with me during this course.

THE END

Exercises

1. For the parabola $y = x^2$, find Pascal's line corresponding to the vertices of a hexagon in order of appearance with the *x*-coordinates at -2, -3, 0, 4, 1, and 2. Also, find the Pascal line corresponding to the 27th permutation of these vertices.

2. Complete the example of the Pascal line for the 43rd permutation of the vertices in the text for the hyperbola.

3. Find the Pascal line for the unit circle for the six vertices $P_1(3/5, 4/5)$, $P_2(1, 0)$, $P_3(-4/5, 3/5)$, $P_4(0, -1)$, $P_5(5/13, -12/13)$, and $P_6(12/13, 5/13)$. Also, find the Pascal line corresponding to the 38th permutation of these vertices.

4. Given the ellipse $\dfrac{x^2}{16} + \dfrac{y^2}{9} = 1$, find the equation of the Pascal line through six rational vertices of the student's choosing. Also, find Pascal's line corresponding to the 14th permutation of these vertices.

Appendix A: Finding More Terms of a Sequence

We get sequences from many sources. Branches of science provide many sequences. For example, a physicist may study the breaking strength of rods of a fixed length made of a certain material having increasing radii. His study generates a sequence of breaking strengths from which he may be able to deduce an equation relating the breaking strengths with the radius of the rod. As another example, suppose a chemist is studying the speed of a chemical reaction under certain heating conditions. She may correlate the reaction times with the temperature and derive an equation relating the two based on her sequence of data. We can also get sequences from other sources as well. For example, an economist may observe how a particular stock behaves under certain market conditions and derive a principle that explains the observed sequence. Of course, she hopes that such a principle is correct, and she may be able to use the principle to predict how other companies would perform under similar market conditions. I had a sequence that I derived from a game I had invented. I computed the first 19 terms by writing a computer program, but I needed the 64th term.

We will discuss four methods that find more terms of a given sequence when the first few terms are given.

The first method is to discern a pattern in the sequence. The pattern can involve many things, such as observing runs of some value, terms increasing (decreasing) by a fixed amount, successive terms exhibiting a constant multiplier, or successive terms coming up based solely on their position in the sequence. From that pattern, we can generate a general term as well as the next term or next several terms if we desire. The second method is to use a system of equations to fit a polynomial to the sequence. The third method applies when the initial terms of a sequence are fractions. We treat the numerator and denominator as separate sequences, and then we reconstruct their fractional form with the found sequences. The fourth method involves going on Neil Sloane's Integer Sequences web site. The web address is the following:

www.research.att.com/~njas/sequences

Input the first few integers of your given sequence, and let the system return to you the results. From the output, you can select what seems to you to be the best answer. I submitted my sequence to Neil's website, and it wasn't there. So Neil included my sequence. He assigned it the number A102169. The interested student can look up my sequence to see what it is.

Suppose the given sequence is 1, 4, 9, 16, 25, 36, ...

Method 1: We see by inspection that these are the squares of the natural numbers, so we might say that the general term is n^2.

Method 2: We construct a polynomial to fit the given sequence. We first need to construct the ordered pairs through which the polynomial passes. We typically say that the given sequence is the y-coordinates, and we use the natural numbers as the index to yield the corresponding x-coordinates. Thus, the first given term is 1, so x is 1 and y is 1. Thus, our first ordered pair is (1, 1). The second given term is 4, so x is 2 and y is 4. Thus, our second ordered pair is (2, 4). We repeat this to generate the following list of ordered pairs: {(1, 1), (2, 4), (3, 9), (4, 16), (5, 25), (6, 36)}. We now fit a polynomial through these six points. This yields a fifth degree polynomial. So we have the following:

$$y = Ax^5 + Bx^4 + Cx^3 + Dx^2 + Ex + F$$

Next, we input each of the points into this model equation to generate the below matrix with its rref version. The columns are A, B, C, D, E, F, and the constant.

$$\begin{bmatrix} 1 & 1 & 1 & 1 & 1 & 1 & 1 \\ 32 & 16 & 8 & 4 & 2 & 1 & 4 \\ 243 & 81 & 27 & 9 & 3 & 1 & 9 \\ 1024 & 256 & 64 & 16 & 4 & 1 & 16 \\ 3125 & 625 & 125 & 25 & 5 & 1 & 25 \\ 9216 & 1536 & 256 & 36 & 6 & 1 & 36 \end{bmatrix} \xrightarrow{rref} \begin{bmatrix} 1 & 0 & 0 & 0 & 0 & 0 & 0 \\ 0 & 1 & 0 & 0 & 0 & 0 & 0 \\ 0 & 0 & 1 & 0 & 0 & 0 & 0 \\ 0 & 0 & 0 & 1 & 0 & 0 & 1 \\ 0 & 0 & 0 & 0 & 1 & 0 & 0 \\ 0 & 0 & 0 & 0 & 0 & 1 & 0 \end{bmatrix}$$

Well, this just tells us that our guess of the general term of x^2 was right. However, other times we may see a different answer come out of the polynomial analysis.

Method 3: The denominator is fixed at one because all of these given elements are integers. Hence, we gain nothing from viewing the sequence from this perspective.

Method 4: The student will just have to log on to Neil's site to experience it for himself or herself.

Our next example is a more difficult sequence than the first. Suppose we have the first four terms 3/5, 19/3, 5/11, 29/7, ... What might the next term in the sequence be?

Our first method says to discern a pattern. I am not clever enough to see one here. Thus, we move to method 2. With four terms, we construct a cubic polynomial. The general cubic is $y = Ax^3 + Bx^2 + Cx + D$. We toss in our terms and construct the matrix.

314

$$\begin{bmatrix} 1 & 1 & 1 & 1 & | & 3/5 \\ 8 & 4 & 2 & 1 & | & 19/3 \\ 27 & 9 & 3 & 1 & | & 5/11 \\ 64 & 16 & 4 & 1 & | & 29/7 \end{bmatrix} \xrightarrow{rref} \begin{bmatrix} 1 & 0 & 0 & 0 & | & 1359/385 \\ 0 & 1 & 0 & 0 & | & -31168/1155 \\ 0 & 0 & 1 & 0 & | & 71587/1155 \\ 0 & 0 & 0 & 1 & | & -14601/385 \end{bmatrix}$$

See, I told you this one was more difficult. Thus, we construct our cubic polynomial.

$$y = \frac{1359}{385}x^3 - \frac{31168}{1155}x^2 + \frac{71587}{1155}x - \frac{14601}{385}$$

To compute our next term, we plug in $x = 5$ into this cubic equation to get $\left(5, \frac{44557}{1155}\right)$. Thus, our guess for what the next term might be is 44557/1155.

The next method says to treat the numerators as an integer sequence and the denominators as an integer sequence. Doing this produces the following two cubics:

$$y_{num} = \frac{34}{3}x^3 - 83x^2 + \frac{557}{3}x - 111$$

$$y_{den} = \frac{-11}{3}x^3 + 27x^2 - \frac{172}{3}x + 39$$

We compute $y_{num}(5) = 159$ and $y_{den}(5) = -31$. Thus, our guess for the next term is $\left(5, \frac{-159}{31}\right)$.

If we do not like any of our guesses at the next term, we can try Neil's site. However, we need to have integers to input. Thus, we multiply each of the fractions by the LCD of the four fractions presented. The LCD is 3x5x7x11=1155. Our sequence then becomes 693, 7315, 525, 4785,... This is the sequence that we would input into Neil's site.

I suppose a word about lotteries is in order. A sequence of lottery numbers is not subject to analysis because they are random. This assumes that the lottery is run correctly, of course, and is not fraudulent. I recall in Brazil a number of years ago that the main politician there had won every local and state lottery for years. That is a fraudulent lottery. Thus, with a fair lottery, it is impossible to predict the next lottery ticket. So a sequence of lottery numbers obtained from past winning tickets is not analyzable. Of course, it is possible to subject it to formal analysis anyway, but the predictive power of such a trend analysis is completely unreliable. One could generate a polynomial or enter in the integers into Neil's site. However, the purpose of analyzing a sequence is to discover the law governing the nature of the sequence. Then, armed with that law (i.e. equation), we can predict the next term or any term we please. The lottery does not operate according to a mathematical or physical law, so any candidate equation found to describe it is wrong.

I was into the lotteries for a while, and I began studying it. I wrote some simulation programs and found to my dismay (contradicting the lottery expert's proclamations that I was reading) that prior history of a number was irrelevant to how often that number would come up again. Thus, I could not compute what the next number would be.

I know that pretty much all of those lottery "experts" out there proclaim that there are patterns in the lottery. What they are seeing is the initial clumpiness of the lottery numbers as they produce a history. This clumpiness is evident for small histories, say 500, drawings. After that, the quantity of each lottery number begins to even out into equally sized piles. For example, if only three 2's had shown up in the first three hundred drawings, the quantity of 2's would catch up to the other numbers by the time 1000 drawings had been completed. Understand, this does not mean that a 2 is *due* in the next drawing. Of course, lotteries typically do not have histories of 1000 or more drawings. However, I saw this happen with my simulations. I would have my program simulate 10,000 drawings or even more, and the sizes of the piles would become closer and closer together. This demonstrates the *Law of Large Numbers* in statistics. This law says that the observed frequencies of outcomes approaches the theoretical frequencies of outcomes as the number of observations (i.e. drawings) increases without bound.

The *Law of Large Numbers* explains why that clumpiness goes away as the number of drawings increases. Thus, the lottery history displays patterns only for small histories of drawings much like flipping a coin ten times yielding seven heads and three tails. If that coin were flipped 10,000 times, it would be close to 5,000 heads and 5,000 tails. The 7 heads and 3 tails "pattern" would not continue for 10,000 flips producing 7,000 heads and 3,000 tails. The lottery expert does not discuss the *Law of Large Numbers*, which means that he or she is either ignorant or lying. In either case, it is wise not to listen to such a person.

I provide this brief analysis as a public service announcement. One day, I may put my lottery software on www.sourceforge.net (the free software foundation) to make it available for anyone. It does not claim to find trends in the lottery tickets because it recognizes that there are none. Rather, it allows a person to put in a list of numbers from which to generate all the possible tickets meeting certain trimming criteria that the user selects. One may wish to have all the tickets without any trimming as well. One may also do other things with this software. It comes with a large help file.

Appendix B: Source Code for PythTrip Program

```c
#include <stdio.h>
#include <stdlib.h>
void MN(void);
void FS(void);
void DN(void);
int primitive(unsigned long int i, unsigned long int j, unsigned long int k);
unsigned long int gcd(unsigned long int x, unsigned long int y);
int main (void)
{
char c='/0', cc='/0';
printf("PYTHTRIP version 1.0 Copyright (c) 2005 by Tony Berard\n\n");
printf("\
   Pythagorean Triples are ordered triples of numbers (a, b, c) that satisfy\n\
the Pythagorean Formula a^2+b^2=c^2. This formula is for a right triangle, and\n\
it says that the sum of the squares of the legs of the right triangle is equal\n\
to the square of the hypotenuse of the triangle.\n\n\
   This program will generate Pythagorean Triples from three distinct modes of\n\
operation.\n\n");
printf("Press the Enter key to continue.\n");
scanf("%c", &c);
Check: ;
printf("1) Simple m, n Formula mode.\n2) Fibonacci Sequence mode.\n\
3) Dye-Nickalls Sequence mode.\n4) Exit.\n\n");
printf("Press 1, 2, 3, or 4 followed by the Enter key.\n");
scanf("%c", &c);
scanf("%c", &cc); //To clear the Enter key character from the input buffer.
if(c == '1') { MN(); printf("\a\nDouble-click the SimpForm.txt file to read.\n"); }
if(c == '2') { FS(); printf("\a\nDouble-click the PTFibSeq.txt file to read.\n"); }
if(c == '3') { DN(); printf("\a\nDouble-click the DNAlgthm.txt file to read.\n"); }
if(c == '4') return 0;
if(c != '1' && c != '2' && c != '3') goto Check;
printf("\nPress the Enter key to continue.\n");
scanf("%c", &c);
goto Check;
return 0;
}
void MN(void)
{
unsigned long int minm=0, maxm=0, minn=0, maxn=0;
char c = '/0', cc='\0';
int i=0,j=0,k=1, flagp=0, flags=0;
FILE *OUP;
OUP=fopen("SimpForm.txt", "w");
printf("\n\
   This is the Simple m, n Formula mode. Given any two positive integers m and\n\
```

n, with m > n, the ordered triple (m^2-n^2, 2mn, m^2+n^2) is a Pythagorean Tri-\n\
ple. The program needs the min and max m and n that the user desires. The pro-\n\
gram will compute all possible Pythagorean Triples within those range specifi-\n\
cations and will weed out all cases where m <= n. A warning is in order\n\
here.\n\
Do not specify ranges that are too wide or else you will get an overabundance\n\
of program output. These m,n expressions have been known for millennia.\n\n");

```c
printf("What is the min m?\n");
scanf("%u", &minm);
printf("What is the max m?\n");
scanf("%u", &maxm);
printf("What is the min n?\n");
scanf("%u", &minn);
printf("What is the max n?\n");
scanf("%u", &maxn);
scanf("%c", &cc);
printf("\nDo you want only the primitive Pythagorean Triples?\n1) Yes.\n2) No.\n\n");
printf("Press 1 or 2 followed by the Enter key.\n");
scanf("%c", &c);
scanf("%c", &cc); //To clear the Enter key character from the input buffer.
if(c == '1') flagp = 1;
printf("\n\
Would you like the program to flag (with an asterisk) the special Pythagorean\n\
Triples, which are the ones with a difference in the sides of just one?\n");
printf("1) Yes.\n2) No.\n\n");
printf("Press 1 or 2 followed by the Enter key.\n");
scanf("%c", &c);
scanf("%c", &cc); // To clear the Enter key character from the input buffer.
if(c == '1') flags = 1;
for(i=minm;i<maxm;i++) {
 for(j=minn;j<maxn;j++) {
  if(i>j) {
   if((primitive(i*i-j*j, 2*i*j, i*i+j*j) && flagp) || (flagp==0)) {
    if(flags && (i*i-j*j-2*i*j == 1 || 2*i*j-i*i+j*j == 1 || i*i+j*j-2*i*j == 1))
     fprintf(OUP, "*");
    if(i*i-j*j<2*i*j) fprintf(OUP,"%d) [%u, %u, %u]", k, i*i-j*j, 2*i*j, i*i+j*j);
    else fprintf(OUP,"%d) [%u, %u, %u]", k, 2*i*j, i*i-j*j, i*i+j*j);
    if((float)k/2 - k/2 < 0.01) fprintf(OUP, "\n");
    else fprintf(OUP, "      ");
    k++;
   } // End if((primitive
  } // End if(i>j
 } // End for(j=minn
} // End for(i=minm
fclose(OUP);
return;
}
void FS(void)
```

```c
{
unsigned long int a=0, b=0, num=0, i=0;
char c = '/0', cc='\0';
FILE *OUP;
unsigned long int *Fib;
printf("\n\
    This is the method to generate Pythagorean Triples from the famous Fibonacci\n\
Sequence. Most often, this sequence usually starts with 0 followed by 1;\n\
however, any two integers can start a Fibonacci sequence. The next term in the\n\
sequence is found by adding the previous two terms. Thus, with the sequence\n\
starting with 0, 1, it becomes 0, 1, 1, 2, 3, 5, 8, 13, 21, ... An example of a\n\
Fibonacci sequence that is different is 1, 3, 4, 7, 11, 18, 29, 47, ... There\n\
is an interesting link between a Fibonacci sequence and Pythagorean Triples.\n\n");
printf("\
    To make a Pythagorean Triangle, take any four consecutive numbers in a\n\
Fibonacci sequence such as 3, 4, 7, 11. The product of the outer two numbers is\n\
the short leg in the Pythagorean Triple. In this example, 3*11 is 33. Twice the\n\
product of the middle two numbers yields the long leg in the Pythagorean\n\
Triple. In this example, 2*4*7 is 56.  The sum of the squares of the middle two\n\
numbers yields the third side--the hypotenuse. In our example, 4^2+7^2 is 65.\n\
Thus, 33^2+56^2 is supposed to be 65^2, and it is! The method works for any\n\
Fibonacci sequence with any four consecutive terms.\n\n");
printf("Press the Enter key to continue.\n");
scanf("%c", &cc);
printf("\
The program will get your starting two values for the Fibonacci sequence, and\n\
it will ask for the number of Pythagorean Triples to compute. A general warning\n\
is in order here. Do not ask the program for thousands of triples because this\n\
will crash the computer. Think of how much memory such a file would have. Do\n\
not start a Fibonacci sequence with numbers above about 20 thousand because the\n\
multiplying and squaring of these large numbers will exceed the integer arith-\n\
metic capacity of the C programming language.\n\n");
printf("What is the first number of your Fibonacci sequence?\n");
scanf("%u", &a);
scanf("%c", &cc);
printf("\nWhat is the second number of your Fibonacci sequence?\n");
scanf("%u", &b);
scanf("%c", &cc);
printf("\
\nHow many Pythagorean Triples would you like the program to compute from this\n\
```

```c
Fibonacci sequence? Warning: Above about 20 to 25 terms will yield
intermediate\n\
computations that exceed the roughly 4 billion ceiling of the C language, de-
\n\
pending on the initial two terms you specify.\n");
scanf("%u", &num);
scanf("%c", &cc);
Fib = (unsigned long int *) malloc((num + 3)*sizeof(unsigned long int));
if(Fib == NULL) {
 fprintf(stderr, "\aThe computer\'s memory is not sufficient to run this
program.\n");
 fprintf(stderr, "Try closing other applications.\n");
 fprintf(stderr, "Press the Enter key to close this program window.\n");
 fscanf(stdin, "%c", &cc);
 exit(1);
}
OUP=fopen("PTFibSeq.txt","w");
Fib[0]=a;
Fib[1]=b;
fprintf(OUP,"The Fibonacci Sequence is the following:\n");
for(i=2;i<num+3;i++)   Fib[i] = Fib[i-1] + Fib[i-2];
for(i=0;i<num+3;i++) { fprintf(OUP, "%u", Fib[i]); if (i<num+2) fprintf(OUP, ",
"); }
fprintf(OUP, "\n\n");
for(i=0;i<num;i++) {
 if(Fib[i]*Fib[i+3] < 2*Fib[i+1]*Fib[i+2])
 fprintf(OUP, "%u) [ %u, %u, %u ]", i+1, Fib[i]*Fib[i+3],
         2*Fib[i+1]*Fib[i+2], Fib[i+1]*Fib[i+1] + Fib[i+2]*Fib[i+2]);
 else fprintf(OUP, "%u) [ %u, %u, %u ]", i+1, 2*Fib[i+1]*Fib[i+2],
         Fib[i]*Fib[i+3], Fib[i+1]*Fib[i+1] + Fib[i+2]*Fib[i+2]);
 if(i==2*(unsigned long int)(i/1.99)) fprintf(OUP, "      ");
 else fprintf(OUP, "\n");
}
fclose(OUP);
free(Fib);
return;
}
void DN(void)
{
unsigned long int sl=0, ll=0, h=0, tsl=0, tll=0, th=0;
int a=0, b=0;
int i=0, num=0;
char c='\0';
FILE *OUP;
printf("\n\
   This is the method to generate Pythagorean triples with a constant gap\n\
between either the two legs or between the long leg and the hypotenuse. This\n\
method is described in the paper A new algorithm for generating Pythagorean\n\
triples. This paper was published in The Mathematical Gazette (1998); volume
82\n\
(March), pages 86-91 by R. H. Dye and R. W. D. Nickalls. The authors state
in\n\
the paper that as far as they know, these algorithms have not been described\n\
before. Thus, this program lets the user explore a recent mathematical
find.\n\n");
printf("Press the Enter key to continue.\n");
```

```c
scanf("%c", &c);
printf("\n\
    To use this method, the program requires a Pythagorean Triple as a seed that\n\
gets the process rolling. Then, successive values are computed being mindful of\n\
the gap specified by the user, which will be between the short leg and the long\n\
leg or between the long leg and the hypotenuse.\n\n");
Check1: ;
printf("\
What is the short leg of the Pythagorean Triple?\n");
scanf("%u", &sl);
printf("\
What is the long leg of the Pythagorean Triple?\n");
scanf("%u", &ll);
if(ll<=sl) {
 printf("\a\nThe long leg must be longer than the short leg!\n\n");
 goto Check1;
}
printf("\
What is the hypotenuse of the Pythagorean Triple?\n");
scanf("%u", &h);
if(sl*sl + ll*ll != h*h) {
 printf("\a\nThis is not a Pythagorean Triple!\n");
 goto Check1;
}
Check2: ;
printf("\nWhere would you prefer the gap to be preserved?\n1) Between the short\
 and long legs\n2) Between the long leg and the hypotenuse.\n");
scanf("%d", &i);
if(i==1) a=ll-sl;
if(i==2) b=h-ll;
if(i != 1 && i != 2) goto Check2;
printf("\n\
\a**A warning is in order:\n\
    It is OK to specify a fairly large number of triples with the preserving of\n\
the gap between the long leg and the hypotenuse. But, do not specify too many\n\
triples with the preserving of the gap between the short and long leg because\n\
the roughly 4 billion ceiling of the C programming language will be exceeded.\n\n\
    How many triples would you like the program to generate?\n");
scanf("%d", &num);
OUP=fopen("DNAlgthm.txt", "w");
if(a) {
 for(i=0;i<num;i++) {
  fprintf(OUP, "%d) [ %u, %u, %u ]", i+1, sl, ll, h);
  if(i == 2*(int)(i/1.99)) fprintf(OUP, "       ");
  else fprintf(OUP, "\n");
  tsl=sl;
  th=h;
  sl = 3*tsl + 2*th + a;
```

322

```
    ll = sl + a; // Not really a recurrence relation.
    h = 4*tsl + 3*th + 2*a;
  } //End for loop.
}
if(b) {
  for(i=0;i<num;i++) {
    fprintf(OUP, "%d) [ %u, %u, %u ]", i+1, sl, ll, h);
    if(i == 2*(int)(i/1.99)) fprintf(OUP, "        ");
    else fprintf(OUP, "\n");
    tsl=sl;
    tll=ll;
    th=h;
    sl=tsl+2*b;
    ll=2*tsl+tll+2*b;
    h=2*tsl+h+2*b;
  } //End for loop.
}
fclose(OUP);
return;
}
int primitive(unsigned long int i, unsigned long int j, unsigned long int k)
{
unsigned long int x, y, z;
x=gcd(i,j);
y=gcd(j,k);
z=gcd(x,y);
if(z>1) return 0; // It's not a primitive triple.
else return 1; // It is a primitive triple.
}
/*   The gcd() function computes the gcd of two ints. Amazingly, this very
necessary math function is not to be found in any of the C library header
functions. Therefore, I had to write this gcd() function. It may be a little
crude, but it gets the job done. In a C textbook, it says that the library
functions are good to use because they have been written by professional pro-
grammers. Well, I am not a pro, but my gcd function works OK. It takes as arg-
ments two ints. It returns an integer value equal to the gcd of the two ints.
*/
unsigned long int gcd(unsigned long int x, unsigned long int y)
{
unsigned long int smaller, gcdval = 1, i;
long double j, l;
unsigned long int k, m;
if (x <= y) smaller = x;
if (x > y) smaller = y;
for(i = 2; i <= smaller; i++) {
  j = (long double)x / i;
  k = x / i;
  l = (long double)y / i;
  m = y / i;
  while(j-k<0.00000000000001 && l-m<0.00000000000001 && x != 1 && y != 1){
    x = x / i;
    y = y / i;
    gcdval = gcdval * i;
    j = (long double)x / i;
    k = x / i;
    l = (long double)y / i;
```

```
        m = y / i;
      } // End while.
   } // End for.
return gcdval;
} // End gcd().
```

Appendix C: Help Me Write the Solutions Manual (or: I Need a Coauthor)

I need some help to complete this course to make it a viable college math course. I need a solutions manual for this edition. I would like to collaborate with a coauthor on the second edition of this book. The first edition will have the solutions manual posted online. The website will be www.plcs.yolasite.com. There's a link there for a pdf file of the solutions manual at its current state of development. I will put other stuff there too over time.

Anyone who submits from 3 to 5 problems from this textbook will be considered. Please follow these eight steps for your submission:

1. State the chapter name and problem number with "solved by your first name and the initial of your last name in city, XX." where XX is the two letter abbreviation of your state.

2. Any diagrams you create need to be done with the program Graph 4.3. A free copy can be obtained at www.sourceforge.net or at www.padowan.dk.

3. Submit your work in a Microsoft Word file. Use the Microsoft Equation editor to write the math portions of your solutions.

4. I do not want you just to solve the problems that you select. Rather, provide a clear narrative that explains the steps in plain English. Students will read this narrative to get their bearings in the math. Place this narrative near the actual math steps in a nice format.

5. If you find an error in one of my problem set ups—either through an under determined or over determined system (or some other error)—fix the problem, and solve that. That kind of thing will really impress me.

6. I have a few omissions in my results, and I say so in the text. If you solve those gaps, I would be really impressed with that as well, although that won't go into the solutions manual, obviously. That would go into a corrected second edition.

7. Write the statement "I am offering these solutions to the public domain and forfeit any of my copyright rights." As public domain, I can then post your solutions on the Internet for free to anyone who wants to download it. I will only include your information from number 1— your first name and the initial of your last name in city, XX. I will also include the statement in this number.

8. Email me at plcs@earthlink.net. Alternatively, you can email me at plcsfirstedition@gmail.com. Please, make sure your work is correct. In the email itself, tell me your information and credentials. Put your solutions in the attachment as an MS-Word file. If you want to submit more than the three to five suggested above, by all means do so. I just need that as a minimum to gauge your level of writing, which includes formatting and all that. Also, if you just want to submit a few solutions, but you do not want to be considered for a coauthor, let me know that in your email. I understand that some people have very limited amounts of time that they can devote. Tell me you are a good Samaritan, and I will know not to consider you for the coauthor position. Thank you.

I will combine all of the unique selections and best renditions of multiply done problems as the results come in as a pdf file done by Word. At some point, I will ask the best writer in my estimation to be my coauthor. Basically, I would like to write the second edition, and I would like my coauthor to write the solutions manual. That way, I get the proceeds from the textbook, and the coauthor gets the proceeds from the solutions manual.

Also, if you just want to offer some suggestions for improvement or any comment related to this work, please contact me. I will post the good ones with my responses on the website www.plcs.yolasite.com. As stated on the back cover, if you find this course to be a worthwhile addition to the college learning experience, please contact your math department head or dean to ask them if they can include this course in your curriculum. Thank you so much.

I have one final comment. This course will actually need a lot of supporting material, such as a study guide, standardized test bank, an online facility for assignments and help for coursework, and other manuals relating to using a CAS to do these problems. Thus, this is far more work than what I can do as a single author, so I need some help. Anything that you, as a student or instructor, do to help put this course on the map would be most appreciated.

Index

30-60-90 triangle, 67, 242
45-45-90 triangle, 67
abscissas, 146
absolute value, 19, 52, 70, 132, 159, 181, 254, 258, 271- 273, 279, 280, 292, 302-304
acute angle, 62, 63, 240
acute triangle, 214, 232
Adam's circle, 249, 252
altitude, 67, 78, 98, 111, 152, 156, 214, 218, 221, 231- 233, 241- 243, 248
angle bisector, 215, 217, 219, 227- 230, 240, 242, 250, 251
anticomplementary triangle, 287
antimedial triangle, 220, 232
antiparallel, 239-242, 248, 249
Apollonian distance, 181
Apollonius, 13, 22, 23, 180, 181, 185, 190, 238
arbelos, 169-173, 175, 184, 187-189
Archimedean radius, 172, 173, 176, 188
Archimedes' quadruplets, 173-175, 189
Archimedes' rectangle, 170, 171, 188
Archimedes' twins, 172, 174, 188, 189
asymptote, 3, 23, 24, 161-168, 186, 238, 239, 255, 303
asymptote line, 23-25, 31, 238, 239, 255, 303
Bevan circle, 221
Bevan point, 221
Bevan, Benjamin, 221
bisector, 160, 190, 215, 217, 219, 224, 227-230, 239, 240, 242, 243, 250, 251
Brianchon, 309
Brocard circle, 231
Brocard line, 231, 236
Brocard points, 231
Brocard, Pierre, 231
Butterworth, John, 221
Carnot, Lazare, 217
Carnot's theorem, 217, 218, 226
Cartesian plane, 92, 211, 236
CAS, 37, 76, 101, 175, 199, 201, 326
Cayley, 312
center-radius form, 92, 93, 102, 103, 215
centroid, 77, 152, 213, 214, 219, 220, 222, 223, 236, 242, 247, 250- 252, 260
Ceva, Giovanni, 212
Ceva's theorem, 212, 213

cevian, 212, 213, 217, 219, 233, 250-252

chord, 20, 22, 24, 31, 60, 128, 136, 137, 142-144, 187, 190, 207-210, 249, 257, 260, 261, 294, 297

circumcircle, 78, 105-108, 152, 212-214, 218, 221, 241, 242, 250, 251, 260, 287, 289

circumradius, 216, 217, 220

cleavance center, 219

cleaver, 211, 219

Cohen, 310

collinear, 20, 33, 34, 68, 73, 92, 93, 182, 187, 190, 205, 209, 211, 212, 214, 217, 218, 220, 250, 260, 292, 302, 309

common pole, 203-207, 209

complementary angles, 62

complete quadrilateral, 152, 155-160

concyclic points, 3, 63, 222, 237, 239, 242, 248

conic section diameter, 203, 207

conjugate axis, 239, 251

conjugate diameter, 22, 23, 208-210

conjugate hyperbola, 24

consistent, 100, 158

convex, 149, 151, 248, 308

Conway circle, 249, 252

cross-join, 32-35

cyclic quadrilateral, 152, 155, 157, 160, 241

cyclocevian conjugates, 250-252

Fermat, 7, 222-224, 242, 244, 246

de Longchamp's point, 220, 225

demarcation line, 269-271, 276

denesting, 3, 37, 51, 53, 54, 60, 159, 175

Desargues, 3, 6, 73-76, 102, 104, 177, 214, 218, 248, 260, 289, 309

diagonal, 64, 67, 71, 146, 148, 155-157, 248

diametrically opposed, 65, 67, 102, 104, 111, 156, 213, 218, 260

directrix, 20-22, 25, 31, 70, 72, 132, 144

discriminant71, 119, 121, 137, 142, 167, 179, 183, 234, 254, 255, 257, 270, 273, 278, 280, 281, 304

distance from a point to a line, 66, 249

divisor, 10, 38, 39, 43, 161

domain, 25, 125, 164, 166, 297

dual, 138, 309, 310

dyadic fraction, 42, 43, 57

Dye-Nickalls, 83-86, 90, 318, 321

eccentricity, 20-25, 70, 256, 257, 259

equilateral triangle, 63, 67, 145, 218, 221-223, 242-245

Euclid, 13, 79, 88, 260

Euler, 214-217, 221, 225, 226, 288, 289

Euler-Gergonne-Soddy triangle, 225

Evans point, 217, 225

excenter, 219, 221

excentral triangle, 221

excircle, 215, 216, 219

extended midpoint formula, 71, 100, 101, 176

external tangents, 204, 209

extouch triangle, 219

eyeball theorem, 3, 260, 261, 263, 264

Fagnano's problem, 214

Farey sequence, 45-47, 58, 260

Feuerbach, 214, 216, 225

Fibonacci sequence, 82, 83, 89, 91, 318, 320, 321

Fletcher point, 225

focal chord, 20, 22, 24, 60, 136, 137, 142, 144, 257, 260, 261, 294, 297

focal diameter, 20, 21

focal parameter, 20

focus, 20-22, 24, 25, 31, 60, 70, 72, 119, 124, 126, 127, 129, 131, 132, 134, 136, 138, 140, 142-144, 202, 237, 251, 265, 266, 288, 290-292, 295-297, 299-301, 303, 306

Free Software Foundation, 316

Frégier, 3, 180, 187, 191, 192, 201

Frost, Robert, 266

Fuhrmann, 220, 248

GCF, 37-40, 56, 86

general tangent line equation, 181, 186, 187, 190

Gergonne, 177, 191, 217-220, 225, 236, 248, 249, 251

Gob point, 218

Gödel, 310

Grebe, 227

Hadwiger-Finsler inequality, 221

Harvey, 0, 237, 239, 251

Heron (or Heronian), 69, 70, 72, 91, 95, 97-100, 103, 211, 226, 228, 251

hexagon, 64, 71, 248, 308-310, 312

hypotenuse, 10, 50, 67, 79, 81-85, 89, 97, 269, 270, 318, 320-322

identity, 11, 34, 60, 74, 75, 87-89, 92, 109, 110, 112, 138, 152, 160, 227, 230, 231, 235, 246, 289

implied discriminant, 257

incenter, 177, 178, 186, 214, 215, 217-221, 225, 228, 230, 249, 250 incircle, 178, 215-217, 219, 249, 251

inradius, 178, 216, 217, 219-221, 249

instantaneous rate of change, 191

internal tangents, 203, 204, 209

intersection method, 16, 19

intouch triangle, 219

inverse, 219

isogonal conjugate, 250, 252

isogonal line, 249, 250

isotomic conjugate, 232, 233, 235, 236, 250

isotomic points, 233, 234

isotomic transversal, 68, 72

Johnson, 3, 285-289

join, 32-35

Kenmotu point, 225

Kiepert's hyperbola, 225

Kiepert's parabola, 225

Kimberling, 211, 261

Kirkman, 312

kite, 156, 178

Kosnita point, 250

L Huillier, 227

latus rectum, 20-24, 31, 134, 136, 290, 292, 306

Law of Large Numbers, 316

LCD/LCM, 38, 56, 111, 135, 315,

Lemoine, 215, 227, 232, 236, 249, 252

Lester, 222, 242, 246, 247, 251

linear asymptote, 161, 164, 165

linear transformation, 95, 102

lottery, 315, 316

maltitude, 152-155, 160

max d line or dmax line, 270-272, 276

medial triangle, 77, 213, 218, 219, 220, 232

median, 77, 212, 215, 227, 229, 230

median, 44, 45, 47, 58

midpoint formula extension, 65

midpoint quadrilateral theorem, 149, 151, 261

Midy's theorem, 40-42, 57, 260

mirror, 68, 196, 229, 250

Mittenpunkt, 219

modular arithmetic, 131

Monge line, 205, 206

Morley's triangle, 225

multiplicity, 188, 195, 269, 285

Nagel, 218-220, 236

Napoleon, 223, 224

negative reciprocal, 10, 97, 108, 146, 147, 159, 188, 195, 199, 200, 209, 229, 241, 242

Newton, 156

Nickalls, R. W. D., 83-86, 90, 318, 321

nine-point circle (nine-point center), 78, 102, 104, 111, 112, 214, 216, 218, 220, 222, 225, 242, 246, 250, 251, 260, 288, 289

Nobb's points, 217

nonlinear asymptote, 165

normal, 61, 128, 169, 186, 190-192, 194, 195, 197, 200-202, 237-239, 251

oblique asymptote, 161, 164, 165

obtuse angle, 62, 63, 70, 232, 240

ordinate, 271

orthic axis, 78, 214, 218, 260

orthic triangle, 78, 102, 104, 214, 218, 233, 236

orthocenter, 78, 111, 156, 157, 160, 214, 218, 220, 222, 232, 233, 235, 236, 250-252, 284, 287, 289

orthocentroidal circle, 214, 242, 247, 251

over determined system, 74, 155, 158, 235, 325

Pappus, 3, 32-37, 121, 260, 310

parabolic asymptote, 3, 161, 164-166

parallelogram, 22, 145-151, 156

part over whole, 100

Pascal, 3, 308-312

pedal triangle, 218, 250, 260

Pedoe's inequality, 221

Pell's equation, 28, 30

PEMDAS, 14, 37

perfect square, 28, 50, 79, 137, 138, 143, 229, 261

perpendicular bisector, 160, 190, 224, 243

perspector (perspectrix), 73-78, 177, 217, 218, 231, 248, 260, 288, 289, 309

plotting a triangle, 92, 226

Plücker, 157, 159, 160, 246, 312

polar line (pole), 3, 203-207, 209, 262

polygon, 61, 63, 64, 152, 170

Power, Frank, 173

powerful number, 28, 30

Prasolov, 78, 102, 211, 218, 260

projective geometry, 308-310

Ptolemy's theorem, 152, 155, 160, 241, 248

Putnam Exam, 128

quadratic equation (formula), 15, 17, 117, 123, 130, 135, 139, 162, 164, 166, 179, 184, 188, 199, 229, 230, 234, 266, 285, 289, 291, 300

quotient, 37, 38, 40, 43, 44, 161

rate of change, 191

ratio, 14, 21, 22, 25-27, 70, 100, 213

rational function, 161-164, 166-168, 257

rational number, 37, 42, 46-48, 50, 51, 54, 120, 137

rational point, 3, 37, 47-51, 58, 59, 79, 92, 94-97, 99, 103, 153, 154, 159, 188, 230, 260, 284, 288

rational roots theorem, 16, 190, 238

rationalize (the denominator), 14, 111, 142, 267

rectangular hyperbola, 24, 61, 225, 283, 284

recursive equation, 27-30, 82, 84, 85, 146

reflect, 65, 67-69, 71, 72, 78, 102, 104, 152, 153, 196, 198, 211-213, 218, 227-230, 233, 250, 260, 287

Regiomontanus, 182

regular, 63, 64, 71, 151, 308, 309

remainder, 10, 38-40, 43, 161

repeating decimal, 42-44

retrocenter, 232, 233, 235, 236

rhombus, 64, 99, 156

root method, 19

rref, 93, 102, 107, 314

Salmon, 312

Schiffler point, 215

Schwatt lines, 231, 236

semi-perimeter, 69, 176, 216, 219

shape and location discussion, 116, 129

sign analysis (sign test),15, 16 257

Simson, 3, 105, 108-112, 218, 260

slant asymptote, 161, 164, 165

Sloane, 25, 313

slope function, 61, 169, 191, 237, 238

Soddy line, 225

Spieker, 219, 220

splitter, 211, 219

square root function, 86, 127, 181

square root method, 108, 229

statue problem, 182, 183, 185, 186, 189, 190

Steiner, 212, 213, 312

subtangent, 181

sum of squares, 81, 93

supplementary angles, 62, 63, 71, 170

surd, 51, 52, 54, 56

symmedial triangle, 231, 236

symmedian line (point), 3, 69, 215, 222, 227, 230-232, 236, 241, 242, 249, 250, 252

symmetry (asymmetry), 20, 31, 66, 67, 100, 181, 182, 196-198, 230, 263

system of parallel chords, 22, 209, 210

tangential triangle, 218

Tarry point, 213

Taylor circle, 248, 252

terminating decimal, 42-44, 57, 58

Terquen, 214

Thales, 64, 170, 188, 201

Thébault, 145, 146, 148-151

transitive property, 44

translation (and rotation) of axes, 124, 225

transversal, 63, 68, 72, 171, 240, 241

transverse axis, 239, 251

triangle center, 211, 215, 217, 219, 220, 261

Triangle Reflection Point theorem, 68, 72

triangular number, 28-30

trigonometry, 5, 8-10, 12, 87, 94, 155, 186, 215, 225

trisected perimeter point, 213

trivial (nontrivial), 76, 114, 115

Tucker, 227, 248, 252

two-variable function (equation), 16, 137, 142, 143, 183

unified definition of conic section, 70, 72

unit circle, 94-97, 99, 102-104, 153, 154, 188, 234, 288, 312

van Aubel, 150, 151, 236

Vecten, 224, 225

vertical angles, 62, 63, 70

vertical asymptote

Viviani, 218, 226

Wallace-Simson line, 3, 105, 108-112, 218, 260

Weill point, 219

Weitzenböck's inequality, 221

Wernau points, 223

Wiles, 7, 222

www.research.att.com/~njas/sequences, 25, 313
www.sourceforge.net, 316, 325
Yff point, 217

www.ingramcontent.com/pod-product-compliance
Lightning Source LLC
Chambersburg PA
CBHW081106170526
45165CB00008B/2339